AF302367

C. Hoffmeister   G. Richter   W. Wenzel

Variable Stars

Translated by S. Dunlop

With 170 Figures and 64 Tables

Springer-Verlag
Berlin Heidelberg New York Tokyo

Professor Dr. rer. nat. Cuno Hoffmeister †

Dr. rer. nat. Gerold Richter
Stellv. Abteilungsleiter am Zentralinstitut für Astrophysik
der Akademie der Wissenschaften der DDR
DDR-6400 Sonneberg-Neufang, Sternwarte Sonneberg

Dr. rer. nat. Wolfgang Wenzel
Abteilungsleiter am Zentralinstitut für Astrophysik
der Akademie der Wissenschaften der DDR
DDR-6400 Sonneberg-Neufang, Sternwarte Sonneberg

Translator: Storm Dunlop
140 Stocks Lane, East Wittering, nr. Chichester, West Sussex PO 020 8 NT, England

Cover picture: George V. Kelvin
In *Symbiotic Stars* by Minas Kafatos and Andrew G. Michalitsianos.
Scientific American, July 1984, p. 78
© Scientific American

Title of the original German edition: *Veränderliche Sterne*, 2. Auflage
published by Johann Ambrosius Barth, Leipzig, and Springer-Verlag, Berlin Heidelberg
New York Tokyo
© Johann Ambrosius Barth, Leipzig
Translation by permission of Johann Ambrosius Barth, DDR-7010 Leipzig, Salomonstr. 18 b

---

ISBN-13: 978-3-642-82271-1   e-ISBN-13: 978-3-642-82269-8
DOI: 10.1007/978-3-642-82269-8

Library of Congress Cataloging in Publication Data. Hoffmeister, C. (Cuno), 1892–1968.
Variable Stars. Translation of: Veränderliche Sterne. Bibliography: p. Includes index. 1. Stars,
Variable. I. Richter, Gerold. II. Wenzel, W. (Wolfgang), 1929–. III. Title. QB835.H5913
1985   523.8'44   85–7928

This work is subject to copyright. All rights are reserved, whether the whole or part of the
material is concerned, specifically those of translation, reprinting, reuse of illustrations, broad-
casting, reproduction by photocopying machine or similar means, and storage in data banks.
Under § 54 of the German Copyright Law where copies are made for other than private use, a
fee is payable to "Verwertungsgesellschaft Wort", Munich.

© Springer-Verlag Berlin Heidelberg 1985
Softcover reprint of the hardcover 1st edition   1985

The use of registered names, trademarks, etc. in this publication does not imply, even in the
absence of a specific statement, that such names are exempt from the relevant protective laws
and regulations and therefore free for general use.

Media conversion: Daten- und Lichtsatz-Service, Würzburg

2153/3130-543210

Preface to the English Edition

This book is a translation, by Mr. S. Dunlop, FRAS, of the second German edition of *Veränderliche Sterne* published by Johann Ambrosius Barth Verlag, Leipzig, DDR, and Springer-Verlag, Berlin Heidelberg New York Tokyo. We have used the opportunity to make improvements and changes in a few places and to add details of new results and discoveries. The foreword to the second German edition remains no less valid for this one, including the changes. We would like to express our thanks to Mr. Dunlop for his expert and sympathetic translation and for the many improvements he suggested.

December 1984

G. RICHTER and W. WENZEL

Preface to the Second German Edition

The suggestion that a second edition of this book should be prepared came equally from the readers, the publishers, and from the editors of the first edition. By a lucky coincidence the completion of the manuscript and the composition of these lines comes at the time of the 90th anniversary of the birth of Cuno Hoffmeister, the author of the first edition, and consequently this current work may be dedicated to his memory.

What Cuno Hoffmeister wrote at the beginning of the preface to the first edition about the difficulties in the selection of references, naturally did not apply just to him at that time alone, but is even more applicable to us nowadays. Indeed there has been a sharp increase in the total number of comprehensive reviews of individual, specialized subjects, which have been the subject of symposia and colloquia, not least through the activities of the International Astronomical Union, for example. The authors of this second edition frequently found it necessary to fall back upon these and to cite them as references. Despite this, in many cases the choice of references remains subjective, and quite frequently, and occasionally intentionally, has led to an emphasis on individual fields of study and interests. We would ask our colleagues, both at home and abroad, to forgive us if they are cited neither sufficiently nor at all, because we have made no attempt to strive for complete coverage.

The book is intended for the same class of readers as the first edition. What has occasionally been felt to be a deficiency in the latter – too little explanation of the physical background – we have tried to improve by modest means. This has also become rather easier, as in the meantime many new discoveries have clarified the situation. The result is that practically all sections of the original have had to be revised, and new classes of variables have been added. In some cases, we have knowingly devoted rather more space than in more "conventional" sections to the direct description of modern fields of research.

The division of the book is, as far as it concerns the classification of variable stars, generally arranged according to the basic causes of the changes in brightness. The first edition's intention (but which did not succeed even there), of taking the evolutionary state, or the age of the stars as a guide, soon proved to be impracticable. The causes of this are of an astrophysical nature: similar physical processes, pulsation for example, can occur in completely different, stellar evolutionary states. Frequent repetition would have been required if such an arrangement had been followed.

This second edition originated in the ordinary course of our work at the Sonneberg Observatory. We heartily thank all our colleagues whose contributions have helped us in our task.

February 1982   G. RICHTER and W. WENZEL

From the Preface to the First German Edition

It is a well-known fact to anyone involved in variable star research that changes in the nature of scientific publication in the last few decades have caused us many difficult problems. It is not only the sheer quantity of material which floods in, threatening to overwhelm us, and which has induced reviewers to merely touch upon their subjects, but it is also the fact that with the increasing overlap between fields of research, one has to be conversant with, and take into consideration, far more than was ever previously necessary.

However, this statement should not be taken to imply that those engaged in this field have failed to keep either their colleagues working in other areas of research, or the general public, fully informed. On the contrary, a series of excellent individual contributions on a wide range of topics has appeared, which, taken overall, would have given a good picture of the state of research, if they had treated the field as a whole. However, this has not been the case. Moreover, such a form of presentation does not allow the importance of individual specific findings, and the links between the various separate areas to be properly examined. The contributions of LEDOUX and WALRAVEN in *Handbuch der Physik* (1958), and of BEYER in *Landolt-Börnstein* (1965) may be considered as next best to a comprehensive survey, although in both of these cases some parts of the material have been handled by different authors.

In Chapter 1 a short explanation is given of how the knowledge of variable stars has reached a considerable degree of refinement, due to the realization that variability is not some special state, but is instead a normal phase in stellar evolution, from which information may be drawn about the ages of the stars and the structure of the Galaxy. The overall presentation given here has been written with this view of stellar evolution in mind, all the more so in that it thus gives starting points for further research. Mention may be made of the possibilities of determining by statistical methods the duration of the various unstable stages, and of the numerous relationships between membership of particular populations and the structure of the Galaxy. The possibilities are now known, the methods have been tried and in part proven, so in this respect the next aim must be to provide the necessary statistical material. We are possibly now at a decisive point in research in this field, which is yet a further reason for giving an account of what has been discovered, and of how it has been achieved.

The basic aim of this book is to give a survey, not just for the specialist, but also for the suitably-informed amateur. Some knowledge is assumed of the most important laws governing the behaviour of gases, and of a few basic concepts in atomic physics, as well as a little about spherical trigonometry and co-ordinate systems. However, for the benefit of those readers who are not fully conversant with this field and such ideas, it was considered advisable to include a reasonably comprehensive section on "Fundamental Concepts", which covers various matters from the definition of what constitutes a star, to galactic structure and stellar evolution. Despite the restricted description given, this should make this book suitable for a wider range of readers.

Because of the vast amount of literature on the subject, it is not possible to avoid a certain amount of unconsciously arbitrary selection of what should be cited as references. For this reason the book includes a list of reviews, which not only includes both older and more recent publications, but also the numerous later reference works that often have very full lists of citations.

This is perhaps an appropriate place to say something about my own introduction to the study of variable stars, especially as this touches upon one or more historical matters, which might otherwise easily become lost. As a young amateur, who had to leave school at the age of 16, I first had variable stars brought to my attention by PHILIPP FAUTH, whose advice I had requested regarding a useful application for my 52-mm telescope. R. LEHNERT, at that time assistant to ANDING at Gotha and later at Bamberg, shortly afterwards gave me a specific recommendation to take up this study. He advised me to observe Algol stars, and I made myself familiar with the Argelander Method. A visit to Bamberg in June 1914 was a landmark. E. ZINNER, then HARTWIG's assistant, gave me a programme of variables that had been discovered in the preceding years, but which were as yet unconfirmed, at least as far as determination of class and period were concerned. This exercise was carried out with good results, and in observing RT CrB I discovered the variability of a neighbouring star. It proved to be a rapid eclipsing star, now known as RW CrB. As ZINNER had to leave Bamberg at the outbreak of the First World War and work in the army meteorological service, I offered my help to Professor HARTWIG and went to Bamberg in April 1915. The period of almost 4 years up to the end of 1918 were my true apprenticeship. HARTWIG had worked for some years with WINNECKE in Strassburg, and WINNECKE had been with ARGELANDER in Bonn from 1856 to 1858.

My main task at Bamberg was to assist with the "Geschichte und Literatur der Veränderliche Sterne" (GuL). The compilation was being done by G. MÜLLER at Potsdam and I had to compare the information in the literature with the card-index independently prepared at Bamberg, to add data that were missing, and to clarify discrepancies. In addition, HARTWIG handed over to me WINNECKE's original observational records, with the task of determining epochs and of cataloguing these in the GuL.

The work on the GuL was also of fundamental importance for me. I was annoyed by the numerous cases, to which nothing was ascribed other than "Class and period unknown", especially as amongst these there were relatively bright, easily-observable stars. This encouraged me to observe these stars with the 160-mm comet seeker, and in some cases, with the 260-mm refractor as well. In many of the objects I was able to determine the class at the very least, and in some, the elements as well, and early enough for them to be included in the first edition of the GuL.

My work at Bamberg ended in 1918 with the return of ZINNER. I had taken care, however, in view of my foreseeable return to a job in commerce, to ensure that I would be able to continue observations at Sonneberg, in that with the income from some writing I had purchased a few instruments and observational reference works. It was from these beginnings that the Sonneberg Observatory started. As its early history has been described elsewhere (BRANDT 1967) this note will suffice here.

But one important fact remains to be mentioned. I continued my work on variable stars during my studies at Jena University (1920 to 1924), at least throughout the vacations, and gradually came to recognize where the subject's main fields would lie in the future. Up to then I had restricted myself to the determination of class and elements, but now another aspect appeared: the problems of statistic. What percentage of stars is variable? How is this fraction divided into the various classes? What is their distribution over the celestial sphere and in space? Through the then-existing "Notgemeinschaft der Deutschen Wissenschaft"[1] I obtained a 170/1200 mm Zeiss triplet from the estate of MIETHE at Charlottenburg. After trials, I was convinced that it was very capable of producing the photographs needed to discover new variables. I began this undertaking in 1926, which in the following years developed into the Sonneberg Fields project for investigating variable star statistics.

The outcome of this project, with its successes and failures has been touched upon, albeit briefly, in the historical review just mentioned. Many of the experimental results obtained from it are described in this book. I hope that other observers and investigators will obtain as much pleasure from their interest in one of the most fascinating areas of modern astrophysics, as it has brought to me.

December 1967

C. HOFFMEISTER

---

[1] An organization founded in 1920 to provide financial and material assistance, and encouragement, to German scientists, otherwise handicapped by the country's impoverished condition following the end of World War I. – Trans.

Contents

1. General Introduction

1.1 Introduction and General Summary

Compared with other branches of astronomy, the study of variable stars is a relatively new field. Its scientific basis was established about the middle of the 19th century. The first relevant observations were those of the supernovae in 1054, 1572, and 1604, although these objects were regarded as forming a special case, and were only recognized in the 20th century as representing a sub-group of variable stars as a whole.

A General Definition of Variability

A variable is a star that changes its brightness. This definition needs some qualification however, as in this general form every star would be included, variations in luminosity being connected with normal evolutionary changes that occur over a period of the order of $10^6 - 10^9$ years. Furthermore, many stars are variable on the scale of a hundredth of a magnitude, and finally, stars which might be taken to be constant, such as the Sun, are variables when observed at extreme wavelengths – in X-rays and the ultraviolet on the one hand, and at centimetre and metre wavelengths on the other. Consequently the definition should be qualified in three ways: firstly, the variability must take place within a period of decades at the most; secondly, it must be observable in the so-called "optical" regions, that is to say in the visual or photographic regions, the latter being extended to include the near infrared. Thirdly, the amplitude must be such as to be perceptible to the eye, either directly or by means of photography, and consequently be of at least 2 or 3 tenths of a magnitude. This last point needs further clarification, as stars with amplitudes < 0.1 mag can certainly be considered as variables. In general however, they will be discussed separately from those with greater amplitudes as they, with certain exceptions, form physically related groups and are only amenable to photoelectric observation.

Historical Survey

The *first stage* of scientific research into variable stars can be seen in the appearance of ARGELANDER's appeal to amateur astronomers that was published in SCHUMACHER's Yearbook for 1844. Of the 133 pages of this discussion, 48 are devoted to variable stars. The author reviewed what was then known about stellar variability and described the step method now named after him, and which put everyone in the position of being able to make brightness estimates to a high degree of accuracy. This induced him to urge all amateur astronomers to undertake co-operative work.

Table 1 gives a list of the stars known in 1844 as showing variability.

Table 1. Variables known in 1844

Star	Discoverer	Year of Discovery
o Ceti (Mira)	HOLWARDA	1639
β Persei (Algol)	MONTANARI	1669
χ Cygni	KIRCH	1687
R Hydrae	MARALDI	1704
R Leonis	KOCH	1782
η Aquilae	PIGOTT	1784
β Lyrae	GOODRICKE	1784
δ Cephei	GOODRICKE	1784
α Herculis	W. HERSCHEL	1795
R Coronae Borealis	PIGOTT	1795
R Scuti	PIGOTT	1795
R Virginis	HARDING	1809
R Aquarii	HARDING	1810
ε Aurigae	FRITSCH	1821
R Serpentis	HARDING	1826
S Serpentis	HARDING	1828
R Cancri	SCHWERD	1829
α Orionis	J. HERSCHEL	1836

ARGELANDER included in his list α Cassiopeiae and α Hydrae, two red stars that were suspected of variability, but which are now classed as invariable. R Cancri was missing from ARGELANDER's list, although its variability was mentioned by SCHWERD (under Hora VIII) in his notes on the "Akademische Sternkarten". Obviously this had been overlooked as the star first appeared on ARGELANDER's programme in 1850. Furthermore the discovery of variability in ε Aurigae in 1821 remained unknown to ARGELANDER until 1847 when HEIS independently detected it.

An English translation of the part of ARGELANDER's "appeal" concerned with variables appeared in 1912 in the American magazine "Popular Astronomy" (CANNON 1912). The translator, ANNIE J. CANNON, noted in her foreword that in ARGELANDER's time the total number of variables known was 18, as given in his list, whereas by 1912 it was 4000, and that observers were to be found not only "in Aachen, Breslau and Bonn", but in almost every country in Europe, in practically every state of the U.S.A., in Japan, South America, Australia, Egypt, and South Africa. From this one may conclude that ARGELANDER's suggestions had been widely adopted, assisted by yet another circumstance: the introduction of photography, which, above all, was responsible for the rapid rise in the number of variables known. Foremost in this field was the Harvard College Observatory in Cambridge, Mass., where from about 1890 onwards there was a systematic search for variables on photographic plates, whilst in the same year the Observatory's southern station started work at Arequipa in Peru. The dense star clouds of the southern Milky Way and of the Magellanic Clouds brought a rich harvest of new discoveries. Interest in variable stars also increased in conjunction with the development of astrophysics, in particular of stellar spectroscopy. This marked the *second stage* in research on variable stars.

The *third stage* came well into the 20th century. Only one branch of research was successfully developed further in the meantime, that of eclipsing stars, which however, by their very nature do not pose the same sort of problem as the basic causes of their variability are mainly optical and mechanical. In the case of the "physical variables", up to the turn of the century and even later, no advances had been made towards an explanation of the phenomena. So there was a tendency to regard variables as rare exceptions. The impetus towards a development of a theory was certainly given by EMDEN's book *Gaskugeln* on gas spheres (1907), and in the following decades it was the work of EDDINGTON on the internal structure of stars, together with the development of atomic physics, that allowed the problem of stellar variability to be seen in a new light.

The pulsation theory arrived, first advanced as a hypothesis for discussion by SHAPLEY (1914), and later set on a mathematical basis by EDDINGTON (1918). The full details of its historical development cannot be given here, but reference may be made to the book by SCHILLER (1923). However the names of MOULTON, LUDENDORFF, BOTTLINGER and GUTHNICK must be mentioned in addition to those already given. The pulsation theory was a very great step forward, and with its confirmation by later research can today be regarded as the true explanation, applicable to δ Cephei stars and related types.

Less lucky were other attempted explanations. The novae may be cited as the most conspicuous example of this. Neither the concept that a star, plunging at high speed into a cloud of interstellar gas and dust, could flare up in the way a meteor does in the Earth's atmosphere, nor the other hypothesis that saw the cause as being due to the impact of a planet, were to be accepted later. These explanations were premature because at that time far too little was known about the processes producing energy within stars. Even today the problems posed by novae and similar types remain incompletely resolved, as do those of stars at the beginning of their evolutionary tracks.

This brings us to the *fourth stage* of our knowledge of variable stars. A very important diagrammatic representation of stellar state, the Hertzsprung-Russell Diagram (see Sect. 1.2), shows the relationship between stellar types with differing physical properties such as temperature, density, and luminosity. It was suspected very early on that an evolutionary development was the basic cause of the observed distribution. But it was the growing insight into stellar energy production by means of nuclear processes that allowed this idea to be given a scientific basis. The fact that the various types of variable stars are found in very distinct regions of the H-R Diagram was ascribed to the evolutionary state of the stars concerned. A new concept of fundamental importance was established when it was realized that variables, with perhaps some exceptions, are not abnormal stars, but merely mark normal stages of development through which every star must pass if it has similar initial characteristics. In addition various types of variability were distributed in the various parts of the Milky Way system – the nucleus, the disk, the spiral arms or the halo – and conversely, their presence, or absence, can serve to reveal the characteristics of the component parts of the Galaxy, as these parts are also of differing ages. The procedure is similar to that used in geology, where the occurrence of given "characteristic fossils" enables the age of the particular strata to be determined.

The discovery of the relationship between stellar variability and general stellar evolution, as well as the association with parts of the Galaxy of differing compositions and ages, has caused the study of variable stars to become highly topical. Current research is directed along two differing lines. In the first, typical or particularly remarkable individual examples are examined in very great detail, both to determine their physical characteristics, and – by means of the knowledge gained by all these investigations – to explain the phenonema observed, which in the majority of classes is not yet completely successful. Secondly, the overall statistical findings are applied to try to obtain a better understanding of the structure of the Galaxy as a function of both space and time. But this work is of even greater significance as the most accurate method of determining cosmic distances in the region of the "Local Group" of galaxies – the Magellanic Clouds, the Andromeda Galaxy, the spiral galaxy in Triangulum (M33), and certain others – rests on the "Period-Luminosity Relation" found in pulsating variables. Even if one reaches out beyond the boundaries of the "Local Group", to distances at which no individual stars are visible with even the largest telescopes, the calibration of the distance scale still depends primarily upon the use of the period-luminosity relation within the more restricted region.

1.2 Explanation of Some Fundamental Concepts

What is a Star?

From a physical point of view a star is a *gas sphere*. This assertion can be readily understood when it is noted that the temperature at the surface of the Sun amounts to 5800 K and that at this temperature all known elements have been converted into the gaseous state. In the interior of the Sun and of other stars the temperatures are far higher than this.

In a gaseous body gas and radiation laws can be applied. These primarily determine the relation between pressure, temperature and density. By observations the surface temperature, the total mass and the volume can be determined, and from these, the mean density. By means of the gas laws and one or two simplifying assumptions one can determine the structure of the body, that is to say in what manner density and temperature change from the surface to the core. Energy production takes place primarily within the core of a star, and predominantly gives rise to very short-wave radiation. This radiation gradually diffuses towards the exterior, and is finally radiated away into space from the surface. Assuming that no disturbance of the energy-production process takes place, the star adopts an equilibrium state, where exactly the same amount of energy is lost by radiation at the surface as is produced in the interior. It should be noted that two opposing forces act upon every element of the star's volume, the one consisting of gas and radiation pressures, which seek to expand the star, and the other gravity, which acts to contract it. As changes in the rate of energy production resulting from stellar evolution only vary slowly, apart from certain critical transition phases, a star can remain for thousands of millions of years in a state of *stable equilibrium*. The Sun is an example of this.

An increase in energy production will expand the star, and a decrease will reduce its radius, always to such an extent that the balance between energy production and radiation loss is re-established. It will be readily appreciated that this also implies a change in the surface temperature.

The physical state of a star is described by the values of various properties, or *parameters*: mass, radius, mean density, luminosity, effective temperature, spectrum, chemical composition, mean rate of energy production, and gravitational acceleration at the surface. Nevertheless, with a given chemical composition, only one of these factors, for example the mass, is normally independently variable; all the others are then derived from it (Vogt-Russell Theorem). In addition, in many stars convection, rotation, and magnetic field strength also play a part.

Stellar Radiation

It is known from physical experiments that gases generally produce an emission-line spectrum, such as in the case of hydrogen for example, in the visible region, the lines of the Balmer series (Hα, Hβ, etc.). However, this rule is not applicable to stars even though they are gaseous bodies. A pure emission spectrum can only occur if the atoms of the gas can vibrate freely, that is when the density is so low that collisions between the atoms do not readily occur. But in stars, apart from the outermost layers, the gas density is so high that the radiation is like that of a solid or liquid body, so that a *continuum* results. By establishing either the peak intensity of the continuum, or its whole range of values, the stellar surface temperature may be determined by means of Planck's law, or the Stefan-Boltzmann law, respectively. Somewhat differing values are obtained according to the method and the spectral range employed. Reference should be made to suitable textbooks for details of the differences between "effective", "radiation", and "colour" temperatures. What is important is that the laws mentioned are only truly valid for a black body; that is to say a theoretical body that has the property of absorbing all radiation falling upon it. No naturally-occurring body strictly fulfils this condition, but stars behave in such a manner that the term "black-body radiation" is a useful first approximation.

The fact that the spectrum of a star is largely determined by its distribution of *absorption and emission lines* is important for our discussion. These lines are only ever produced in the outer layers, that is in the stellar atmosphere. The solar Fraunhofer lines are formed in the chromosphere or reversing layer, which has a very limited vertical extent, and which at total solar eclipses shows the flash-spectrum in emission. The absorption lines thus only ever give information about the chemical composition of the stellar atmosphere; moreover, the absorption spectrum is very strongly dependent upon the temperature: the lower the temperature of the stellar surface, the richer the spectrum becomes in absorption lines.

Even at the Sun's temperature many elements in the chromosphere are ionized, calcium for example, and chemical compounds are only found in the atmospheres of red stars; e.g. the absorption bands of TiO in long-period variables. In the deep interior of a star there are essentially no chemical elements, but only a gas, which consists of elementary particles (protons, neutrons, and electrons), deuterons and helium nuclei formed from them, and radiation quanta, and in which, in the later evolutionary stages, heavier nuclei are also present.

Emission lines are always formed outside a star and indicate the existence of an *extended gaseous shell*, whose luminosity is primarily excited by radiation from the star. These shells are very important for the explanation of the phenomena encountered in variable stars. If the shell exhibits relatively low temperatures ($\approx 15\,000 - 20\,000$ K), then emission lines with low excitation potentials are found – the Balmer series of hydrogen and the lines of neutral helium (He I). At high temperatures (50 000 K and more) we find a high-excitation spectrum with lines of ionized Helium (He II = He$^+$) and of singly- or even multiply-ionized oxygen, carbon, nitrogen and iron (O II, III,..., C II, III,..., N II, III,..., Fe II, III,...), and other lines. If the density of the gas is exceptionally low, then in addition to the "permitted" lines, so-called "forbidden lines" appear, which in the description of spectra are enclosed in square brackets, e.g. "lines of [Fe II]".

The measurement of the relative intensities of certain of these lines enables conclusions to be drawn about the temperature and density of the gas emitting the lines.

Other components of stellar radiation, such as corpuscular and neutrino fluxes as well as extreme short-wave radiation in the ultraviolet and X-ray regions will only be discussed here in certain exceptional cases.

The coolest stars – if we except some objects which radiate practically only in the infrared – have surface temperatures of about 2000 K. (All the temperatures given here are on the Kelvin scale, which differs from the Celsius scale in that its zero point does not lie at the freezing point of water, but at "absolute zero", $-273°$C. Consequently the relationship is: Kelvin temperature = Celsius temperature $+ 273°$. In the case of the high temperatures found in stars the difference is mostly of minor importance.)

Apparent Brightness of Stars – Magnitudes

Even in ancient times the stars were classified according to their brightness, that is according to their "apparent" brightness to terrestrial observers. It was an arbitrarily chosen arrangement; the brightest stars were assigned to the first class and the next brightest, such as those of, say, Ursa Major, to the second. Five or six such groups were thus obtained down to the limit of visibility to the naked eye. The magnitude system used today was adjusted to fit the designations already in use, as far as it was possible, by POGSON in 1856.

The perception of stellar brightness by an observer follows FECHNER's *physiological law*: the eye perceives similar differences where, in fact, similar ratios exist. That is to say that the perceived intensity corresponds to an arithmetical series, if the actual intensity of the stimulus follows a geometrical series. The usage was retained of designating magnitude, m, by a value which increases with a decrease in brightness. POGSON found that on average a star of magnitude m was about 2 to 3 times as bright as a star of magnitude $m + 1$, and he chose for his system the constant ratio of 2.512:1. This figure is so based that its logarithm has the exact value of 0.4. The logarithm of the intensity ratio between two stars, differing by n magnitudes, is thus $0.4\,n$. The general formula is

$$I_1/I_2 = 10^{-0.4(m_1 - m_2)}$$

where the subscripts 1 and 2 refer to the two stars. A difference of 5 magnitudes corresponds to an intensity ratio of 100:1, and 10 magnitudes to 10 000:1. For practical reasons the scale has been extended beyond the first magnitude towards brighter stars, by the introduction of magnitudes 0, -1, -2, etc. A specific magnitude will be denoted here by the use of a superscript m, and a difference in magnitudes by "mag".

By way of illustration, let us consider the following question: What is the relative intensity of the Sun and Sirius? The Sun has a magnitude of $-26^{m}\!.78$ visually, and Sirius $-1^{m}\!.44$. The Sun is thus 25.34 mag brighter than Sirius. The logarithm of this ratio is $25.34 \times 0.4 = 10.136$, which numerically is $1.3677 \times 10^{10} = 13.677$ thousand million.

A light source with a strength of 1 International Candle appears, at a distance of 1120 metres, to have the brightness of a first magnitude star.

The sign] before a magnitude value means "brighter than", and the sign ["fainter than". This arrangement is expedient because stellar magnitudes increase with decreasing brightness, so that the use of the signs $>$ and $<$ can readily give rise to confusion.

Magnitude Values in Various Wavelength Regions

Stellar magnitudes mostly show differences if they are measured in different wavelength regions, and this is further complicated by the differing *sensitivities* of the receptors used. For the eye the maximum response is at approximately 540 nm (m_v) and for a non-sensitized photographic plate at about 430 nm (m_{pg}). Both values can vary from case to case: for photographic plates as a result of sensitization, where due to an extension at the long-wavelength end of the spectral region covered, an increase in the total response is obtained.

In the case of the eye, it can vary due to differences between individuals. A consequence of this is that a red star is, in general, brighter visually than photographically; with blue stars the situation is reversed. Magnitudes that are obtained photographically by means of suitable sensitization and filters to correspond with the normal sensitivity of the eye are classed as "photovisual" (m_{pv}). By comparing plates taken in different spectral regions it is then easy to determine the colour of a star, which is often of great importance in the case of variables. (Bolometric magnitudes relate to the total radiation, that is radiation in every wavelength region of the electromagnetic spectrum. As the Earth's atmosphere is only transparent in certain spectral regions, the determination of bolometric magnitudes involves some difficulties.)

The magnitude difference between the photographic and visual magnitudes mentioned, $m_{pg} - m_v$, is called the *International Colour Index* (CI). It is positive for yellow and red stars, and negative for extremely hot blue stars. In practice many other spectral bands are used, for example those designated U (ultraviolet), B (blue), V (visual), G (green), R (red), and I (near infrared). The U-B-V system in particular, and also the R-G-U system, serve to distinguish the colour of a star and the distribution of energy within its spectrum. Following the infrared band I in the medium and far infrared are the further bands J, K, L, M, N, and Q (the latter extending to a wavelength of 22 μm).

Absolute Magnitudes, Spectral and Luminosity Classes

Astrophysics is largely concerned with comparing stars with one another on the basis of their luminosity. Differences in this respect would be directly observable if all stars were at the same distance from us. As this is not the case, we must reduce the apparent magnitudes to an arbitrary and appropriate unit distance. For this the *distance of 10 parsecs* = 32.6 light-years has been chosen. Hence if we read that the Sun has an absolute visual magnitude of $4\overset{M}{.}71$, this implies that at the unit distance of 10 parsecs the Sun would appear to have the same magnitude as a star of $4\overset{m}{.}71$. Absolute magnitudes are indicated by the use of the symbol M. We shall employ a superscript M in a similar manner to the superscript m.

More than 99% of all stars belong to the *main sequence of stellar spectra*, which is classified by means of the letters O, B, A, F, G, K, and M. It is in effect a *temperature scale*. At one end are the hottest stars with surface temperatures of 100 000 K and absorption lines of ionized helium, He II. Class B is characterized by lines of neutral helium, He I, and by the first appearance of hydrogen absorption (the Balmer series), which is predominant in type A, but at F is more and more replaced by the lines of ionized calcium, Ca II. Metallic lines, including those of Fe, appear in G spectra, alongside the very strong Ca II lines and the very much weakened H lines, and are preponderant in class K. With the transition to class M molecular bands appear, particularly that of TiO, and these are predominant in the deep red M-type stars.

In addition there exist various *branches*. Included among the hot stars at the beginning of the sequence are class P for the emission spectra of planetary nebulae, Q for novae, and W for Wolf-Rayet stars with broad emission lines. A branch of the cool stars is formed by carbon stars of class C (formerly designated classes R and N), where the TiO bands of class M are replaced by those of CN (cyanogen), CO (carbon monoxide), and C_2. Another variant is that of class S with bands of zirconium oxide, ZrO.

Table 2. Relationship between spectral class and colour index (*upper portion*: luminosity class V, *lower*: class III)

Spectrum	CI	$B-V$	$U-B$
B0	$- 0^m\!.21$	$- 0^m\!.30$	$- 1^m\!.08$
B5	$- 0.14$	$- 0.18$	$- 0.58$
A0	0.00	$- 0.02$	$- 0.02$
A5	$+ 0.21$	$+ 0.15$	$+ 0.09$
F0	$+ 0.38$	$+ 0.29$	$+ 0.02$
dF5	$+ 0.49$	$+ 0.42$	$- 0.01$
dG0	$+ 0.59$	$+ 0.58$	$+ 0.05$
dG5	$+ 0.74$	$+ 0.68$	$+ 0.21$
dK0	$+ 0.93$	$+ 0.81$	$+ 0.48$
dK5	$+ 1.22$	$+ 1.15$	$+ 1.08$
dM0	$+ 1.52$	$+ 1.40$	$+ 1.23$
gF5	$+ 0.51$	$+ 0.42$	$+ 0.07$
gG0	$+ 0.77$	$+ 0.66$	$+ 0.27$
gG5	$+ 1.00$	$+ 0.81$	$+ 0.50$
gK0	$+ 1.23$	$+ 0.99$	$+ 0.85$
gK5	$+ 1.67$	$+ 1.50$	$+ 1.80$
gM0	$+ 1.86$	$+ 1.54$	$+ 1.84$

Table 2 shows the relationship between spectral class and colour index, being sub-divided from class F5 into dwarf (d) and giant (g) stars.

Colour index is so defined that an A0 star has the same brightness at both visual and photographic wavelengths.

In the first decade of the 20th century it was discovered that red stars form two completely different groups, between which no directly transitional forms exist. These are the giants of very low density, such as α Boo and α Tau, and the numerous, much fainter stars of relatively high density. This discovery was indeed the point of departure for the Hertzsprung-Russell Diagram (H-R Diagram), named after its authors RUSSELL and HERTZSPRUNG. It can be described as follows: If stars are plotted as points on a two-dimensional diagram, with the x-axis indicating the spectral classes from B to M or N, and the y-axis showing absolute magnitude (luminosity), then the stars are by no means evenly distributed. They fall firstly in an approximately diagonal line from upper left to lower right, the *main sequence*, and secondly in a less sharply-bounded region to the right of the upper part of the main sequence, the *giant branch*, and thirdly in very sparsely populated groups both above and below the main sequence and the giant branch. The characteristic groups may be seen in Fig. 1. Instead of the spectral class one can also use effective temperature or colour index as argument; in the latter case the resulting plot is known as a colour-luminosity diagram.

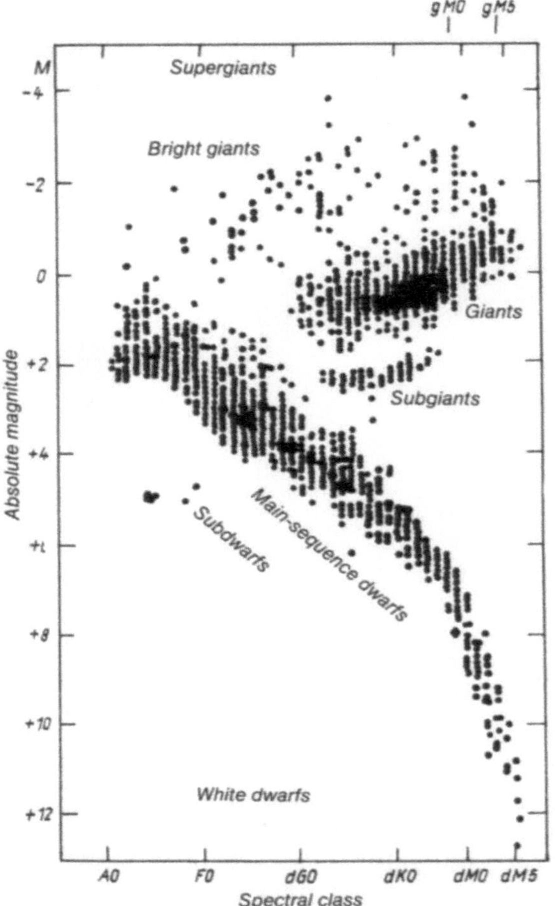

Fig. 1. Hertzsprung-Russell diagram (excluding B stars) based on spectroscopically determined absolute magnitudes (Mt. Wilson)

As both very bright and very faint stars can have the same temperature, particularly amongst red objects, the absolute magnitude (luminosity) is to a great extent a criterion of the effective surface area of the star. Luminosity classes may therefore be devised, and these are indicated by a Roman numeral following the spectral class as follows:

Ia, Ib ... supergiants
IIa, IIb ... bright giants
   III ... normal giants
   IV ... subgiants
   V ... main sequence dwarfs

The luminosity is determined from the presence and sharpness of certain spectral lines. In the physical sense it is the energy flux from the star, which may be given in Watts or erg/s. Luminosity L and absolute bolometric magnitude M are related mathematically by the equation:

$$L = 3.9 \times 10^{26} \times 10^{-0.4(M - 4.72)} \text{ W}.$$

Distances

Astronomers are wrongly supposed to use particularly large, so-called "astronomical" numbers. These large numbers only appear if the layman insists that cosmic distances be expressed in the units which he normally uses, for example in kilometres. Astronomers, however, mostly follow the practice of choosing a unit of measurement appropriate to the quantity being measured. In the Solar System the unit is the mean distance between the Earth and the Sun, 149.598×10^6 km. This is known as the *Astronomical Unit* (a.u.). Thus the distance of Venus from the Sun may be expressed as 0.723 a.u., that of Jupiter as 5.203 a.u., of Saturn 9.546 a.u., and of Neptune 30.09 a.u.

The unit for measurements of the Galaxy is the *parsec* (pc). The name is essentially an abbreviation for "parallax second". It is the distance which corresponds to a parallax of one second of arc, that is to say the distance at which the radius of the Earth's orbit, the astronomical unit, subtends the angle of $1''$.

The light-year is more commonly used. It is the distance covered by light in one year, the velocity of light being 299 793 km/s. In school or general public lectures it always has to be pointed out that the light-year, despite its name, is a measurement of length, similar to "an hour's walk" or "an hour's drive".

The relationship between the various units is as follows:

1 pc = 206 265 a.u. = 3.2617 light-years
1 light-year   = 63 239 a.u.

For distances in our galaxy we often use the next greater unit, the "kiloparsec" (kpc). 1 kpc = 1000 pc.

A still greater unit, the megaparsec (Mpc), is employed for extragalactic distances. 1 Mpc = 1000 kpc = 10^6 pc = 3.26 million light-years.

There is the following relationship between apparent magnitude m, absolute magnitude M, and distance r (in pc):

$$m - M = 5 \times (\log r - 1) + A$$

The parameter $m - M$ is known as the "distance modulus". In this equation A (in mag) represents a possible contribution from interstellar extinction (see below).

The Galaxy, Stellar Populations

The Sun and its planets belong to our general stellar system, also known as the Milky Way System or the Galaxy, which consists of 150–200 thousand million stars. As a whole it has a flattened, lens-like form, similar to that of a *spiral nebula*, with a nucleus rich in stars and spiral arms extending from it. The greatest diameter in the main plane amounts to about 30 kpc = 100 000 light-years. The Solar System is situated close to the main plane, about 8 kpc from the nucleus, which lies in the direction of the constellation of Sagittarius. The visible band of the Milky Way is a good indication of the position of the main plane of the Galaxy. The system so far described is surrounded by an almost spherical, low-density shell, the galactic halo, having a diameter of at least 50 kpc. It includes RR Lyrae stars, amongst others, as well as the globular clusters.

In 1952 the concept of stellar populations was introduced by BAADE. He noted that in the Galaxy the mixture of stars in the spiral arms (called Population I) was of a completely different nature to that found surrounding the nucleus, as observed in the bright star clouds of the Milky Way in Sagittarius (Population II). Young Population I stars are found in the spiral arms, and Extreme Population II stars include those of the nucleus, the globular clusters, and the galactic halo. Nowadays we also recognize certain intermediate classes. Population I includes essentially all the younger stars, and also the regions in which stellar formation is still taking place, whilst Population II consists of the old stars.

In investigations into the structure of the Galaxy special co-ordinates are used: the galactic latitude b of an object on the celestial sphere measures the angular distance, perpendicular to the central line of the Milky Way (the galactic equator). Galactic longitude l is measured along the galactic equator from the direction of the galactic centre. So, for example, the centre of the constellation Scutum has the approximate co-ordinates $l = 23°$, $b = +2°$. In this book we shall use as a basis, and without any special symbols, the revised system used since 1959, in which the position of the galactic equator and centre have been determined by modern methods.

There is yet another spatial co-ordinate system, in which R indicates the distance of an object from the galactic centre, when its position is projected onto the galactic plane, and where z is the perpendicular distance of the object from that plane. In the case of the Sun, for example, $R \approx 8\,\text{kpc}$, $z \approx 15\,\text{pc}$.

Interstellar Material

Apart from stars the Galaxy contains considerable quantities (about 5% by mass) of gas and dust. The dust is present in the luminous reflection nebulae and in the dark clouds that are visible, at least in part, to the naked eye, but it is principally located in the spiral arms so that in most places it is impossible to see through the system in the main plane. For us the galactic nucleus itself is concealed behind dense dark clouds. Another large field of dark clouds lies in the constellations of Taurus and Orion. Hydrogen is present everywhere in interstellar space, but is only at a higher density in the spiral arms, where it is mainly found in conjunction with the dust clouds. As regards the total mass it is the interstellar gas which predominates with 99%, as against 1% for the dust. It consists of about 60% hydrogen, 38% helium, and 2% heavier elements (sodium, calcium, iron, etc.).

In certain regions, generally very limited in extent, the spectral lines from molecular clouds are detectable, and which include combinations that are unstable under terrestrial conditions (e.g. OH).

The gas becomes visible in those places where it is excited by energetic stellar radiation, as in gaseous nebulae like the Orion Nebula. However, even neutral hydrogen H I radiates an emission line at 21 cm in the radio region. Approximately 10% of hydrogen clouds are ionized by stellar radiation and these are known as H II regions.

The interstellar material, in particular the dust, which consists of very fine particles, causes difficulties in all work on stellar distribution, due to the effect of interstellar extinction, and which is very hard to measure. Moreover, the extinction reduces the light of stars behind the dust clouds by an amount which is wavelength-dependent, so that a reddening is produced. The colour index of such stars is thus larger, by an amount that is known as the colour excess, than is appropriate for their spectral class.

Stellar Evolution

Quite early on it was suspected that the H-R Diagram might be an evolutionary pattern. Like LOCKYER, it was assumed that stars formed from nebulae and began their life as red giants, becoming hotter and denser and moving onto the main sequence, and finally becoming extinct as red dwarfs. Evolution would thus have largely taken place along the main sequence towards the cool dwarfs. It was for this reason that the spectral classes of the latter are still known today as "late", in contrast to the "early" classes O to A. At that time, of course, the energy sources of stars were unknown. It was the development of atomic physics, in particular research into nuclear fusion processes, that produced a deeper insight into the possible paths which stellar evolution could follow.

The course of development accepted nowadays, on strong grounds, is the opposite to that just mentioned. A *protostar* is a sphere of gas and dust which is contracting gravitationally – that is to say under the influence of its own gravitational attraction – and which consists predominantly of hydrogen. With the contraction the potential energy is partially converted into heat, that is, into the particles' kinetic energy. As a consequence the interior of the protostar is so strongly heated that finally, at about 10 million K, the first *nuclear reactions* can commence. These are the proton-proton process which produces a deuterium nucleus, and the proton-helium process, where 4 protons are converted into one He^4 nucleus. In both cases a small portion of the mass is "radiated away". It reappears in the form of γ-radiation, which slowly diffuses from the interior towards the outside, by means of repeated absorption and re-emission, and which finally leaves the outer layers in the form of UV, visible, and heat radiation. Those young stars are unstable, which results in their being *variable*. With the end of contraction the star arrives at a stable position on the main sequence. Here it can remain for some millions, or some thousands of millions of years, slowly becoming brighter and consequently moving upwards and to the left on the H-R diagram. The rate of evolution is dependent upon the mass of the actual star under consideration.

A new phase of instability is reached when all the hydrogen is exhausted in the centre of the star, resulting in a purely helium core. The nuclear energy processes are consequently interrupted, and the star contracts. In doing so it heats up to such an extent that a new nuclear fusion reaction becomes possible, converting He^4 into carbon C^{12}, and partly also into oxygen O^{16}, and which liberates a great amount of energy. The star leaves the main sequence and becomes a giant; as such it temporarily passes through the phases in which free oscillations (pulsations) are characteristic. Whilst the helium-carbon process, which requires a temperature of approximately 100 million K, operates in the interior, the hydrogen-helium process continues in the layers further out that are still rich in hydrogen. It should be noted that the time taken for this evolution is strongly dependent upon the initial mass of the star. The more massive a star is, the quicker it evolves, and if in model calculations we increase the mass of a star from 1 to a few solar masses, the duration, in particular the length of time spent on the main sequence, may be shortened by a factor of 100.

Up to the helium-"burning" phase, calculations of stellar models can be carried out in a very satisfactory manner. However, we know very little about developments beyond this point. The *final phase* of stellar evolution is reached when all sources of energy are exhausted. Naturally, the radiation pressure, which together with gas pressure opposed the force of gravity, then disappears. The gas pressure alone is not able to resist the gravitational forces, and the star collapses. A final phase of stellar evolution which is observed is that of the "white dwarf", a star whose material is largely "degenerate". This implies that the nuclear particles are so densely packed that mean densities of more than 10^5 g/cm^3 are produced. If a star has a mass of 1.5 to 3 solar masses towards the end of its evolution, then when all its sources of nuclear energy are exhausted it finally collapses into a "neutron star", whose central density can be greater than 10^{14} g/cm^3.

Both of these end products of stellar evolution are probably to be found in the "eruptive binaries".

1.3 Light-Curves and Periods

Basic Concepts

A variable star is classed as *periodic* when its behaviour, in particular the maxima and minima of its brightness, recur at nearly equal intervals. The word "nearly" should be taken to mean that small irregularities will be accepted. If these should exceed a certain amount, let us say departures of the order of a third of the period, or else if the light-curve has changed with time, so that maxima and minima cannot be determined, even if later the regular changes are re-established, then the star is classed as *semi-regular*. If no periodicity can be discovered then the star is taken as being *irregular*, and this includes those cases in which waves of considerable magnitude occur, but without discernable regularity. A light-curve is obtained by plotting observed magnitude values against time, and by laying a curve through the points obtained. This assumes that an adequate number of observations is available. In the case of rapid, regular variables, one can combine observations from different cycles, by using the known period, and thus obtain a mean light-curve. For this the phase of each observation must be established, that is the time from the nearest preceding maximum or minimum, and which is conveniently expressed as a fraction of the period.

The *symbols* normally employed and their significance are as follows:

$$\left.\begin{array}{l} M_{\mathrm{E}} = \text{Time of Maximum} \\ m_{\mathrm{E}} = \text{Time of Minimum} \end{array}\right\} \text{with Epoch number E } (\mathrm{E} = 0, 1, 2 \ldots)$$
$$P \quad = \text{Period}$$

The following discussion applies to work on those stars where the maxima are more marked than the minima. The opposite case (e.g. with eclipsing stars) is of course identical, if "minimum" is read for "maximum".

The elements (or formula) for the calculation of maxima are:

$$M_{\mathrm{E}} = M_0 + P \times \mathrm{E},$$

where M_0 is the initial maximum. The subscript E is frequently disregarded.

An indication of the asymmetry of the light-curve is given by:

$$\varepsilon = (M - m)/P,$$

where $M - m$ is the duration of the rise; this value is only relevant to curves having a general sine-wave form, and is thus not applicable to eclipsing stars, where D indicates the duration of the eclipse and d the duration of a constant magnitude at minimum.

Amplitude: The difference in brightness between maximum and minimum.

The phase mentioned above, ϕ, is calculated for any time t according to the equation

$$\phi = \frac{t - M_0}{P} - \mathrm{E}(t).$$

Phase Calculations

With the last of the equations given above it is possible to calculate the interval (expressed in units of the period P) between an observation carried out at a time t and the preceding maximum, where the latter is given by the epoch number $\mathrm{E}(t)$. The equation serves to reduce the possibly very numerous observations to a single epoch in order to obtain a *mean light-curve*. This procedure is applicable to regular variables, including eclipsing binaries.

In all these calculations we use the so-called Julian Day, which is progressively numbered, and where the time of observation is expressed as a decimal

fraction of a day, usually to three or more significant places. This is described
in a later section. The reader who is not familiar with the concept might like
to study this first, before proceeding.

If, for example, we wish to calculate the phase of the RR Lyrae star
VX Aps, which has the elements:

$$M = 243\,4239.361 + 0\overset{\mathrm{d}}{.}484578 \times \mathrm{E}$$

at time $t = 243\,4540.550$, then we form the expression

$$\frac{4540.550 - 4239.361}{0.484578} = 621.5491,$$

and thus obtain phase 0.549 (of cycle 621; but the cycle is generally without
significance). This form of calculation – with variable t – may be carried out
very easily on an electronic calculator.

Determination of the Elements

The determination of the elements of a periodic variable, that is of the initial
maximum M_0 and the period P, can take place in the following manner given
a number of observed maxima or minima. We write down the series of dates
in a column, perhaps with a note of their probable reliability, and take the
differences between consecutive values. All these differences are of the nature
of nP (where $n = 1, 2, 3, ...$). In Mira stars we quite frequently find
$n = 1$, that is the period length, appearing directly in the differences. With
rapidly-varying stars the determination of the correct period is more difficult,
but we can succeed if we begin with the smallest differences. In the case of
RR Lyrae stars we know that in the vast majority of objects the period lies
between $0\overset{\mathrm{d}}{.}3$ and $0\overset{\mathrm{d}}{.}6$. The procedure is then to choose two well-established,
small differences, and by means of a slide-rule, to discover which reasonable
value for the period will satisfy them. If this gives too many possible values
then further differences must be used and an effort made to determine which
is the right value. In any case the aim is to obtain the best possible *approximation* of P, which can then be further refined, until to some extent it fits the
whole series of epochs. For M_0 we may accept the first of the dates given. We
thus obtain an approximate formula, with which we may calculate the appropriate value closest to each of the observed epochs. By taking the *observed
minus calculated* $(O - C)$ differences, and plotting these graphically as ordinates against the epoch number, we may use this to make further adjust-
ments. The representation is called an $O - C$ diagram. The procedure is best
clarified by an example (Table 3).

We should note that in the case of a Mira-type star the observed maxima
may be uncertain by 10–20 days, owing not only to errors and any un-
favourable distribution of observations, but also to variation in the form of
the light-curve from cycle to cycle. The 5 differences observed suggest that the
epoch numbers given in column 3 may be accepted. The true period is

Table 3. $(O-C)$ values in the Mira star AU Oph

Obs. Maxima	Differences		E	C_1	$O-C_1$	C_2	$O-C_2$
	(Days)	(Epochs) n					
241 6631			0	6631	0^{d}	6622	$+\ 9^{\mathrm{d}}$
	3292	10					
9923			10	9931	$-\ 8$	9934	-11
	3305	10					
242 3228			20	3231	$-\ 3$	3246	-18
	2677	8					
5905			28	5871	$+34$	5896	$+\ 9$
	310	1					
6215			29	6201	$+14$	6227	-12
	695	2					
6910			31	6861	$+49$	6889	$+21$

therefore likely to be somewhat longer than 300 days. Taking 330^{d} as a first approximation, then we obtain the calculated dates given in column C_1. From the $O-C_1$ values we can see that to obtain the best possible agreement the initial epoch would have to be a few days earlier, and the period increased by 1 or 2 days. The determination of the most probable values can be carried out by calculation, but a simple graphical procedure will suffice here, as shown in Fig. 2. The straight line is laid down by eye, so that the deviations of the observations are minimized as much as possible, and are to some extent equally divided between plus and minus values. In the present case the uncertainty is rather large; however, this is implicit in the nature of the case, chiefly as only 6 observed maxima are available. After this latest adjustment the initial epoch M_0 has an $O-C$ value of $+9^{\mathrm{d}}$, and thus becomes 6622. The slope of the line amounts to 36 units over 31 cycles, i.e. $+1.2$ units per cycle. This is the correction to the period, which thus has a value of $331\overset{\mathrm{d}}{.}2$. The improved elements are then

$$M = 241\ 6622 + 331\overset{\mathrm{d}}{.}2 \times \mathrm{E}$$

From this the calculated values of C_2 and $O-C_2$ can be obtained, and it can be seen from the diagram that the latter are the distances of the points from the adjusting line.

GAUSS established that the line of best fit is most rigorously given by

$$\Sigma\,(p \times (O-C)^2) = \mathrm{Min.,}$$

that is to say that the sum of the squares of the errors should have the smallest possible value, p being a weighting factor. Previously we have simply taken $p=1$. If however, the estimates of maxima are known to be of differing reliability, it is advisable to assign the weight of 2 or 3 to the most accurate, according to their quality, the value of which has to be assessed by the person doing the calculations.

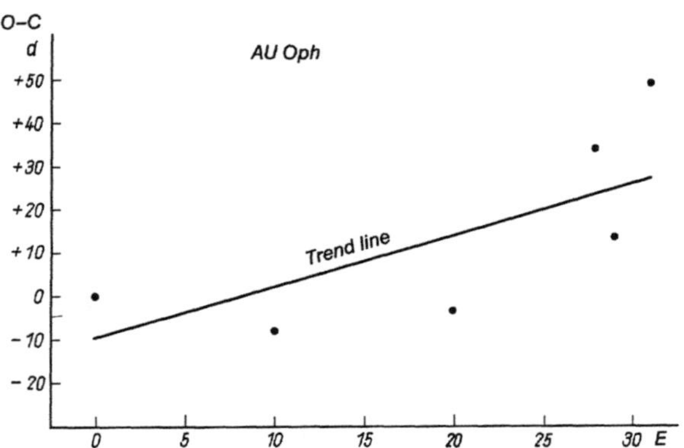

Fig. 2. An example of the graphical method of $O-C$ curve-fitting (see Table 3)

It is immaterial what approximate value of the period we start with, so long as it is reasonably close to the true value. With a long series of observations one may take a well-established maximum $\tilde{M}$ towards the end of the series, and take the preliminary mean value given by

$$P_{\mathrm{m}} = (\tilde{M} - M_0)/\tilde{E}.$$

The graphical method described has the advantage that it is easy to understand. Many observers, however, prefer to use a method where the corrections are calculated mathematically. This is generally carried out by GAUSS' *method of least-squares*, just mentioned, as far as stars with invariable periods are concerned, as we have previously assumed. Details of the procedure may be found in the appropriate literature. Sophisticated electronic calculators generally have a "linear regression" program that can be successfully employed for this, and which offers an easy method of improving periods and initial maxima (or minina).

Variable Periods

A theoretical assumption in the use of the method of least squares is that the $O - C$ values may be treated as random errors. For short series of observations, and for the initial determination of elements we may generally make this assumption. But with longer series of observations the question must nevertheless arise of whether the period is constant. A variation will be shown on the $O - C$ diagram by the fact that the best fit is obtained by the use of a curve rather than a straight line. For a long time is was thought that cyclic changes in period lengths were the rule, so that a *sine term* was added to the linear equation:

$$M = M_0 + P \times E + k \times \sin(\alpha \times E + \phi)$$

where the constant k is the semi-amplitude in days of the amount by which the maxima and minima are early or late when compared with the linear

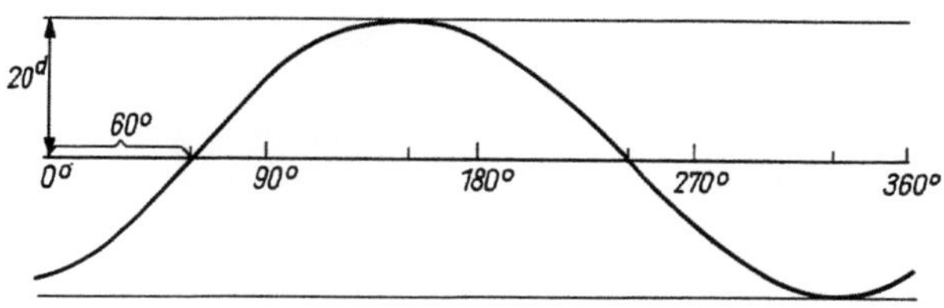

Fig. 3. Graphical representation of the sine term $+20^d \times \sin(3°E + 300°)$

formula, and $2\pi/\alpha = P_1$ is the period of the sine term in units of P; ϕ is a constant, which determines the phase of the sine-wave at the epoch $E = 0$.

If we drop the requirement that E should be an integer, and regard it as steadily increasing, designating it by E', then for any time T the following is valid:

$$T = M_0 + P \times E' + k \times \sin(\alpha \times E' + \phi),$$

and differentiating with respect to E', we obtain the instantaneous value P' for the period:

$$P' = P + k \times \alpha \times \cos(\alpha \times E' + \phi).$$

An example may make this clear. Let the elements be

$$M = M_0 + P \times E + 20^d \times \sin(3° \times E + 300°).$$

The period of the sine term is $360/3 = 120$ epochs, which means that in 120 cycles the value of P runs through all its possible values. With $k = 20^d$ and $\alpha = 2\pi \times 3/360 = 0.052358$ (radians) we obtain $k \times \alpha = \pm 1\overset{d}{.}047$ for the range of the variation in the period.

Figure 3 shows the effect of the sine term $+ 20^d \times \sin(3° \times E + 300°)$ on the position of the maxima, where, as just noted, the length of the longer cycle is $360° \triangleq 120\,P$ and the angle $\phi = 300°$ gives the phase at the epoch $E = 0$.

The use of sine terms has lost a lot of its significance, particularly in relation to long-period variables, in the last few decades, since it was recognized that the period changes are mostly of a different nature. There are also cases of progressive increase or decrease in periods. This type of change can be taken into account by the use of additional terms incorporating *powers of E*:

$$M = M_0 + P \times E + k_1 \times E^2 + k_2 \times E^3,$$

where k_1 and k_2 are positive or negative constants expressed in days.

1.4 Julian Date and the Expression of Dates and Times

As may be realized from the foregoing discussion, in order to carry out any calculations, days must be consecutively numbered. This aim is met by the Julian Period, which was introduced in 1581 by JOSEPH JUSTUS SCALIGER.

In astronomical yearbooks we can find appropriate tables from which the Julian Date may be easily calculated if we know the number for January 0. For example we may require the Julian Date (JD) for 1968 May 29. From a table we find that 1968 January 0 = 2439856 and add 31 + 29 + 31 + 30 + 29 to obtain: 1968 May 29 = 2440006. A statement of the unit used (day) is usually omitted, for example in the elements of variables. Table 4 will enable the JD to be calculated for several years; although in practice one either uses the tables which may be found in various publications, or else calculates tables for oneself, giving the figure for each day of a particular year.

Table 4. Julian Dates for January 0

1961	243	7300	1981	244	4605
1962	243	7665	1982	244	4970
1963	243	8030	1983	244	5335
1964	243	8395	1984	244	5700
1965	243	8761	1985	244	6066
1966	243	9126	1986	244	6431
1967	243	9491	1987	244	6796
1968	243	9856	1988	244	7161
1969	244	0222	1989	244	7527
1970	244	0587	1990	244	7892
1971	244	0952	1991	244	8257
1972	244	1317	1992	244	8622
1973	244	1683	1993	244	8988
1974	244	2048	1994	244	9353
1975	244	2413	1995	244	9718
1976	244	2778	1996	245	0083
1977	244	3144	1997	245	0449
1978	244	3509	1998	245	0814
1979	244	3874	1999	245	1179
1980	244	4239	2000	245	1544

It is important to realize that the Julian Day is reckoned from *noon to noon, Universal Time*, and does not begin at midnight. Universal Time (UT) is the mean solar time at the zero meridian (i.e. Greenwich Mean Time, GMT). Calculation from noon to noon was introduced to avoid a change of date at midnight UT, and as a consequence the Julian Date begins 12 hours later than the corresponding calendar date. 1968 May 29 is JD 2440006; this day lasts from 1968 May 29, 12^h to 1968 May 30, 12^h UT.

The unfamiliar reader will probably take this as being an unnecessary complication. So we should explain how this came about. For a very long time astronomers had reckoned the day as beginning at noon, for the reasons just given, and all almanacs were calculated on this basis. But the editors of the leading British almanac, the Nautical Almanac, decided, chiefly at the insistence of the Admiralty, that from the beginning of 1925 the day would begin at midnight. As for reasons of economy of labour there was an exchange of material between the publishers of almanacs, with also a considerable division of the work, the other publications – the American Ephemeris,

Connaissance des Temps, Berliner Astronomisches Jahrbuch, etc. – were forced to introduce the same changes. An exception was only made in the case of the Julian Date, after international discussion, as a change would have meant a break in continuity, thus defeating the whole object of the arrangement.

The time of day is given in *decimals of a day*, generally to three or more decimal places for short-period variables, and always in Universal Time. Tables are given in the various handbooks for the conversion of hours and minutes into decimals of a day. Table 5 may help to prevent errors.

Table 5. Examples of decimal days

Date	UT	JD	
1968 May 29	12^h	244	0006.000
	20		0006.333
	21		0006.375
	22		0006.417
	23		0006.458
May 30	0		0006.500
	1		0006.542
	2	244	0006.583

With rapid variables, that is those with periods less than a few days, it is necessary to apply corrections to the observed times to allow for the finite speed of light. If we consider a variable that is situated close to the ecliptic, then at opposition the light from the star will reach the Earth about 8 minutes earlier than it will reach the Sun; at quadrature there will be no difference, and towards conjunction a delay will occur. As is well-known, it was by means of this effect that OLAF RÖMER determined the speed of light from the eclipses of Jupiter's satellites. Because of this the observed dates of maxima and minima, or even better, the individual observations, are reduced to the position of the Sun. This so-called *light-time correction* depends upon the star's position relative to the ecliptic; it has a value of zero at the poles of the ecliptic. The formula by which it may be calculated is:

$$\text{light-time correction} = -0^d\!0057 \times R \times \cos\beta \times \cos(L - \lambda)$$

where R is the Earth's radius vector, which only varies very slightly from the value of 1, L is the longitude of the Sun, and λ, β are the ecliptic co-ordinates of the star. The dates corrected for light-time are called *heliocentric dates*, and are identified by the use of the solar symbol $\odot$. Tables of light-time correction have been published by PRAGER (1932).

1.5 Nomenclature of Variable Stars

The division of the heavens into *constellations* has been partly handed down from antiquity, and has been retained by modern science. At the beginning of the 17th century BAYER (1572 to 1625) designated the brightest stars in each constellation by the use of Greek letters, and if these did not suffice, with Roman ones. Numbers for the fainter stars were added later in the star catalogues of HEVELIUS, FLAMSTEED and others, so that on modern charts of the brighter stars one finds designations from very varied sources alongside one another. When variables have BAYER letters, no special names are required; some examples of this are δ Cephei, η Carinae, α Herculis, β Lyrae, α Orionis, and β Persei, amongst others.

A new phase in general stellar designations began with the large Durchmusterung Catalogues, which appeared after 1850. Approximate positions and magnitudes were given for some hundreds of thousands of stars down to about 10^m, and they were identified by numbers within each degree-wide declination zone. These catalogues, the northern and southern Bonner Durchmusterungen, the Cordoba Durchmusterung and the Cape Photographic Durchmusterung – the word "Durchmusterung" has come to be international – and lastly the Henry Draper Catalogue of Stellar Spectra, published by Harvard Observatory, are of the greatest importance for research into variable stars, as we shall see when we come to the description of observations and their analysis. In discussing a variable it is essential to state whether it is given in one of these catalogues. However for naming variables these catalogue numbers were out of the question.

The production of the Bonner Durchmusterung led to many new discoveries and so its initiator, ARGELANDER, found himself forced to introduce a *special system of nomenclature*. The lower-case letters and the first part of the alphabet in capital letters had already been allocated, but capitals towards the end of the alphabet remained unused. As ARGELANDER believed variability to be a rare phenomenon, he thought that no-one would ever discover more than 9 variables in any constellation. He therefore recommended that the variables in each constellation should be designated by the letters R, S, T, U, V, W, X, Y, and Z, together with the genitive of the Latin name of the constellation. This system has continued in use to this day, although it soon had to be extended. In Table 6 we give the constellation names, together with the Latin genitives, and the IAU-approved abbreviations. Examples of typical designations are thus R Aquilae, U Cephei, and X Leonis Minoris, abbreviated R Aql, U Cep, X LMi.

It was soon obvious that the 9 letters provided by ARGELANDER would be by no means sufficient, particularly when after the introduction of photography the total number of new discoveries rose rapidly. So the double letters RR, RS, RT, ..., to ZZ were used. When this potential was exhausted, a new series was started with AA...AZ, BB...BZ, etc. (omitting J), which ended with QQ...QZ, as the combinations from RR onwards had already been used. Inversions, such as BA, are not permitted. Consequently there were 334 letter designations available for each constellation.

In the meantime the Dutch astronomer NIJLAND had proposed the adoption of a single system of nomenclature. Under each constellation, variables would be designated by V 1, V 2, etc. R Aql would thus have been called V 1 Aql, and RR have become V 10. Despite the suitability of this proposal,

Table 6. The constellations

Latin name	Genitive	Abbreviation	English name
Andromeda	Andromedae	And	Andromeda
Antlia	Antliae	Ant	Air Pump
Apus	Apodis	Aps	Bird of Paradise
Aquarius	Aquarii	Aqr	Water Bearer
Aquila	Aquilae	Aql	Eagle
Ara	Arae	Ara	Altar
Aries	Arietis	Ari	Ram
Auriga	Aurigae	Aur	Charioteer
Boötes	Boötis	Boo	Herdsman
Caelum	Caeli	Cae	Chisel
Camelopardalis	Camelopardalis	Cam	Giraffe
Cancer	Cancri	Cnc	Crab
Canes Venatici	Canum Venaticorum	CVn	Hunting Dogs
Canis Major	Canis Majoris	CMa	Big Dog
Canis Minor	Canis Minoris	CMi	Little Dog
Capricornus	Capricorni	Cap	Goat
Carina	Carinae	Car	Keel
Cassiopeia	Cassiopeiae	Cas	Cassiopeia
Centaurus	Centauri	Cen	Centaur
Cepheus	Cephei	Cep	Cepheus
Cetus	Ceti	Cet	Whale
Chamaeleon	Chamaeleontis	Cha	Chameleon
Circinus	Circini	Cir	Compass
Columba	Columbae	Col	Dove
Coma Berenices	Comae Berenices	Com	Berenice's Hair
Corona Austrina	Coronae Austrinae	CrA	Southern Crown
Corona Borealis	Coronae Borealis	CrB	Northern Crown
Corvus	Corvi	Crv	Crow
Crater	Crateris	Crt	Cup
Crux	Crucis	Cru	Cross
Cygnus	Cygni	Cyg	Swan
Delphinus	Delphini	Del	Dolphin
Dorado	Doradus	Dor	Dorado
Draco	Draconis	Dra	Dragon
Equuleus	Equulei	Equ	Little Horse
Eridanus	Eridani	Eri	River Eridanus
Fornax	Fornacis	For	Furnace
Gemini	Geminorum	Gem	Twins
Grus	Gruis	Gru	Crane
Hercules	Herculis	Her	Hercules
Horologium	Horologii	Hor	Clock
Hydra	Hydrae	Hya	Hydra (water monster)
Hydrus	Hydri	Hyi	Sea serpent
Indus	Indi	Ind	Indian
Lacerta	Lacertae	Lac	Lizard
Leo	Leonis	Leo	Lion
Leo Minor	Leonis Minoris	LMi	Little Lion
Lepus	Leporis	Lep	Hare
Libra	Librae	Lib	Scales
Lupus	Lupi	Lup	Wolf
Lynx	Lyncis	Lyn	Lynx
Lyra	Lyrae	Lyr	Lyre
Mensa	Mensae	Men	Table (Mountain)
Microscopium	Microscopii	Mic	Microscope

Table 6. (continued)

Latin name	Genitive	Abbreviation	English name
Monoceros	Monocerotis	Mon	Unicorn
Musca	Muscae	Mus	Fly
Norma	Normae	Nor	Level (square)
Octans	Octantis	Oct	Octant
Ophiuchus	Ophiuchi	Oph	Ophiuchus (serpent bearer)
Orion	Orionis	Ori	Orion
Pavo	Pavonis	Pav	Peacock
Pegasus	Pegasi	Peg	Pegasus (winged horse)
Perseus	Persei	Per	Perseus
Phoenix	Phoenicis	Phe	Phoenix
Pictor	Pictoris	Pic	Easel
Pisces	Piscium	Psc	Fishes
Piscis Austrinus	Piscis Austrini	PsA	Southern Fish
Puppis	Puppis	Pup	Ship's Stern
Pyxis	Pyxidis	Pyx	Ship's Compass
Reticulum	Reticuli	Ret	Net
Sagitta	Sagittae	Sge	Arrow
Sagittarius	Sagittarii	Sgr	Archer
Scorpius	Scorpii	Sco	Scorpion
Sculptor	Sculptoris	Scl	Sculptor
Scutum	Scuti	Sct	Shield
Serpens	Serpentis	Ser	Serpent Serpens caput (serpent's head) Serpens cauda (serpent's tail)
Sextans	Sextantis	Sex	Sextant
Taurus	Tauri	Tau	Bull
Telescopium	Telescopii	Tel	Telescope
Triangulum	Trianguli	Tri	Triangle
Triangulum Australe	Trianguli Australis	TrA	Southern Triangle
Tucana	Tucanae	Tuc	Toucan
Ursa Major	Ursae Majoris	UMa	Great Bear
Ursa Minor	Ursae Minoris	UMi	Little Bear
Vela	Velorum	Vel	Ship's Sails
Virgo	Virginis	Vir	Virgin
Volans	Volantis	Vol	Flying Fish
Vulpecula	Vulpeculae	Vul	Little Fox

there were thought to be good grounds for rejecting it. First, the letter designations had been used for decades in the literature, and secondly, specific classes of variability had become linked with certain designations, such as U Gem, SS Cyg, and RR Lyr, so that it did not seem sensible to alter these to numbers. However, a concession was made in favour of NIJLAND's proposal: when in any constellation – Sagittarius was the first – QZ was reached, and the letter designations were thus exhausted, the next variables were designated V 335, V 336, etc., so that the last number at any time showed the number of variables known in that constellation up to that date, with the exception of some bright variables with their own designations. In most of the Milky Way constellations the figure of 334 has now been exceeded. Some examples of the highest totals obtaining at present (KHOLOPOV et al. 1981) may be mentioned: V 1359 Aql, V 828 Cen, V 1761 Cyg, V 2127 Oph, V 1084 Ori, V 927 Sco, and V 4069 Sgr.

As variables are very numerous in the star clouds of the Milky Way, an exact delineation of *constellation boundaries* was essential. The Astronomische Gesellschaft, which, up to the first World War, was recognized as an international organization, had already decided in 1867 that all work relating to constellation boundaries should be based upon the Uranometria Nova, an atlas published by ARGELANDER in 1843. In the case of variables, however, there were considerable difficulties. First of all the boundaries in the Uranometria Nova were irregularly curved lines, and at the very small scale of the charts there were a number of doubtful cases. Secondly it was discovered that different copies of the charts did not agree exactly as the stars and constellation boundaries had been printed from various copper plates. In consequence the International Astronomical Union decided upon a new definition of the boundaries, with the specification, firstly that these should all be circles of Right Ascension and Declination, and secondly that they should be so placed that alterations to the existing classification of variables by constellation would be avoided. For the southern sky the first requirement had already been fulfilled by the Uranometria Argentina published by K. GOULD in 1877, which is for the equinox of 1875.0. The task was therefore to extend this system to the northern sky. The result is the Délimitation Scientifique des Constellations, Cambridge 1930, the work of DELPORTE at Uccle near Brussels. Tables are given with the charts listing all boundaries by Right Ascension and Declination. For the sake of uniformity the equinox of 1875.0 had to be used, and this point must be borne in mind in using the work.

A partial departure is the naming of variable stars in globular clusters and in the two Magellanic Clouds.

The nomenclature described here is applied to stars when the fact of their variability is confirmed, and something is known about its nature. Up to the time of the second World War the editors of the Astronomische Nachrichten undertook the allocation of a preliminary designation to all stars shown to be variable. This took the form of a *running number* in conjunction with the year, starting anew each year, for example 377.1943 Sge. Independently of this, some institutions, where systematic searches for variables were undertaken, introduced their own schemes, beginning with Harvard Observatory, which chose, from the start of the work, to give the designation HV with a running number, and without naming the constellation. Since the provision of preliminary designations has not been resumed by the Astronomische Nachrichten, and no successor has been found, a number of discoverers have adopted a similar procedure. The matter will be touched upon again in the section on the discovery of variable stars.

2. Pulsating Variables

2.1 Classical Pulsating Stars

2.1.1 Historical Summary and Classification

As mentioned earlier, the first two variables of the type discussed here were discovered in 1784 by PIGOTT (η Aql) and GOODRICKE (δ Cep). At that time no-one could have had the slightest suspicion of the significance which this type of variable would later assume for astronomy as a whole. Certainly the great regularity of the variations in brightness was recognized very early on, but more than a century had to pass before the discoveries were made that the radial velocities (BELOPOLSKY), and, clearly the effective temperature (K. SCHWARZSCHILD), changed in synchronism with the variations in brightness. Although radial pulsations of a homogeneous star had been discussed theoretically by A. RITTER as early as 1879, a rather artificial, double-star theory was still being held up to the time of the work by SHAPLEY (1914) and EDDINGTON (1918), also mentioned in the introductory chapter. One must remember that the famous Period/Luminosity relation for δ Cephei stars had already been discovered by LEAVITT in 1912. This relationship, through which these stars became the most accurate and most certain means of distance determination for nearby extragalactic objects, could thus be applied even before a proper theory for its interpretation was in existence.

Attempts were also made at that time, albeit without success, to explain the changes in brightness by means of oscillations in the shape of the stars, that is, by a mechanism which we encounter again in modern theories of non-radial pulsations (Sect. 2.3).

Unfortunately the way in which these stars are classified is not completely consistent. In English-language usage the expression "Cepheid" is frequently used to embrace all the stellar types which we shall discuss in this current section (2.1), differentiation being achieved by the use of explanatory terms, such as "Population I Cepheid". The selfsame type may also be found referred to as "δ Cephei stars" or even more explicitly as "classical δ Cephei stars". As has happened in this case, observers tend to quote the names of prototype objects for the purposes of classification. In this book we avoid the use of the term "Cepheid" as it has already been applied to the members of a meteor shower.

2.1.2 δ Cephei and W Virginis Stars

Definitions, Statistics and Light-Curves

The stars in these two groups (Figs. 4 and 5) are characterized by periodic light variations with periods lying between 1 and about 70 days, periods below 2 days and over 50 days being rare. They are occasionally described as "long-period δ Cephei stars" in contrast to the "short-period" objects – the RR Lyrae stars (Sect. 2.1.3). The amplitudes are fairly large, mostly lying between 1 and 2 mag. Values as low as 0.1 mag are found, even if only rarely. These well-defined classes are distinguished amongst the physical variables as being those that exhibit the least irregularities in period lengths and in the shape of the light-curves. Moreover, from a determination of the period the *Period/Luminosity relationship* allows the absolute magnitudes to be obtained. Note that in general the determination of periods poses no problems, assuming that a suitable number of either photographic plates, taken over an appropriate period of time, or of other photometric measurements is available.

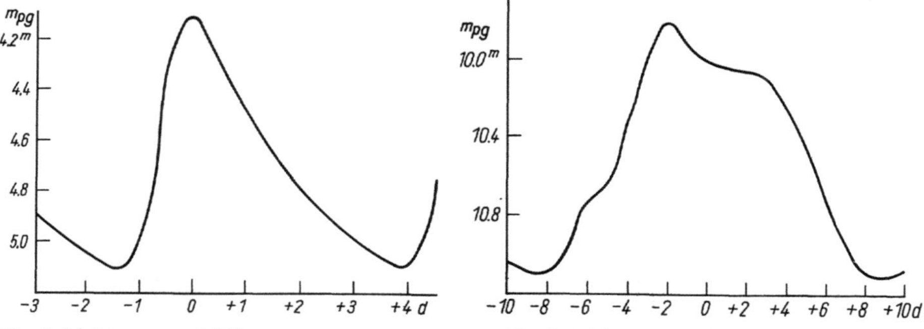

Fig. 4. Light-curve of δ Cep   Fig. 5. Light-curve of W Vir

It is a fortunate coincidence that both the Magellanic Clouds are rich in δ Cephei stars. As they are systems which are distant, and not very large, it is possible to consider all the individual stars as being at the same distance from us, so that their apparent magnitudes should only have to be corrected by the distance modulus $m - M$, to obtain their absolute magnitudes. The relationship was discovered by Miss LEAVITT (1912) from 25 stars in the Small Magellanic Cloud. From the distances of the few bright δ Cephei stars (δ Cep, η Aql, and a few others – the fundamental work by SHAPLEY only incorporated 11 stars) their absolute magnitudes were determined, thus calibrating the Period/Luminosity relationship, which can now in principle be used for reliable distance determinations for galactic and extragalactic objects. However, the calibration was subject to many inaccuracies and errors. Thus it was necessary, for example, for the values of extragalactic distances to be doubled in 1952, as before then the absolute magnitudes of the δ Cephei stars had been set too low. One notes that by combining radial velocity and proper motion determinations, SHAPLEY arrived at a mean parallax for the 11 stars of only 0''.0034; the small size of this value is a sign of its uncertainty. Subsequently the Period/Luminosity relation (Fig. 6) has been precisely established following many different studies, and has been adjusted by the incorporation of further parameters (chemical composition and effective temperature) as a result of the increasing accuracy of observations. In the fundamental work by SANDAGE and TAMMANN (e.g. 1969) the 13 galactic δ Cephei stars listed in Table 7 were used for the calibration. Their absolute magnitudes are well known from their situation in clusters and associations, or in a dust cloud of known distance (SU Cas) (RACINE 1968).

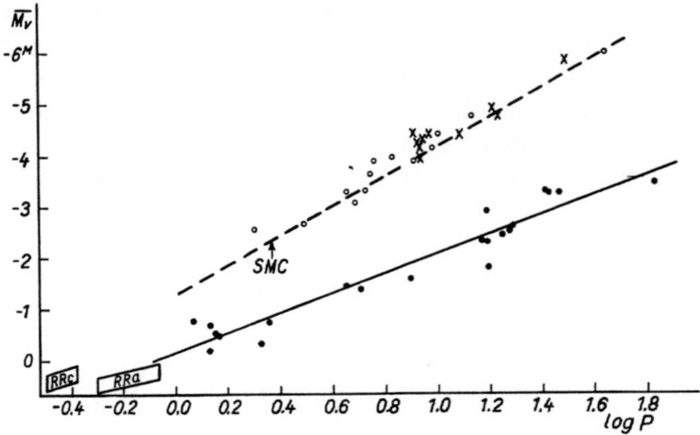

Fig. 6. Period-Luminosity relationship of δ Cephei stars (*above*) and W Virginis stars (*below*). ○ variables in galactic clusters, × variables in the Large Magellanic Cloud. The broken line shows the relationship found in the Small Magellanic Cloud, while the lower line is an average from the points plotted. The position of RR Lyrae stars in the globular cluster ω Centauri is shown schematically for comparison purposes. After DICKENS and CAREY (1967), with additions from Table 7 (where $\overline{M}_v$ is defined)

Table 7. The δ Cephei stars used for the luminosity calibration

Star	Cluster	P	$\overline{M}_v$
SU Cas	–	$1^d.95$	$- 2^M.54$
EV Sct	NGC 6664	3.09	− 2.62
CE Cas b	NGC 7790	4.48	− 3.205
CF Cas	NGC 7790	4.87	− 3.075
CE Cas a	NGC 7790	5.14	− 3.275
UY Per	h, χ Per	5.36	− 3.54
VY Per	h, χ Per	5.53	− 3.91
U Sgr	M 25	6.74	− 3.93
DL Cas	NGC 129	8.00	− 3.84
S Nor	NGC 6087	9.75	− 4.03
VX Per	h, χ Per	10.89	− 4.34
SZ Cas	h, χ Per	13.62	− 4.71
RS Pup	Pup III	41.38	− 5.95

($\overline{M}_v$ is the mean magnitude obtained from the average intensity of the luminosity curve)

In the General Catalogue of Variable Stars (GCVS) by KUKARKIN et al. (1969, 1971, 1974, 1976), 396 δ Cephei stars are listed whose classification can be regarded as definite, and which, with a few exceptions, belong to the Milky Way system. Those exceptions are primarily a few objects in the outermost regions of the Small Magellanic Cloud, which physically belong to it (e.g. GESSNER 1981a). As δ Cephei stars, being typical members of Population I, are mostly located within a small distance of the galactic plane, their discovery is hindered, here and there, by interstellar dust clouds, despite their high luminosity.

Alongside this group of the *classical* δ Cephei stars, which belongs to the true Population I, there exists the second group, the *W Virginis stars*, which are distinguished by their amplitudes, light-curves, spectral characteristics, and radial velocity curves. These are often referred to in the literature as "Population II Cepheids" despite the fact that it has been known for a long time (e.g. WOOLLEY 1966 and RICHTER 1967a) that a majority of W Virginis stars belong to the disk population. From the catalogue mentioned above we find 102 stars belonging to this group. It should be mentioned that in the case of approximately 270 further cases (partly indicated by question marks in the catalogue), no distinction can be made as to whether they belong to one group or the other, mainly owing to the lack of suitable observations.

A distinct Period/Luminosity relation exists for the W Virginis stars; its slope is less than that of the δ Cephei stars, and the absolute magnitudes are on average more than 1 mag fainter (see Fig. 6). The determination is most successful in the case of globular clusters, and, as for the δ Cephei stars, the question of whether the Period/Luminosity relation for W Virginis variables is consistent in different stellar systems (or clusters) cannot yet be answered precisely.

The distribution of stars into the various *period ranges* that is shown in Table 8 is taken from a work by PETIT (1960). He included 525 cases. This figure is slightly larger than the total of the two numbers that we have just quoted for 1976. This is due to a fair number of objects that were previously falsely classified, and which have since been excluded. It also shows that in any case later searches did not add a very considerable number of new variables of these two types. This latter fact is a result of the stars' high luminosity; they had been fairly exhaustively discovered out to great distances, at least as long as they were not situated behind large, dark cloud complexes.

Table 8 covers a range of P from $\approx 1^{\mathrm{d}}-100^{\mathrm{d}}$, and the number of stars at the extremes is so small that those few cases which are erroneously included do not alter the statistics. A maximum for the δ Cephei stars can be recognized at $\log P = 0.75$, $P = 5^{\mathrm{d}}.6$, with a secondary maximum at $\log P = 1.1$, $P = 12^{\mathrm{d}}-13^{\mathrm{d}}$, at a point where the W Virginis stars exhibit their primary maximum, whilst these latter stars show a flat secondary peak at $P = 2^{\mathrm{d}}.2$. The δ Cephei star with the shortest period currently known in our galaxy is V 473 Lyr (BREGER 1981) with $P = 1^{\mathrm{d}}.49$, although it does show certain peculiarities. Among the W Virginis stars BX Del appears to show one of the shortest periods ($1^{\mathrm{d}}.09$); see the comprehensive description of the

Table 8. Period distribution of galactic δ Cephei and W Virginis stars

$\log P$	δ Cephei type	W Virginis type
0.0	0	2
0.05	0	3
0.15	1	6
0.25	3	5
0.35	6	10
0.45	22	4
0.55	42	5
0.65	63	1
0.75	72	3
0.85	58	1
0.95	25	4
1.05	29	8
1.15	30	25
1.25	18	16
1.35	13	12
1.45	6	11
1.55	6	6
1.65	3	2
1.75	0	2
1.85	1	1
1.95	0	0
Total	398	127

($\log P$ represents the mean of 0.1-mag-wide $\log P$ intervals; P is in days)

photometric behaviour – the form and variability of the light-curve – by FUHRMANN (1982) in this connection. In this lower-period range there are also, for example V 553 Cen (a carbon star with $P = 2^{d}06$), RT TrA ($1^{d}95$), SW Tau ($1^{d}58$) and BL Her ($1^{d}31$). This last star is frequently regarded as the prototype of a small sub-group of W Virginis stars, in particular when stars of similar period-length in globular clusters are under discussion (Chap. 5).

In an investigation into a "photometric classification of pulsating variables with periods between one and three days", DIETHELM (1981, 1983) obtained, from material which unfortunately only consisted of 28 galactic pulsating stars, the following photometric criteria for the types mentioned:

RR Lyrae stars: Smooth light-curve in V, only a small hump prior to the beginning of the rise, which is steep ($\leq 0.26\,P$). The amplitude in B is greater than in U.

W Virginis stars: Hump on the rise around $0.2\,P$ before maximum.

BL Herculis stars: Hump on the decline about $0.25\,P$ ($\pm 0.1\,P$) after maximum.

δ Cephei stars: Smooth light-curve with gradual rise (duration $\approx 0.3\,P$); $P > 2^{d}3$.

Not included in this investigation were the variables with light-curves approximately similar to sine-waves (duration of rise $\gtrsim 0.4\,P$), – e.g. BP Cir:

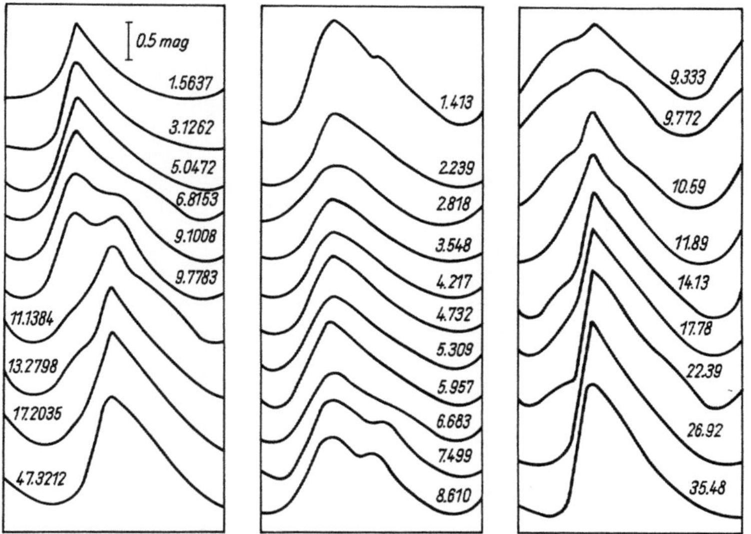

Fig. 7. Relationship between period-length and shape of the light-curve in δ Cephei stars. *Left*: Magellanic Clouds; *centre* and *right*: the Galaxy (after PAYNE-GAPOSCHKIN)

$P = 2^{d}40$, amplitude $= 0.33$ mag in V, rise $\approx 0.43\,P$, see KURTZ 1979 – and the double-period stars (see below).

It must be stressed that in this period range much observational work still remains to be undertaken.

The *amplitudes* are dependent upon the periods. At periods of 2^{d} to 3^{d} the visual amplitude is about 0.5 mag, and the photographic (in the B region), 1.0 mag; at $P = 40^{d}$ to 50^{d} the values are 1.2 mag (vis.) and 1.7 mag (pg.). In the ultraviolet the amplitudes are even greater, in the case of δ Cephei about 3.4 times the visual value.

HERTZSPRUNG (1926) referred to *systematic behaviour in the light-curves* (Fig. 7) of the δ Cephei stars. Among the shortest periods the curves are smooth. Between $P = 6^{d}5$ and 9^{d} one often observes a wave on the descending branch, the phase of which decreases with increasing period. At $P = 10^{d}$ the hump coincides with the maximum, and at greater periods appears on the rise, particularly on the lower portion between 14^{d} and 15^{d}. When $P > 15^{d}$ the curve is again generally smooth. Amongst the bright representatives, η Aql ($P = 7^{d}177$) has a significant wave on the decline.

An extensive and consistent series of photoelectric observations in 5 colours of about 150 δ Cephei stars in the southern Milky Way was carried out by PEL (1976), and discussed and compared in subsequent publications with model parameters obtained theoretically.

KWEE (1968) studied the light-curves of general field W Virginis stars. He found that secondary waves and humps are widely encountered, in particular at periods between 1 and 3 days, and also on the decline between 13 and 19 days (Fig. 8). In general, sudden period changes appeared to be much more

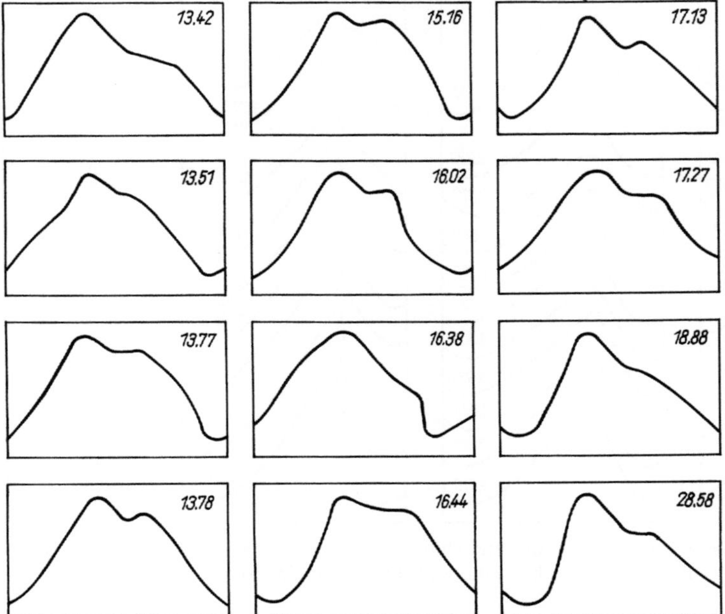

Fig. 8. Light-curves of 12 galactic W Virginis stars of differing period-length (after
PAYNE-GAPOSCHKIN)

frequently encountered among W Virginis stars than among δ Cephei objects.

A completely abnormal, and up to now unique, case that attracted astronomers' attention in 1966, is that of the W Virginis star RU Cam, where
$P = 22\overset{d}{.}26$. At the beginning of 1966 FERNIE and DEMERS (1966) announced
that RU Cam had seemingly ceased its variations. From the Sonneberg
Observatory's photographic sky patrol plates, HUTH (1966) found the surprising behaviour shown in Fig. 9. At the same time that author investigated
the long-term changes in the period, which varied between $22\overset{d}{.}055$ and
$22\overset{d}{.}187$. The spectrum is variable, K0 at maximum and R2 at minimum;
consequently we are dealing with a carbon star, and the corresponding
spectral classification is CO_1 to $C3_2e$. The existence of carbon at the surface
can be explained for such old stars, where in the interior some He has already
been converted into C, by the mass loss of the originally outermost layers.
However, the temporary suppression of the pulsations – the amplitude increased again from 1967 onwards – for the time being remains a mystery.
According to observations by WALLERSTEIN and CRAMPTON (1967) the variation in the radial velocity also came to a standstill.

Physical Properties

The δ Cephei stars are *supergiants*, in general of luminosity class Ib; on the
Hertzsprung-Russell Diagram they occupy a narrow, steep strip, slightly

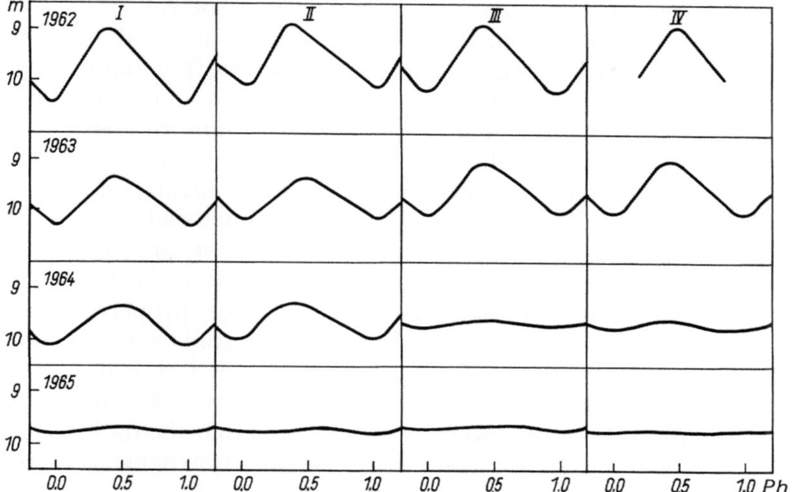

Fig. 9. Three-monthly mean photographic light-curves of RU Cam, after HUTH (1966)

inclined towards the right, the limits of which are approximately absolute magnitudes $M_v = -2$ and -6, and spectral classes F5 and K0. Longer periods are associated with later spectral class and greater colour index. The spectrum and colour of the objects not only show this relationship with period from star to star, but also exhibit a strong phase relationship in individual stars, so that the colour shifts towards the red as the star becomes fainter. Consequently as mentioned above, the photographic amplitudes are, in most cases, considerably greater than the visual ones. We give some examples of the variations in spectral class in Table 9.

Table 9. Spectral variations in some δ Cephei stars

Star	Period	Spectral class
SU Cas	1^{d}94	F5–F7
δ Cep	5.37	F5–G2
η Aql	7.18	F6.5–G2
ζ Gem	10.15	F7–G3
X Cyg	16.38	F7–G8
T Mon	27.01	F7–K1

The mean *radii* of δ Cephei stars are of the order of 5×10^6 to 100×10^6 km, approximately 10 to 150 solar radii. The largest star of this type would thus, if placed at the Sun's position, extend as far as the orbit of Venus. As might be expected, the mean radii show a close correlation with the period, and this may be expressed by the relation $R = 4 \times 10^6 P$ (with units of km and day). For the majority of the brighter δ Cephei stars, radial velocity curves are available in addition to the light-curves. By integrating the radial velocity curve, one obtains the curve of the changes of radius, ΔR, in the form $R - R_{\text{Min}}$, on a linear scale. Alternatively, from the spectral variations one can derive the changes in the effective temperature and thus, by means of the Stefan-Boltzmann Law, the changes in surface luminosity. From the latter, and the apparent magnitude, a curve of the radius changes can be derived in the form R/R_{Min}. The radius R, easily found by combining the two curves, must in principle, when linked with the temperature, give the value of the luminosity, which thus follows quite independently of the distance, apparent brightness or interstellar extinction. Names such as those of BAADE, W. BECKER, BOTTLINGER, VAN HOOF and WESSELINK are associated with this method of treatment, which first served to prove the pulsation theory, and later allowed the determination of the physical properties connected by the relationships just mentioned. Table 10 gives some values for the maximum variation in the radius of δ Cephei stars (RR Lyrae has been included as well); in this particular discussion the subscripts Min and Max apply to the extreme values of the radius. It will be seen that the radius (and diameter) of a star is about 10% greater at its maximum expansion phase than at greatest contraction.

Table 10. Radius variation in pulsating stars

Star	P	$R_{\text{Max}}/R_{\text{Min}}$
RR Lyr	$0\overset{d}{.}57$	1.072
T Vul	4.44	1.152
δ Cep	5.37	1.119
η Aql	7.18	1.091
ζ Gem	10.15	1.085

Figure 10 shows in somewhat schematic form, the variations of some of the properties of δ Cephei stars, as a function of the phase. The light- and radial velocity curves of δ Cephei stars are almost mirror-images of one another. A systematic time difference only appears to exist in the occurrence of the secondary waves of the two curves, but it seems that this can be explained theoretically.

The light changes of pulsating stars can be attributed to two opposing effects. At the stage of maximum contraction the visible surface is reduced – in the ratio of 0.81:1 for a 10% change in radius. If the temperature were to remain constant the intensity of the star's light would therefore be reduced by 19%. However, because of the gas laws, the contraction produces an increase in temperature, by which the reduction in surface area is greatly over-compensated, as in accordance with the Stefan-Boltzmann Law the total radiant energy output (or bolometric luminosity) is proportional to the 4th power of the absolute temperature. We might therefore expect brightness maximum to correspond to the phase of greatest contraction, that is, of minimum radius. This is not quite the case, instead maximum light is about $0.13\,P$ later than the simple theory suggests, and as just mentioned, coincides with the greatest velocity of contraction (see Fig. 10). Attempts have been

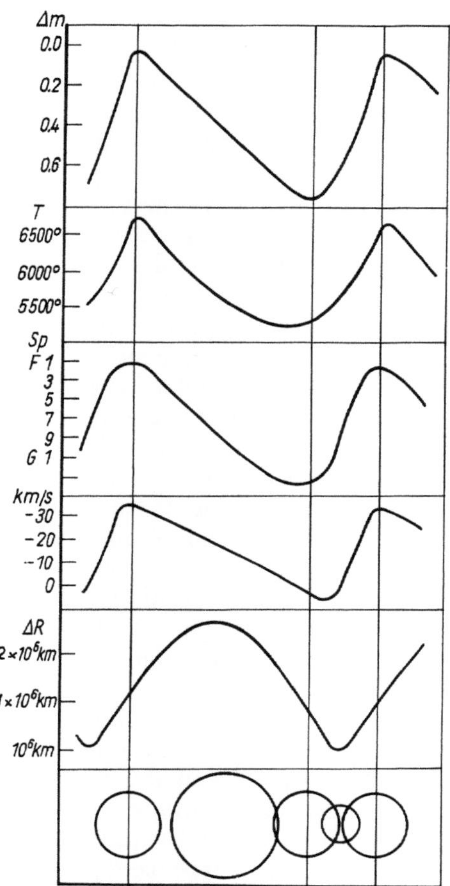

Fig. 10. Variation with time of some properties of δ Cep (from *top*: luminosity, effective temperature, spectral class, radial velocity, radius, and surface area of the star)

made, on various theoretical grounds, to solve this "phase lag problem"; however, it is not seen by the theoreticians as being of great importance.

The *masses* of δ Cephei stars lie between 3 and 16 solar masses, depending upon their periods. The order of magnitude is well established by theoretical model calculations, but various methods give differences, by up to a factor of 3, in the masses obtained. An independent, direct mass determination was possible in the case of BM Cas, an eclipsing system in which one component pulsates. According to THIESSEN (1956) the system consists of a supergiant, of spectral class A5 and absolute magnitude $- 8^{M}.4$ – consequently one of the most luminous stars known – and a δ Cephei star with $P = 27^{d}$, and $M_{bol} = - 6.0$. A mass of 14.3 solar masses was calculated for the latter. This value is not implausible for the period quoted, but it should not be accorded too high a weight as it is now uncertain whether we are really dealing with a δ Cephei star without any abnormalities.

The masses of the W Virginis stars are significantly less and have been quoted as being about 0.55 solar masses (BÖHM-VITENSE et al. 1974).

Origin of Pulsations and Evolutionary State

It is definitely established that it is primarily the outer layers of the star which pulsate. The fundamental cause was first recognized about 1960, after many authors (among them EDDINGTON, ZHEVAKIN and ROSSELAND) had performed important preliminary work. It is not, as was at first thought, that once pulsation has been initiated it continues to regulate the energy production in the deep interior of the star for evermore – nuclear processes being, as is well-known, very temperature-sensitive. It is rather that the *absorption characteristics* of the outer layers maintain the pulsation. The process has sometimes been called the *Kappa Mechanism*, from the Greek letter κ, used as a symbol for the coefficient of absorption of the energy flowing from the stellar interior. A very clear description of the processes involved is given by KIPPENHAHN and WEIGERT (1964, 1965). In the main it is the zone where helium is doubly ionized, lying a few hundred thousand km below the stellar surface, which is responsible for driving the pulsation. Within this zone, helium is increasingly ionized as the temperature rises towards the interior, eventually becoming completely ionized. With a slight compression, such as can always occur through some small disturbance, that is with an increase in pressure and temperature, there is a rise in the absorption of radiation within this zone. This additional energy over-compensates for the normal heat loss found in stable stars (which has a damping effect), and the expansion of the affected layers of gas overshoots the original rest position. This expansion now produces the opposite effect to that just described, and an undamped oscillation results. In principle this can persist as long as the general evolutionary process maintains the dimensions and properties of the excitation zone within the star. Various criteria can be used to determine whether a particular, theoretical, stellar model gives rise to pulsational variability.

The evolutionary state of the δ Cephei stars – and more recently, of the W Virginis and RR Lyrae stars – is fairly well-understood from extensive series of *model calculations*. They are objects in whose centres hydrogen has been completely transformed into helium. Differing interpretations merely vary in the extent to which the triple-alpha process has already enriched the stellar core with carbon. HOFMEISTER et al. (1964) worked from the hypothesis that energy production is primarily taking place at the outermost layer of the "burnt-out" carbon core through further transformation of helium. In their classic work on objects of 7 solar masses these authors discovered, in the models considered, the multiple changes which occur in the overall evolutionary expansion and contraction of the outer layers. These changes have now been well-investigated and are seen on the Hertzsprung-Russell Diagram as multiple reversals in the evolutionary track. Since that time similar evolutionary calculations have been carried out by many authors for other stellar masses – such names as DEMARQUE, IBEN and PACZYNSKI should be mentioned here – and it is evident that stars in a wide range of neighbouring masses behave in a similar way. Nevertheless IBEN (1974) showed in a review of the theoretical position that δ Cephei stars are undoubtedly objects where energy production is taking place within a still-existing helium core. All calculations are in very good agreement with the observational evidence that it is at this evolutionary stage that the tendency to pulsations appears, and that the stars are only found within the region of the H-R Diagram known as the *instability strip*. Massive stars (the δ Cephei stars) reach this region considerably earlier and at a somewhat different position than the objects of just a single solar mass (W Virginis and RR Lyrae stars). The position of the blue edge of the strip is possibly determined by the helium content of the outer layers and the mass of the stars concerned, whilst the red edge is probably related to the occurrence of convection beyond it, which would suppress the mechanism driving the pulsations (IBEN 1974). In passing, it may be noted that there are also non-variable stars within this strip (SCHMIDT 1972). Cox et al. (1973) attempted to explain this by a lack of helium within those regions of the stellar interior which would otherwise possess helium ionization zones.

The relationship between the periods and the mean density $\bar{\varrho}$, is expressed by the equation:

$$P \times \sqrt{\frac{\bar{\varrho}}{\bar{\varrho}_\odot}} = \text{const.} = Q.$$

The *pulsation constant Q* has, in the basic stellar model, a theoretical value of $0^{\text{d}}\!03$, if one assumes that the pulsations are occurring in the fundamental mode. However, an exception to this latter assumption could be those stars with light-curves like sine-waves (the sinusoidal variables) mentioned somewhat earlier, in which the presence of the first overtone has sometimes been inferred (PEL and LUB 1978).

Pulsating Stars with Double Periods

The existence of a group of δ Cephei stars with an abnormally large scatter in their photoelectric light-curves was first recognized by OOSTERHOFF (1957). The analysis of the measurements showed that the observations could be explained by the *superimposition of two oscillations*. If P_0 and P_1 are the two respective periods $(P_0 > P_1)$, then a beat cycle is produced with a period P_b given by:

$$\frac{1}{P_1} - \frac{1}{P_0} = \frac{1}{P_b}.$$

In Table 11 we give, from a review by FAULKNER (1977), 11 typical cases, the total number of which has not been increased recently despite specifically-aimed searches (see for example the work by BARRELL 1982). We exclude here CO Aur, which as an ostensibly semi-regular star (Sect. 2.2.2) is occasionally observed by amateurs, and which according to MANTEGAZZA (1983) is a δ Cephei star pulsating in the first and second overtones. However, this has yet to be confirmed. But for comparison purposes we have included the two well-studied, multi-periodic RR Lyrae stars AC And (FITCH and SZEIDL 1976) and AQ Leo (JERZYKIEWICZ and WENZEL 1977).

Table 11. Multiple-period δ Cephei and RR Lyrae stars

Star	P_0	P_1	P_1/P_0
Y Car	$3\overset{d}{.}6398$	$2\overset{d}{.}5590$	0.703
GZ Car	4.1588	2.933	0.705
TU Cas	2.1392	1.5183	0.710
UZ Cen	3.3344	2.355	0.706
BK Cen	3.1739	2.2366	0.705
VX Pup	3.0117	2.136	0.709
V 367 Sct	6.2930	4.3849	0.697
BQ Ser	4.2707	3.012	0.705
U TrA	2.5684	1.8249	0.710
AP Vel	3.1278	2.1993	0.703
AX Vel	3.6731	2.5928	0.706
AC And	0.7112	0.5251	0.738
AQ Leo	0.5498	0.4101	0.746

The way in which the periods are superimposed may be clarified by a simplified example. Let the primary period be $P_0 = 3\overset{d}{.}000$, upon which a period P_1 is superimposed, which is 1% less, being thus $2\overset{d}{.}970$. At Epoch 0 the maxima of the two periods coincide, giving a steep shape to the resultant light-curve. In each following cycle the maximum of P_0 lags by $0\overset{d}{.}03$ with respect to the maximum of P_1, so that after 49.5 cycles of P_0 ($= 148\overset{d}{.}5$) the minimum of P_0 and the maximum of P_1 coincide (assuming that the light-curves are symmetrical), and after a further $148\overset{d}{.}5$ the conditions at Epoch 0 recur. As a consequence the beat period $P_b = 297^d$. This is also obtained by use of the equation given above.

As Table 11 indicates, the actual conditions are more complicated. The difference in the periods shown is 25–30%. The beat period is thus only a few days and moreover is not an integer. What is particularly important however, is that an exact analysis of the light-curves reveals a so-called non-linear coupling of the two periods. This especially occurs in the two RR Lyrae stars (loc. cit.), but is also the case in U TrA for example (OOSTERHOFF 1957), revealing itself in the fact that further periods also occur and which may be calculated from the equation:

$$1/P_{ij} = |i/P_0 + j/P_1| \quad (i;\, j \text{ integers}).$$

In AQ Leo for example, amongst other periods that given by $i = j = 1$ is present, namely $P_{11} = 0\overset{d}{.}2348$. (See also the discussion of this subject in Sect. 2.1.3 on RR Lyrae stars.)

Model calculations show that the periods P_0 and P_1 may, with a high degree of probability, be identified with the *fundamental and first overtone* radial pulsation modes. The observed P_1/P_0 ratio corresponds fairly exactly with the theoretical value.

The significance of the δ Cephei stars with double periods lies in the fact that theoretical considerations have led to the possibility of masses and radii of the stars concerned being determined from the knowledge of the two periods alone (e.g. PETERSEN 1973). The ratio P_1/P_0 also determines the pulsation constant Q and hence likewise the mean density (e.g. FITCH 1970). (For further details the publications cited should be consulted, where many other references will also be found.) The masses obtained (0.7–1.7 solar masses), and radii (14–23 solar radii) do not agree with the accepted values for normal δ Cephei stars with similar periods (4.5 $\mathfrak{M}_\odot$ and 30 $R_\odot$). The fundamental reason for this well-discussed mass discrepancy (see for example STOBIE 1980 and references cited there) is still uncertain. Possibly the recently discovered (BALONA 1983) binary nature of Y Car (see Table 11) will lead to an independent determination of the mass of this object.

In relation to the *evolutionary stage* of these objects, which are also known as double-mode stars, two alternative suggestions seem at present to be worthy of consideration. STELLINGWERF (1975) suggests the possibility that a stable pulsation might be involved with the two oscillations being equally excited, whereas FITCH for example, in the work already mentioned (1970), maintains that the objects concerned are in the middle of the very rapid process of switching from pulsations in the fundamental mode to those in the first overtone. Only further future observations will be able to clarify the situation.

2.1.3 RR Lyrae Stars

General Description and Statistics

The RR Lyrae stars are essentially distinguished from the δ Cephei stars by their shorter periods, the Population to which they belong (II) and with it their distribution in the Galaxy, their position on the Hertzsprung-Russell Diagram, and their evolutionary state. In review articles they are not infrequently discussed together with the W Virginis stars. Their great abundance in globular clusters is notable, and because of this they are occasionally called *cluster-type variables*. With respect to period-length they lie above the boundary of the δ Scuti stars – which is at about $0^{\rm d}2$ – and below the transition to the δ Cephei and W Virginis stars at about $1^{\rm d}0$. The upper boundary in particular is well-defined by a deep minimum in the period-length frequency distribution: in the gap between periods of $0^{\rm d}9$ and $2^{\rm d}2$ only a very few cases are found. Some of these have already been discussed under the appropriate heading in Sect. 2.1.2 with the δ Cephei and W Virginis stars. DIETHELM (1981, 1983) gives UX Nor ($P = 2^{\rm d}4$) as the RR Lyrae star with the longest period.

The *distribution over all the possible periods* can be seen in Table 12 (after KUKARKIN 1975). A and B indicate globular clusters having stars of average and low metallicity, respectively. Clusters of high metallicity do not contain RR Lyrae variables. Table 12 shows that the period-distributions in the main body of the Galaxy and those in globular clusters of varying metal content exhibit certain distinct, characteristic differences. These indicate that despite the complete uniformity of their variablility, the RR Lyrae stars do not form a homogeneous group, and were originally formed by very different pro-

Table 12. Period distribution of RR Lyrae stars

Period	Percentages		
	Galaxy	Globular clusters A	Globular clusters B
$0^{d}225$	0.8	1.5	0.4
0.275	2.3	5.8	3.2
0.325	4.6	7.7	8.5
0.375	5.6	3.1	27.6
0.425	8.5	5.4	6.8
0.475	19.4	20.0	1.4
0.525	19.6	23.8	3.2
0.575	18.1	17.6	13.1
0.625	11.5	9.6	19.0
0.675	5.7	3.8	9.0
0.725	2.5	1.0	5.9
0.775	0.8	0.4	0.9
0.825	0.3	0.2	0.8
0.875	0.3	0.1	0.2
	100.0	100.0	100.0

(The first column gives the mean of the corresponding period-interval)

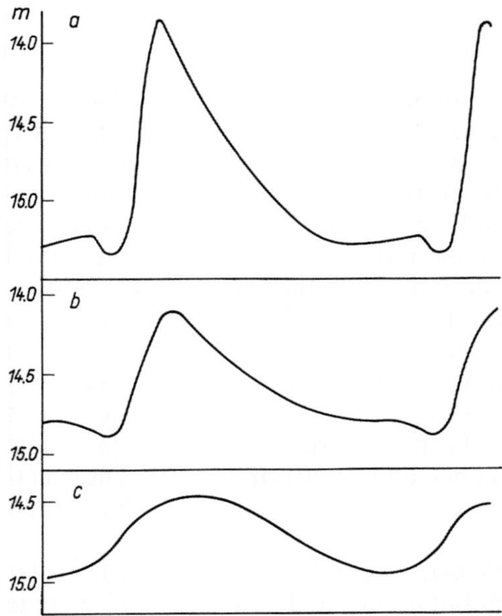

Fig. 11. Principal forms of light-curve in RR Lyrae stars

cesses. This is also shown by a number of other parameters, which will not be discussed here, but for which reference should be made to the literature, for example to the review by KUKARKIN already mentioned.

Various sub-types may be distinguished from the *light-curves*. BAILEY divided the stars into three groups a, b and c, whose curves are shown in

Fig. 11. The mean periods also differ, that for a being $0\overset{d}{.}48$, b $0\overset{d}{.}58$ and c $0\overset{d}{.}32$. As there is a continuous transition between groups a and b, classification is often doubtful. In addition, curves of type a are almost 4 times as numerous, so a distinction is only made nowadays between types RRab and RRc. Less than 10% of the cases in the general galactic population belong to the well-defined RRc class. In contrast, in metal-poor globular clusters variables with periods between 0.3 and 0.4 days form a high peak in the frequency distribution.

As representative of many good series of observations on RR Lyrae stars, we will only mention here the work by LUB (1977), who has published uniform photoelectric light-curves in 6 colours for 90 objects. This material has been subjected to comprehensive analysis.

The catalogues of KUKARKIN et al. (1969, 1971, 1974, 1976) contain more than 5800 RR Lyrae stars belonging to the Galaxy. Amongst these, 50% are RRab variables, and just 6% RRc objects; the sub-class of the remainder is undetermined. It will be seen from this that for every δ Cephei or W Virginis object there are almost 8 known RR Lyrae stars.

Spurious Periods

At this point mention must be made of the problem of spurious periods, which have directly led to quite a few falsely-determined periods amongst faint RR Lyrae stars, but which also occur in other types of periodic variables. In practice this state of affairs arises because after reduction of the observed magnitude values by the procedure described in Sect. 1.3, one can obtain the same phase ϕ for an observation from many completely different periods. Two "periods" P_1 and P_2 give the same phase for a particular observed value when they are related by:

$$\phi(t) = \frac{t - M_0}{P_1} - E_1(t) = \frac{t - M_0}{P_2} - E_2(t)$$

or

$$(t - M_0)\left(\frac{1}{P_1} - \frac{1}{P_2}\right) = E_1 - E_2$$

(E_1 and E_2, being epoch numbers, are integers.) If in an observational series the individual observations are always separated by a time difference T, or complete multiples thereof (owing to "observational windows"), one may also write:

$$t - M_0 = K \times T \quad \text{(where } K \text{ is an integer)}$$

giving for the "correlated periods" the relationship:

$$\frac{1}{P_1} - \frac{1}{P_2} = \frac{E_1 - E_2}{K} \times \frac{1}{T}$$

In RR Lyrae stars frequently

$$T = 1 \text{ sidereal day} = 0\overset{\text{d}}{.}9973$$

– the measurements being continually made at the same hour angle – and

$$|(E_1 - E_2)/K| = 1$$

The true (P) and spurious (P_f) periods are then related by:

$$\left| \frac{1}{P} - \frac{1}{P_f} \right| \approx 1.0027 \text{ d}^{-1}.$$

Occasionally T is equal to a mean solar day – the observations always being made at the same hour of the night – but in this case the difference from a sidereal day is negligible for all practical purposes. T may also equal a synodic month, with observations only being made during the moonless period. Within the limits of the observational accuracy ever available, P and P_f both then produce approximately equivalent mean light-curves, even if the intervals between measurements only approximately satisfy the conditions mentioned. On the basis of observational material alone, spurious periods can only be recognized as such in those cases where observations have been successfully obtained outside the sequence determined by T – in the example mentioned this would be at different hour angles from the customary one.

There are numerous examples in the literature, and spurious periods can considerably upset the statistics relating to sub-groups of RR Lyrae stars, as has been shown for example by PAVLOVSKAYA (1957), WENZEL (1962) and SHUGAROV (see KUKARKIN 1975). By way of making this plain we give in Table 13 a few of the cases encountered.

Table 13. Spurious periods in RR Lyrae stars

Star	Spurious period	True period
V 672 Aql	$0\overset{\text{d}}{.}346$	$0\overset{\text{d}}{.}530$
RV Del	0.332	0.498
XX Hya	0.337	0.508
DD Lyr	0.271	0.373
V 1514 Sgr	0.341	0.519

A very interesting example is that of BG Oct, recently discussed by GESSNER (1981 b), where 60 observations (nonetheless, only obtained over a single summer) can be fitted to either $P_1 = 0\overset{\text{d}}{.}5992$ or else $P_2 = 0\overset{\text{d}}{.}7490$, and where $(E_1 - E_2)/K = 1/3$.

Variability of the Light-Curves and Periods

Most of the RR Lyrae stars repeat their light-curves from cycle to cycle with amazing regularity. It was discovered as early as the beginning of the 20th

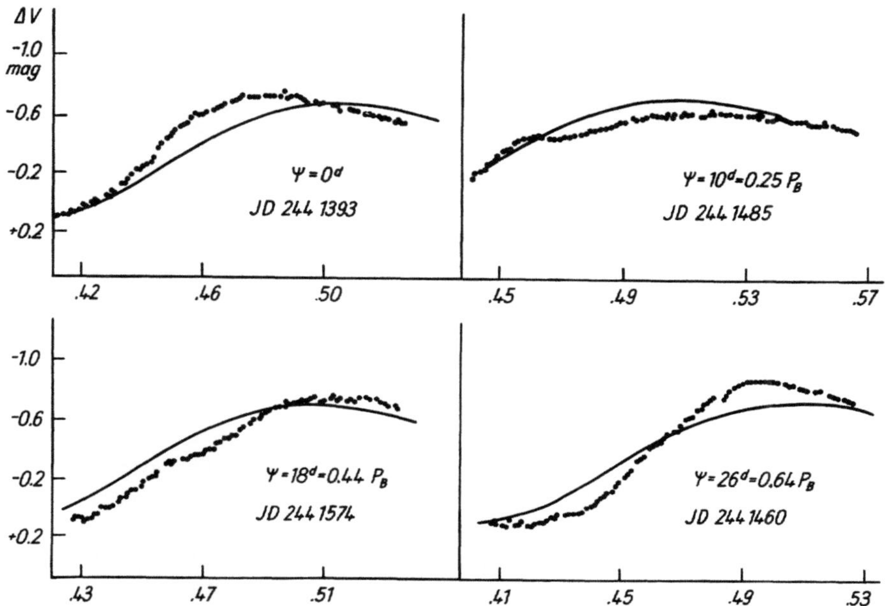

Fig. 12. Blazhko effect in RR Lyrae. Shape of the rise and maximum of the light-curve for four different phases ψ of the Blazhko period ($P_{\mathrm{B}} = 40\overset{d}{.}8$) from photoelectric observations (*points*) made at Budapest. The mean curve is shown by the continuous lines (after SZEIDL 1976)

century, however, that some of these stars display marked changes in the height of maxima, and that the times of maxima cannot be expressed by a linear equation. RR Lyr itself (Fig. 12) shows such a phenomenon (SHAPLEY 1916), which is nowadays named the *Blazhko Effect* after the actual discoverer. In RW Dra, which still serves as a typical example, BLAZHKO found that the secondary changes just mentioned occurred with a period of 41.6 days. SZEIDL, one of the experts on this particular subject, estimates (1976) that about 15–20% of field RRab stars exhibit changes in their light-curves. In a table he lists 26 RRab variables and 3 RRc stars for which the secondary periods (also sometimes called the Blazhko periods, P_{B}) were known at that date. As examples we have extracted from this list, somewhat arbitrarily, those cases that were investigated at Budapest (Table 14).

In the list mentioned the longest Blazhko period is that of RS Boo (537^{d}) and the shortest BV Aqr ($11\overset{d}{.}6$); the latter however is unconfirmed. The concentration between 20 and 40 days, which occurs even in Table 14, is very striking. It appears as if the effect is more pronounced in metal-poor RRab stars and in globular clusters than in the remaining RR Lyrae-type variables. Occasionally the magnitude of the Blazhko phenomenon is itself subject to changes, as in the case of RR Lyr itself, for which DETRE (1969) found a 4-year cycle. In this star the direction of the magnetic field seems to vary with the Blazhko period and its magnitude over the 4-year cycle. According to SZEIDL (1976) this could be "of fundamental importance for the under-

Table 14. RR Lyrae stars with Blazhko effect

Star	P	P_B
RR Gem	0^d397	37^d
SW And	0.442	36.8
RW Dra	0.443	41.7
AR Her	0.470	31.6
SZ Hya	0.537	25.8
RW Cnc	0.547	29.9
TT Cnc	0.563	89
RR Lyr	0.567	40.8
AR Ser	0.575	105
DL Her	0.572	33.6
Z CVn	0.654	22.7
TV Boo	0.313	33.5 RRc

standing of the nature of the Blazhko Phenomenon". In point of fact, COUSENS (1983) in a detailed, quantitative discussion has made it seem plausible that the 41-day Blazhko cycle in RR Lyr can be explained by the combined effects of fundamental radial pulsations, rotation, and magnetic field; the magnetic field producing a significant non-radial, pulsational component (see Sect. 2.3) near the stellar surface. The star is an oblique rotator, not rotating about the axis of the (bipolar) magnetic field, but about another, inclined axis. In the course of this rotation, various aspects of the pulsation are presented to the observer, and a modulation of the fundamental pulsation results. Another proposed explanation (involving double-mode pulsations) will be described later.

A further form of change in the light-curves of RR Lyrae stars has already been mentioned in the discussion of double-period δ Cephei stars. Here we find the simultaneous excitation of the fundamental mode and of the first overtone (AQ Leo), or even of the fundamental, first and second overtones (AC And) and their interference.

Table 15. Components in the light-curve of AQ Leo

i	j	P_{ij}	A_{ij}
1	0	0^d4101	0.2210 mag
0	1	0.5498	0.1124
1	1	0.2348	0.0522
2	0	0.2051	0.0476
1	-1	1.6151	0.0395
2	1	0.1494	0.0216
3	0	0.1367	0.0175
0	2	0.2749	0.0169
1	2	0.1646	0.0116
2	-1	0.3271	0.0115
2	2	0.1174	0.0111
3	1	0.1095	0.0092
1	-2	0.8334	0.0072
4	0	0.1025	0.0059

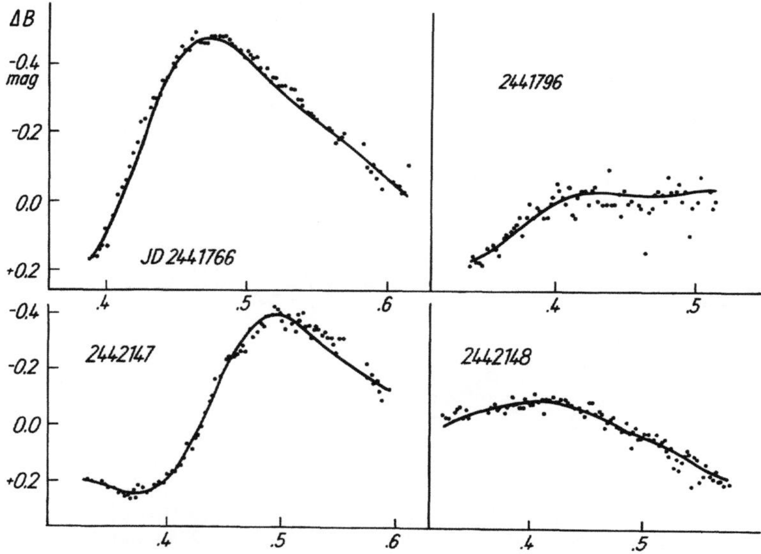

Fig. 13. Comparison of the observed (*points*) and calculated (*line*) B light-curves of the double-period RR Lyrae star AQ Leo over four nights (after JERZYKIEWICZ and WENZEL 1977)

As an illustration of these effects we include the comprehensive Table 15, with rounded values, of the components noted in the light-curve of AQ Leo by JERZYKIEWICZ and WENZEL (1977).

The synthetic light-curve is the sum of the sine waves:

$$m(t) = \bar{m} + \sum_{i,j} A_{ij} \sin\left(2\pi t / P_{ij} + \Phi_{ij}\right),$$

in which not only harmonics of the fundamental and first overtones occur, but also combination periods (both i and $j \neq 0$). P_{ij} and A_{ij} are the parameters given in Table 15. The Φ_{ij} terms, which we will not discuss here in detail determine the mutual phase-lag of the individual waves, and $\bar{m}$ is the mean magnitude taken over a representative period of time.

The agreement with the photoelectric light-curve on which it was based is excellent (Fig. 13). Above all, however, the observations for the subsequent years, which were not used in the analysis, were correctly represented. This suggests that the analysis has properly interpreted the physical state of affairs involved in the *double-mode pulsations*. An investigation by Cox et al. (1983) of those mean light-curves of RR Lyrae stars in the globular cluster M15 which show considerable scatter, also brought to light a whole series of objects apparently having double periods. With $P_0 = 0^{d}55$, the resulting value for $P_1/P_0 = 0.746$ was in agreement with that for AQ Leo, and with the aid of new model calculations a mass of $0.65\mathfrak{M}_{\odot}$ was obtained. In this case there is no mass discrepancy as with the δ Cephei stars (see the corresponding part of Sect. 2.1.2, and the discussion below of the parameters of RR Lyrae stars).

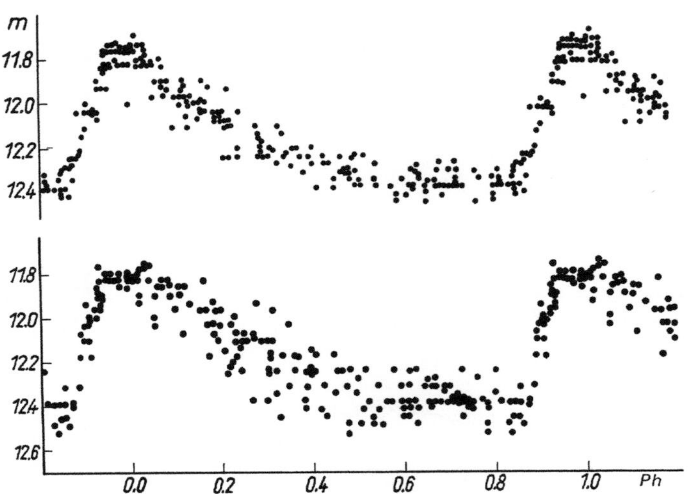

Fig. 14. Mean visual light-curve of the RR Lyrae star LX Lyr according to HOFFMEISTER 1970. *Top*: undisturbed form (JD 243 2791-2835); *below* disturbed form (2682-2780 and 2850-2865)

It should be noted that recently BORKOWSKI (1980), using AR Her as an example, has also tried to explain the Blazhko effect by double-mode pulsations, where apart from the fundamental period $P_0 = 0\overset{d}{.}470$, another oscillation exists with the reciprocal period $1/P_1 = 2/P_0 + 1/P_B$ ($P_B = 31\overset{d}{.}6$, the Blazhko period). P_1 probably then corresponds to the third overtone.

A third class of irregularity in the light variations of RR Lyrae stars to be mentioned, is the changes in the periods, which are limited in amount, but may be either irregular and abrupt, or else secular. The latter cause the $O - C$ curve to have a parabolic shape (see the end of Sect. 1.3). ROSINO (1972), who comprehensively described the relationships found in globular clusters, noted the variability in periods as being of the order of 10^{-10} days per day. Extensive work in this field has been carried out by BELSERENE, WILKENS, SZEIDL, OOSTERHOFF, COUTTS and others. The basic assumption that these secular shifts are connected with the evolution of RR Lyrae stars across the pulsational strips on the H-R Diagram could not be supported, as the period changes are at least an order of magnitude too large for this, and moreover, because both senses occur (increases as well as decreases of period). See additionally Sect. 5.1.2.

All instabilities in the actual light-curve primarily make themselves obvious as increased scatter of the individual observations when a mean light-curve is produced. This also shows that other irregularities in the form of the curve seem to take place, which do not have their source in any of the known causes already mentioned in this section. According to HOFFMEISTER (1970, p. 66), a tendency to increased scatter appears to occur in many stars on the descending branch, around phase 0.3, as exhibited by LX Lyr, for example (Fig. 14). An earlier, comprehensive discussion by HOFFMEISTER (1955) showed that of 30 general field stars investigated, 20 had some form of disturbance to their light-curves. Amongst these there were only 4 or 5 cases with the true Blazhko effect. Z Mic might be regarded as a prototype for such stars with an apparently irregular occurrence of increased scatter. Despite great exertions on the part of observers regarding irregularities in the RR Lyrae stars, much remains to be done. The possibility certainly exists that further observations will demonstrate that all the effects can be accounted for by one comprehensive theory.

Physical Properties

RR Lyrae stars of the RRab subclass have long been regarded as the most homogeneous class of all variables. From this it was inferred that they were very suitable as indicators of Population II regions, and thus for research into the structure of the Galaxy. (Unfortunately, as yet they have no significance for distance determinations in other galaxies, other than the Magellanic Clouds and some dwarf galaxies, as in the Andromeda Galaxy and M33 in Triangulum they lie just below the limit of the largest telescopes.) More recently the indications are increasing, as already mentioned, that physically different groups also exist among the RR Lyrae stars, which may be characterized, for example, by differing metal content and population membership. Even the invariance of their *absolute magnitude* is no longer a definite hypothesis. The average value for the mean absolute visual magnitude may be taken as being $M_v = + 0.6$ with a scatter of $0.3-0.4$ mag, with $M_B = + 1.0$. These values are only weakly period-dependent.

The latter point is also valid for the *spectral classes* determined from the strength of the hydrogen absorption lines. In the vast majority of the RRa stars these change, irrespective of period, from about A7 at maximum light to about F5 at minimum. On the other hand, the spectral class determined from the K line of Ca II shows marked differences from star to star, particularly at minimum, where the scatter amounts to about a spectral class. A leading, classic investigation of this was originated by PRESTON (1959), who introduced

$$\Delta S = 10 [Sp.(\text{H}) - Sp.(\text{Ca II})],$$

derived at minimum magnitude, as a parameter to describe the spectra, and in particular the *metal content*. $\Delta S = 0$ indicates strong Ca II lines and a high metallicity, and $\Delta S = 10$ low values of these properties. The hydrogen type, on the other hand, is primarily an indicator of temperature, as one would expect from a normal spectral classification scheme. PRESTON himself, and numerous later authors (see KUKARKIN 1975), recognized that relatively metal-rich stars belong to the disk population in the Galaxy (having galactic orbits with low inclinations and low eccentricities, small motions relative to the Sun, and a low scatter in spatial velocities). The metal-poor RR Lyrae stars are members of the halo population. The two classes are thus presumed to be of different origin. However, there still exists a large number of unexplained anomalies.

The RRc stars have systematically earlier spectral classes at minimum. However, they show the same sort of scatter in their spectral characteristics.

The variation in the two spectral classes defined above, with respect to the phase of the magnitude changes, can be gathered in all its essentials from Table 16 and Fig. 15 (after PRESTON 1959, Fig. 2). It is least amongst the Ca II type of metal-poor objects.

Table 16. Spectral variations in RR Lyrae stars

Phase	$Sp.$ (H)	$Sp.$ (Ca II)		
		$\Delta S = 0$	6	10
$0^{P}\!\!.8$	F5	F5	A9	A5
0.0	A7	A6	A2	A2
0.1	F0	F1	A5	A3
0.3	F4	F4	A8	A5
0.6	F5	F5	A9	A5

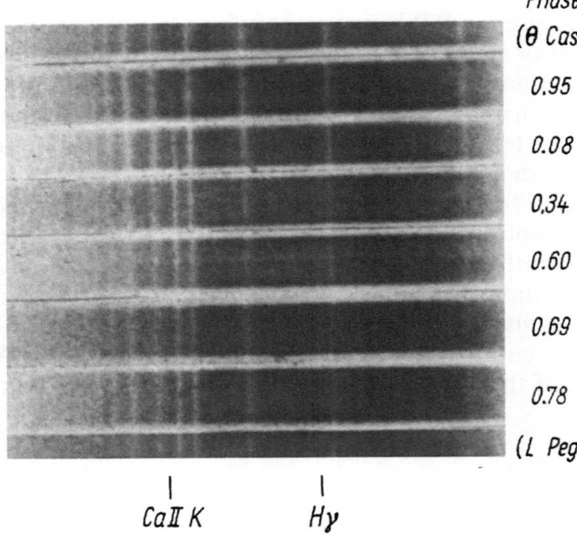

Fig. 15. Six spectrograms of the RR Lyrae star XZ Dra at various phases of the light-curve; ϑ Cas (A7V) and ι Peg (F5V) are shown for comparison (after PRESTON 1959)

Yet another spectroscopic peculiarity to be mentioned is the existence of hydrogen emission lines and line-splitting on the rising branch of the light-curve, first observed (e.g. STRUVE 1947, SANFORD 1949) in the relatively bright star, RR Lyr itself (Fig. 16). This led to the hypothesis of *shock waves* in the atmospheres of RR Lyrae stars and other Population II pulsating variables (see Figs. 17 and 18). Later research has shown, however, that the physical conditions in the outer layers during pulsation are far more complicated.

Radii and *masses* of galactic RR Lyrae stars were derived by WOOLLEY and SAVAGE (1971) by an extension of the Baade-Wesselink method described

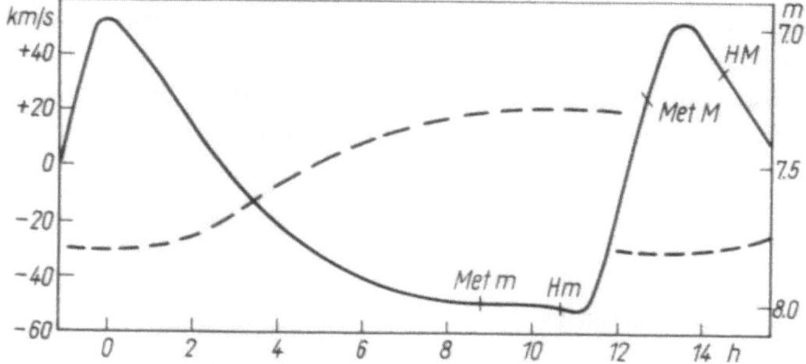

Fig. 16. Mean light-curve (*continuous*) and mean radial-velocity curve (*broken*) of RR Lyr. Note the presence of two velocities about 2 hours before maximum light (splitting of the spectral lines). *Met m* and *Met M*: latest and earliest spectral class based on metal lines; *Hm* and *HM*: the same based on hydrogen lines

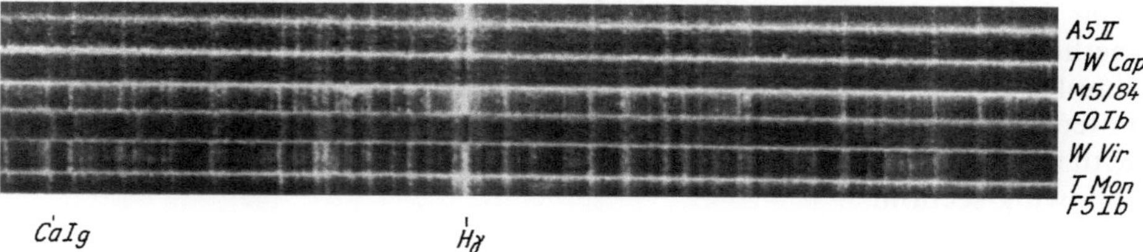

Fig. 17. Spectrograms of W Virginis stars (TW Cap, star 84 in M5, W Vir) and the classical δ Cephei star T Mon; with three comparison spectra of normal supergiants. Note the Hγ emission component in the W Virginis stars; this is lacking in the δ Cephei star (after WALLERSTEIN 1958)

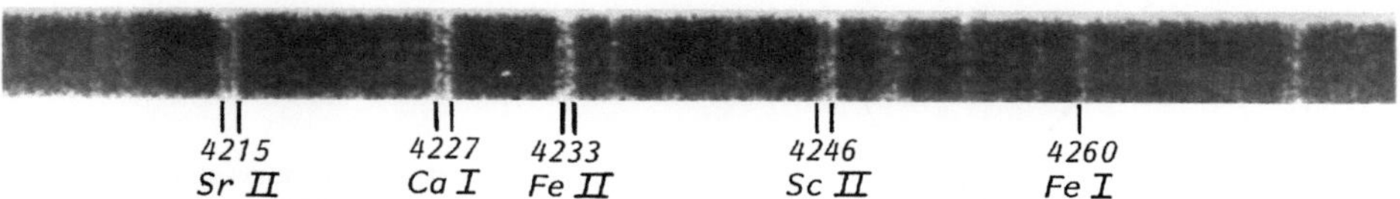

Fig. 18. Part of a spectrogram of W Virginis star 42 in the globular cluster M5, close to maximum light. Note the doubling of several absorption lines, which indicates the existence of two layers, moving in different directions (outwards and inwards) (after WALLERSTEIN 1959)

for the δ Cephei stars. They obtained, for RRab stars with $P > 0^{d}44$ and $M_v = +0.40$, the values $R \approx 5.5 R_\odot$, $\mathfrak{M} \approx 0.5 \mathfrak{M}_\odot$, and for RRc stars with $P > 0^{d}36$ and $M_v = +0.8$, $R \approx 4.5 R_\odot$, $\mathfrak{M} \approx 0.6 \mathfrak{M}_\odot$. From these values it follows that the pulsation constant $Q \approx 0^{d}03$, in general agreement with the theoretical value. The uncertainties in these values, however, allow the hypothesis, upheld by numerous theoreticians, that the RRc variables pulsate in the first overtone; its period amounts to about 3/4 of the fundamental.

VAN HERK (1965) determined the proper motion, parallax and spatial velocity of 210 stars. The mean parallax derived from the proper motions, $0''00097$, corresponds to the *mean absolute magnitudes* $M_{pg} = +0.87$ and

$M_v = +0.68$ for the set of stars considered, agreeing well with the values obtained by other means (see above).

An outline of the evolutionary state of the RR Lyrae variables has already been given in the discussion of the other classical pulsating stars (Sect. 2.1.1).

2.1.4 δ Scuti Stars

Definition, Characteristics and Statistics

The δ Scuti stars are variables of spectral classes A or F with pulsational periods below $0\overset{d}{.}3$. In the period range from about $0\overset{d}{.}2-0\overset{d}{.}3$ we find both δ Scuti and RRc stars, and it is not possible to distinguish between the two groups on the basis of period alone. The amplitudes amount to a few thousandths to a few tenths of a magnitude, a typical value being 0.02 mag. Most of these variables are thus only amenable to photoelectric determination of their light-curves. Figure 19 includes two characteristic light-curves. The method of designation is still controversial. We follow here, for the sake of simplicity, the recommendation by BREGER (e.g. 1979), who is one of the most noted specialists in this field. The publication just mentioned gives a review with numerous references, from which we have extracted some details for the following section.

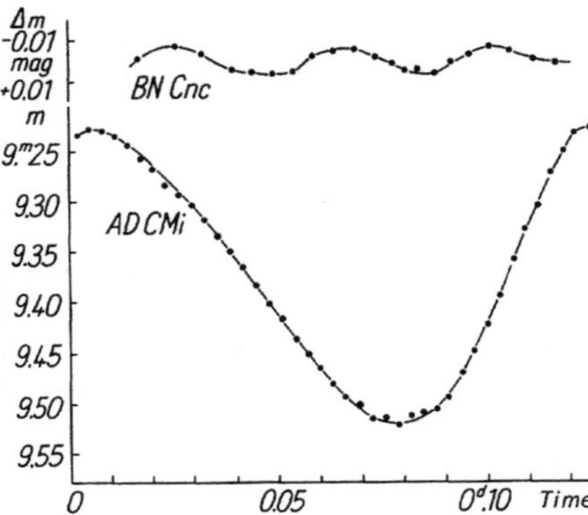

Fig. 19. Photoelectric light-curves (visual band) of the two δ Scuti stars BN Cnc and AD CMi, after BREGER (1975)

The first definite candidate with a large amplitude, CY Aqr, was discovered by HOFFMEISTER (1934) on plates taken by the Sonneberg photographic sky patrol, and recognized as "ultra-short-periodic" by JENSCH (1934), with a period of $0\overset{d}{.}061 = 88$ min. The light-curve (Fig. 20) is only slightly variable and resembles that of a normal RRab star (but not RRc). The visual observation of CY Aqr can be very exciting, as the rise of almost a magnitude takes place in 10 minutes, so that anyone at a telescope can directly see the increase in brightness, and must hurry to complete the magnitude estimates. Nevertheless, the time of the sharp maximum can be determined by such a method to an accuracy of a minute. JENSCH (1936) used this fact, in the context of a school experiment, to measure the speed of light, in just the same way as in

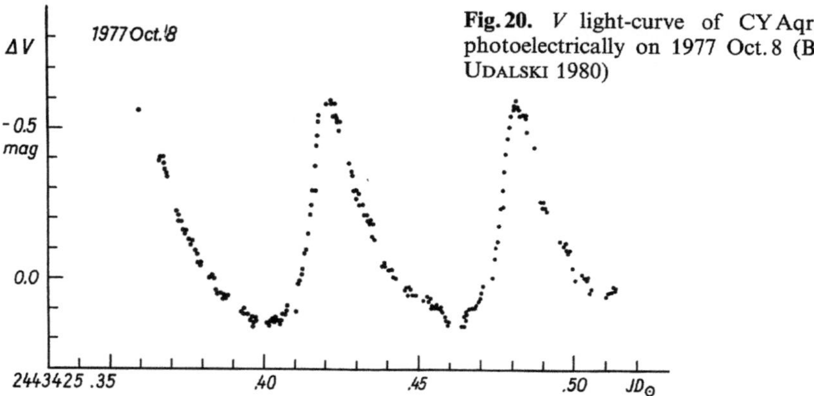

Fig. 20. V light-curve of CY Aqr, observed photoelectrically on 1977 Oct. 8 (BOHUSZ and UDALSKI 1980)

his time, RÖMER managed to do by means of the phenomena of the satellites of Jupiter (see Sect. 1.4).

The designation "ultra-short-periodic variables" was later adopted again by, for example, EGGEN (without the amplitude being taken into account). It can be criticized, however, in view of the fact that pulsating white dwarfs exhibit very much shorter periods (Sect. 2.3.2). The first suggestion that objects like CY Aqr differ from the RR Lyrae stars in their physical characteristics was given by H. J. SMITH (1955). He called them "dwarf cepheids", regarding them as being a miniature form of the δ Cephei stars, and placed them in an intermediate population. However, the expression is misleading in many respects. As a result other authors occasionally substituted the terms "AI Velorum stars" (BESSELL 1969) or "RRs variables" (KUKARKIN et al. 1969) when fairly large amplitudes were present. According to BREGER (1979) and other authors, however, *no physical difference exists between objects with large and small amplitudes*, and as the latter are several orders of magnitude more frequent, the name of the prototype δ Scuti is used for the whole group.

The statistical study of the δ Scuti stars is distorted by many troublesome *selection effects*. Photographic discovery methods are only of importance for those objects with the greater amplitudes, yet they are not effective with long-exposure plates, due to the very short periods. Numerous photoelectric surveys of specific types of stars (for example, among A2V to F0V stars) have raised the number of known members very considerably, but in a very unsystematic manner. The GCVS catalogue and its 3 Supplements (KUKARKIN et al. 1969–1976) list 157 cases (including the questionable ones), while a list by BREGER (1979) contains 129 well-researched bright or particularly interesting δ Scuti stars. The distribution of amplitudes (in the V region) from this list is given in Table 17. Table 18 contains individual details for the 13 stars with amplitudes greater than 0.45 mag, and for δ Sct itself.

By way of a supplement to this, it may be mentioned that the bright stars α Lyr, γ UMi and γ CrB are probably weakly pulsating. Some authors, for example KHOLOPOV (1981), have provisionally included them with the δ Scuti variables, where they may be borderline cases.

Table 17. Amplitude distribution of δ Scuti stars

Amplitude V	Number
$\leq$ 0.05 mag	90
0.051 −0.100	14
0.11 −0.20	3
0.21 −0.30	4
0.31 −0.40	4
0.41 −0.50	4
0.51 −0.60	5
0.61 −0.70	5
	129

Table 18. Some δ Scuti stars

Star	Period	Amplitude V	Spectrum
SX Phe	$0\overset{d}{.}055$	0.51 mag	sd F0
CY Aqr	0.061	0.73	F0
DY Peg	0.073	0.54	A9
AE UMa	0.086	0.7	A9
EH Lib	0.088	0.50	F0
RV Ari	0.093	0.70	A0
AI Vel	0.112	0.67	F2
V 703 Sco	0.115	0.50	F2
SZ Lyn	0.120	0.51	F0
DY Her	0.142	0.49	F4 III
RS Gru	0.147	0.56	A8
VZ Cnc	0.178	0.61	F2 III
BS Aqr	0.198	0.51	F3
δ Sct	0.194	0.29	F3 III − IV

Physical Properties − Occurrence of Pulsations

Most δ Scuti stars belong to Population I, and a number are recorded in open clusters − see the compilation by FROLOV and IRKAEV (1982). A very few variables, however, show low metal content (an indicator of Population II) and are probably located below the normal Population I main sequence on the Hertzsprung-Russell Diagram. (An extreme example is SX Phe, where the luminosity is indicated in Table 18 as sd = subdwarf.) In addition high spatial velocities may point to Population II membership in some cases. There is the suggestion that these should be described as a separate, metal-poor subgroup (BAGLIN et al. 1980).

Absolute magnitudes lie essentially between $M_v = 0$ and $+ 3$ (SX Phe being the exception with $+ 4\overset{M}{.}1$) thus forming a continuation to the lower edge of the pulsation strip occupied by the δ Cephei, W Virginis and RR Lyrae stars. A well-defined Period-Luminosity relationship exists, if the dependence of the luminosity upon the spectral class is taken into account, by means of appropriate corrections. In distinction to the δ Cephei and

RR Lyrae stars, the radial-velocity curve is not always a good mirror-image of the light-curve, and moreover there is also a phase shift between the two of about 1/10 period. The temperature maximum is reached shortly before maximum brightness. After earlier failures, the Baade-Wesselink method (Sect. 2.1.2) of comparing the changes of radius as derived from the light-curve, and from the radial-velocity curve, led to usable radii and masses for δ Scuti stars (see BREGER 1980 for references). In the case of stars with $P = 0\overset{d}{.}14$, these are about $3R_\odot$, and $2\mathfrak{M}_\odot$, and indeed these are independent of whether objects with large (RRs) or small amplitudes (δ Scuti stars in the narrower sense) are under consideration. The pulsational constant resulting from these rough mean values is $Q = 0\overset{d}{.}038$. More exact calculations incorporating concepts from theoretical models confirm that generally *radial pulsations* account for the largest part of the variability. As with other pulsating stars, the source of these pulsations is probably to be found in the *Kappa-mechanism* (Sect. 2.1.2), and indeed in the He^+ ionization zone.

However, it appears that departures from this occur, and that in some individual cases non-radial pulsations are present (see Sect. 2.3). The often-cited work by BREGER (1979) takes as a representative example the variable 1 Mon $=$ V 474 Mon, in which SHOBBROOK and STOBIE (1976), as well as MILLIS (1973) found frequencies of 7.217, 7.346 and 7.475 cycles per day (that is to say periods of $0\overset{d}{.}1386$, $0\overset{d}{.}1361$ and $0\overset{d}{.}1337$). The equality in the differences between any two neighbouring frequencies is typical for such oscillations (see Sect. 2.3). The amplitude in this case amounts to 0.2 mag, and the spectrum is described as F2 IV. V 571 Mon $=$ 21 Mon, V 376 Per and V 1208 Aql $=$ 28 Aql may be similar cases. DZIEMBOWSKI (1974) has also shown theoretically that non-radial modes may be excited in δ Scuti models.

As mentioned above, the δ Scuti stars occupy the lower part of the instability or pulsation strip in the Hertzsprung-Russell Diagram, already explained in connection with the δ Cephei stars. However, only about a third of the stars in this region show measurable variability, and whether this is always of the δ Scuti type remains to be investigated. Not all the factors responsible for the occurrence of pulsation or for its suppression are yet known. Probably rotation plays a part here, because in slowly rotating stars with little "sideways" circulation, He^+ diffuses downwards out of the ionization zone, and "metals" outwards towards the surface. Too low a He^+ content in the ionization zone leads to a breakdown of the *Kappa-mechanism* and to pulsational stability (Am stars). Generally it is considered that "the delicate balance between the complicated processes that tend to mix and those that separate the stellar matter in the very thin layers that excite or damp the oscillations, in some cases will produce a particularly strong excitation" (PETERSEN 1976 in a review of the theoretical aspects).

Multiple Periodicity

The determination of the periods of δ Scuti stars is often very difficult in view of the mostly very small amplitudes, and it is not uncommon for it to be necessary to revise values published earlier. The work is additionally impeded

Table 19. Multiple periods in δ Scuti stars

Star	P_0	P_1	P_2	P_1/P_0	P_2/P_1	P_2/P_0
VZ Cnc		0^d1784	0^d1428		0.8006	
VY Hya	0^d2234	0.1727		0.7732		
δ Sct	0.1983		0.1164			0.6005
V 703 Sco	0.1500	0.1152		0.7683		
V 474 Mon	0.1361		0.0826			0.6069
CC And	0.1249		0.0749			0.5999
AI Vel	0.1116	0.0862		0.7727		
BP Peg	0.1094	0.0845		0.7715		
V 571 Mon	0.0999	0.0750		0.7507		
RV Ari	0.0931	0.0720		0.7726		
AE UMa	0.0860	0.0665		0.7734		
CY Aqr	0.0610	0.0454		0.7443		
SX Phe	0.0550	0,0428		0.7782		

by the frequent occurrence of secondary periodicities, which are super-imposed on the primary oscillation. The phenomena are exactly the same as the effects described for the δ Cephei and RR Lyrae stars, and the analysis of the light-curves is carried out by the same means as for those types. Table 19 contains the cases quoted by FITCH (1976) as definitely showing multiple periodicity, which were taken from a list compiled by FITCH and SZEIDL (1976). The Table may be seen as a continuation of Table 11 towards shorter periods.

The difficulties which occur in the analysis can be seen by the case of V 474 Mon, for example, which was cited above as a non-radial pulsator, but which appears here as having a fundamental radial mode (P_0) and the appropriate second overtone (P_2). In the light-curve of the well-studied, bright star, δ Sct itself, a total of 9 periods were detected, including non-radial modes. In addition, combination periods were present as in the δ Cephei and RR Lyrae stars (FITCH 1976). Moreover, the observed period ratios given in Table 19 agree satisfactorily with those calculated for radial pulsations from theoretical models. These, dependent upon the metal content of the stellar material, lead to $P_1/P_0 = 0.74$ to 0.78 and $P_2/P_1 \approx 0.81$ (e.g. Cox et al. 1979). It appears that not all δ Scuti stars are pulsating in the fundamental mode.

2.2 Slowly-Varying, Pulsating Stars

2.2.1 Mira Stars

The most important group among the slow variables is that of the Mira stars, so-called after the prototype o Ceti, otherwise known as Mira. Their particular characteristic is the *large amplitude* of the essentially continuous changes in magnitude, which means that there is a *very high probability* of their being discovered. It may be assumed that nearly all Mira stars that reach magnitude 11 at maximum are known, and this is confirmed by experience. The Mira stars are red giants and supergiants, and on the H-R Diagram they form a well-defined group at the far right-hand end of the giant branch. However, the population to which they belong is by no means uniform, as the statistical data will show. Most of the spectra show emission lines of hydrogen and often of some other elements. In the closely related semiregulars, which have shorter periods and smaller amplitudes, the emission features are rarer.

Periods and Amplitudes

The demarcation of the true Mira type is somewhat arbitrary. An amplitude of at least 2 mag is taken as being the limit, or according to other authors at least 2.5 mag; the stars with smaller amplitudes are classed with the prototype object $Z\,Aqr$ (SRa, Sect. 2.2.2). We may take the shortest period for the true Mira type as about 90^d. Here nature comes to the aid of the statistician, as stars with periods between 50^d and 90^d nearly always show irregularities, and are thus included in the semiregular class.

The Mira star with the shortest known period ($90^d\!.65$) would seem to be $T\,Cen$ (spectrum K0$-$M4, range $5^m\!.5-9^m\!.0$ vis.). However, in the GCVS it is considered to belong to the "SRa" class (see Sect. 2.2.2).

With respect to period length the Mira class overlaps those of the long-period δ Cephei stars and the semiregulars of the RV Tauri and S Vulpeculae types.

The Mira star with the longest period appears to be $BX\,Mon$ ($P = 1374^d$, $Sp. = $ M4ep, $m_{pg} = 9^m\!.5-13^m\!.4$). Some semiregulars have even longer cycle lengths.

The frequency distribution of period lengths for the Mira stars is given in Tables 20 and 21 (after IKAUNIEKS 1963). The maximum of the distribution is at 278 days.

Table 20. Distribution of Mira stars according to period length, subdivided into the three main spectral classes

Period	Spectral class		
	M	C	S
101^d-150^d	32	–	–
151 $-$200	68	–	–
201 $-$250	148	3	4
251 $-$300	172	3	7
301 $-$350	184	5	11
351 $-$400	113	14	14
401 $-$450	65	17	6
451 $-$500	25	7	5
501 $-$550	12	4	–
551 $-$600	4	–	–
601 $-$650	1	–	–
651 $-$700	2	–	–

Table 21. Distribution of Mira stars and semi-regulars (SR) according to period length

Period	Mira	SR
$\leq$ 50^d	0	23
51^d– 100	4	182
101 – 150	138	355
151 – 200	331	250
201 – 250	653	144
251 – 300	774	96
301 – 350	507	82
351 – 400	308	63
401 – 450	177	16
451 – 500	.62	15
501 – 550	27	9
551 – 600	4	5
601 – 650	4	3
651 – 700	4	4
701 – 750	1	5
751 – 800	0	3
801 – 850	0	3
851 – 900	0	1
901 – 950	0	2
951 –1000	0	1

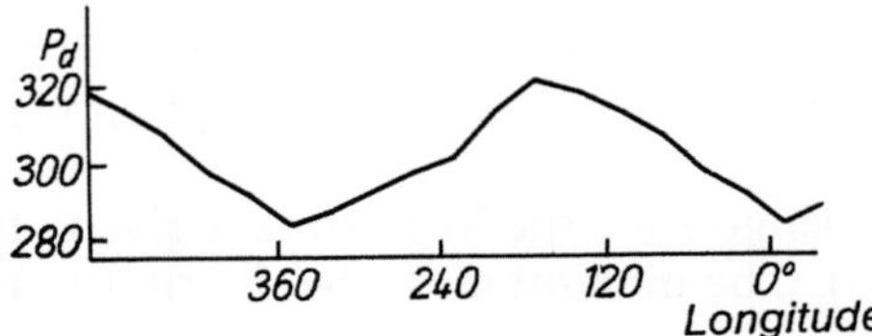

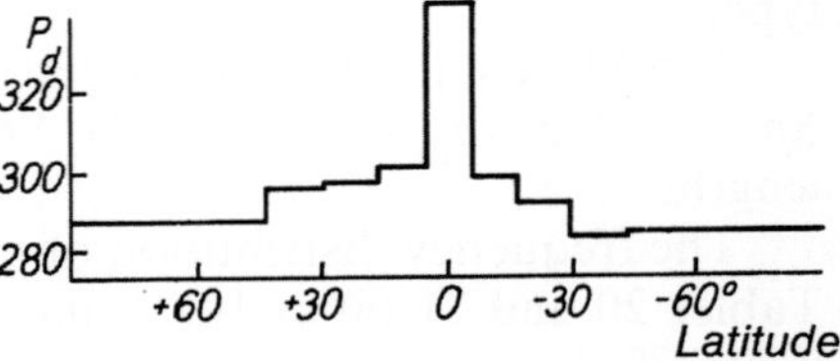

Fig. 21. Distribution of mean period-lengths with respect to (*left*) galactic longitude and (*right*) galactic latitude (after AHNERT)

A correlation between mean period length and position in the Galaxy was first recognized by AHNERT (1939)–Fig. 21. He obtained $P = 299^d$ for 998 Mira stars with maxima brighter than 10^m5; 342^d for 117 stars with galactic latitudes of less than $\pm\,5°$; 242^d for 198 cases in the Harvard G Scorpii field and 259^d from 50 stars in the 67 Ophiuchi field investigated at Sonneberg. This means that the periods in the vicinity of the galactic centre are significantly shorter than the mean value, while those in the regions close to the spiral arms are significantly longer. Consequently the *difference between the two Populations* is also shown by the properties of the Mira stars.

KUKARKIN (1949) found the same effect when he determined the mean value of the periods relative to galactic longitude (Table 22, where longitude 0° corresponds to the direction of the galactic centre). This means that in the neighbourhood of the centre there are numerous Mira stars and that their

Table 22. Relationship between period length and galactic longitude in Mira stars

Galactic longitude	Number	P
30°– 90°	541	282$^\text{d}$6
90 –150	209	307.0
150 –210	144	319.0
210 –270	111	300.4
270 –330	315	284.4
330 – 30	890	256.0

periods are short; at the anticentre there are few Mira stars, and their periods are long.

Recently some authors (MAFFEI 1967 and EVANS 1976) have doubted whether the increasing proportion of short-period Mira stars in the direction of the galactic centre really corresponds to the true situation. EVANS scoured the region around the galactic centre for red variables in three "windows" with red and infrared plates. As a result very many long-period Mira stars were found, which had remained undiscovered on earlier searches with blue plates. This is due to the fact that with increasing period-length the absolute blue magnitude decreases. The long-period Mira stars in the galactic centre at a distance of about 7–9 kpc (Sect. 7.2) are thus below the limit of the telescopes used for the blue plates. However, as the mean spectral class and thus the mean colour index shifts towards the red with increasing period-length (see Fig. 23), these long-period Mira stars are easily found on red and infrared exposures.

Spectra

Mira stars belong overwhelmingly to spectral class M, in particular Me (Fig. 22); in other words they have hydrogen emission lines, and occasionally some others. A much smaller number, however, are divided among spectral classes S, N, R and C (the latter are the carbon stars). The distribution of Mira stars with respect to spectral class is given in Table 20, after IKAUNIEKS (1963). The vast majority of Mira stars, however, have not yet been classified according to their spectrum. A spectral catalogue of Mira stars of classes Me and Se has been prepared by KEENAN (1966). Figure 23 shows the period-spectrum relationship.

Spectra of class M without emission lines among variable stars are predominantly restricted to the semiregular and irregular variables, and to Mira stars with relatively short periods (mean period $\bar{P} = 216^\text{d}$), as against $\bar{P} = 298^\text{d}$ for Me; $\bar{P} = 367^\text{d}$ for Se; $\bar{P} = 379^\text{d}$ for N. (It may be noted here that in the Henry Draper Catalogue the old designations Ma, b and c are used and that the modern Me occurs as Md.)

The observation of the continuum is impeded by the *absorption bands*, normal in M stars, primarily those of TiO. Naturally the strength of these bands varies during the course of the changes in brightness.

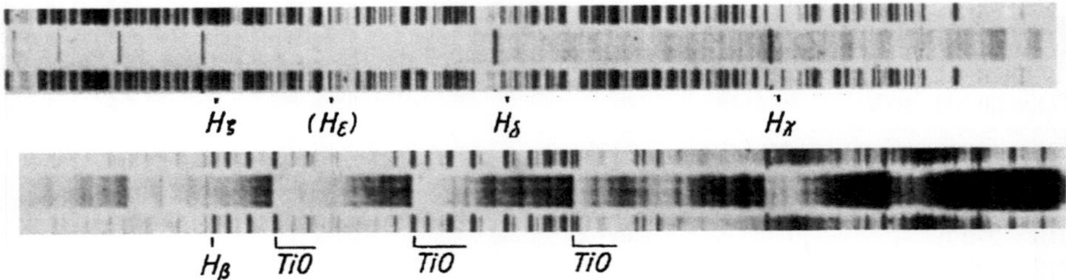

Fig. 22. Spectrogram of Mira; *below* red end of the spectrum, *above* blue. Note the titanium oxide absorption bands and the hydrogen emission (after STRUVE 1954)

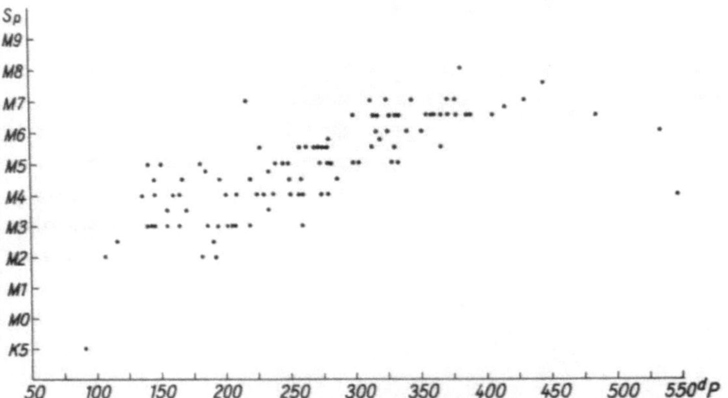

Fig. 23. Period-Spectrum relationship for Mira stars (after KEENAN 1966)

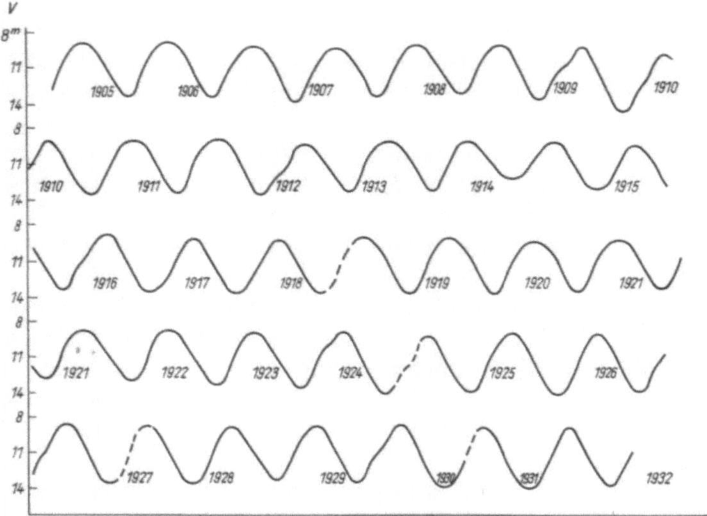

Fig. 24. Light-curve of S Boo (after NIJLAND)

Light-Curves

The height of maximum can vary greatly in one and the same star. An especially impressive example, as it can be seen by the naked eye, is o Cet itself. According to the very thorough investigation by GUTHNICK (1902) of the material available up to that time, the extreme range of magnitude at maximum was between $1^m\!.7$ and $5^m\!.2$. The form of the light-curve changes correspondingly.

The properties of the other bright Mira star, χ Cyg, are similar. The comprehensive investigation by ROSENBERG (1906) gave extreme visual magnitudes at maximum of $3^m\!.3$ and $7^m\!.3$, but only a range of $4^m\!.5$ to $5^m\!.5$ for the vast majority of maxima, and 12^m to 14^m for the minima. Such changes in the light-curve are typical of the Mira stars, even if the ranges of maxima in o Cet and χ Cyg are taken as being extreme cases.

The star S Boo has a relatively regular light-curve (Fig. 24). LUDENDORFF (1928) introduced the following classes to describe the light-curves of the Mira stars.

Type α:	The rise is markedly steeper than the decline. The minimum, with only a few exceptions, is consistently wider than the maximum.
	Subdivisions:
Type α_1:	has curves with nearly, or completely constant phase at minimum of considerable duration (approximately 1/3 to 1/2 of the period) and generally with a very steep rise.
Type α_2:	The minimum no longer shows a constant phase of considerable extent, but is still very wide. The rise is generally very steep.
Type α_3:	Minimum is not so wide as in α_2, but the rise is still extremely steep.
Type α_4:	As α_3, but with a less steep rise.
Type β:	The rise is only slightly steeper than, or the same as, the fall, the light-curve being essentially symmetrical.
	Subdivisions:
Type β_1:	The maximum is sharper than the minimum.
Type β_2:	The maximum is as sharp or as flat as the minimum.
Type β_3:	The maximum is flatter than the minimum.
Type β_4:	The maximum is very wide and shows a constant phase of considerable duration.
Type γ:	Light-curves with waves (humps) on the ascending branch or with double maxima.
	Subdivisions:
Type γ_1:	Waves on the ascending branch.
Type γ_2:	Double maxima.

Examples:

α_1 ... Y Vel		β_1 ... R Boo		γ_1 ... R Aur	
α_3 ... o Cet		β_3 ... X Cam		γ_2 ... R Nor	
α_4 ... R Dra					

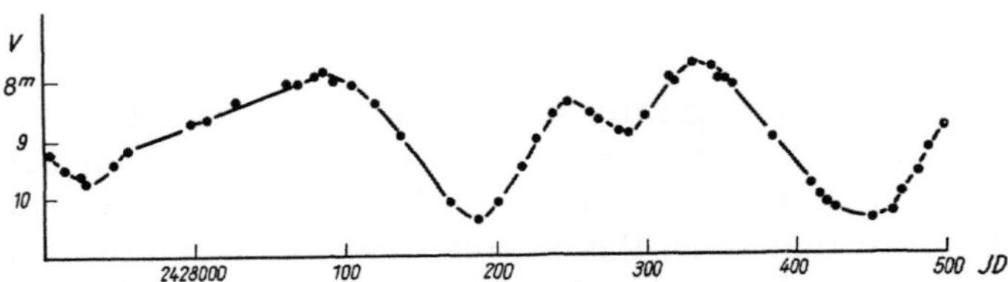

Fig. 25. Light-curve of V Boo (after HOFFMEISTER)

This classification by LUDENDORFF is purely based upon the phenomena observed, and is without any particular significance for the understanding of the physical processes involved. It is, however, of historical interest, and the reader will still encounter this classification scheme in various places in the literature relating to Mira stars.

As already intimated, no-one should expect a star to follow the mean light-curve ascribed to it all the time. Only rarely will two cycles be completely similar. In many stars, however, the departures are extraordinarily large. V Boo may be considered as an example, the mean period of which has been established as $258\overset{d}{.}8$, but where, however, the intervals between successive maxima can amount to $230^d{-}290^d$, while those of the minima only range between 250^d and 270^d. The corresponding great changes in the form of the light-curve often result in the appearance of double or even triple maxima (Fig. 25).

Reference may also be made here to the extensive studies by NIJLAND, which were published from 1930–1938, under the title "Mean Light-curves of Long Period Variables", and in addition to the comprehensive presentation by THOMAS (1932) and to a somewhat later statistical discussion of 357 stars by FEUCHTER (1967).

Period Changes

Those objects whose periods have shown considerable changes over decades are the subject of particular interest. Two cases will be mentioned here, R Aql and R Hya (see also WOOD 1975).

When discovered in 1856, *R Aql* had a period of about 348^d, which in the following 120 years shortened to 284^d. The process is still continuing in a regular manner. TURNER (1920) gave a set of 5 formulae for 5 periods of time, of which we give the first and the last:

$$\text{Epochs } 0{-}20: \quad M = 239\,9173 + 345\overset{d}{.}0\,E$$
$$\text{Epochs } 81{-}86: \quad M = 242\,5729 + 301\overset{d}{.}5\,E$$

A later discussion of the two stars by SCHNELLER (1965) gave the period of R Aql at Epoch E as being:

$$P_E = 348\overset{d}{.}980 - 0\overset{d}{.}554202\,E + 0\overset{d}{.}000552309\,E^2$$

Without the positive E^2 term, the period would finally reduce to zero, which is obviously not possible. Differentiating this equation, we find that the absolute minimum of the light-curve is reached at $P = 210^{\mathrm{d}}$, at epoch $E = 502$. This is approximately 400 years after the initial epoch, and consequently around the year 2250. Naturally, this calculation is largely only by way of an example, as with these stars it is very difficult to give reliable predictions over a long period of time. The second-power term had already been included in the old formulae given by TURNER and MÜLLER. MÜLLER added yet another, periodic term.

In $R\,Hya$, which can reach visual magnitude 4 and has been known as a variable since 1704, a very much less regular course of development has been observed. Then the period-length amounted to 500^{d}. For the period between 1903 and 1962 all 55 maxima have been verified by observation; the mean period was $400\overset{\mathrm{d}}{.}055$. If the overall period of time is divided into 4 parts, then the following individual values are obtained:

$$
\begin{array}{lll}
1903-1923 & P = & 405^{\mathrm{d}} \\
1923-1935 & & 415 \\
1935-1941 & & 400 \\
1941-1962 & & 386
\end{array}
$$

Since then the period-length has not changed significantly. PRAGER has already noted that the period does not vary continuously, but probably in discrete steps. In consequence, as we shall see in the next section, R Hya does not basically differ from normal Mira stars, except in the size of the jumps. T Cep is possibly a similar case.

A possible interpretation of the phenomenon of period-changes in Mira stars as a consequence of *helium-flash* activity has been given by WOOD and ZARRO (1981).

Period Changes in General

There is hardly one Mira star where a constant period will account for the maxima over a long period of time. In some cases the $O - C$ diagrams show very large departures, over a long time the maxima occurring earlier or later than the mean – that is to say, earlier or later than suggested by the mean period calculated from a very long series of observations (Fig. 26). Efforts have been made to explain these departures by means of an additional periodic term in the formula (a sine term), which has mostly resulted in a considerable reduction in the sum of the squared residuals. This was based upon the hypothesis that the changes in brightness are primarily controlled by periodic processes in the stellar interiors, which we know to occur in pulsating stars. It was thus a concept which, even at that time, was not without a certain physical plausibility. In particular MÜLLER in the first edition of "Geschichte und Literatur des Lichtwechsels der Veränderlichen Sterne" treated many Mira stars on this basis. As an extreme example we may give here the elements proposed by GUTHNICK for o Cet:

$$
\begin{aligned}
M = {} & 241\,5574.96 + 331\overset{\mathrm{d}}{.}6926\,E + 9\overset{\mathrm{d}}{.}5\sin(1\overset{\circ}{.}4\,E + 245\overset{\circ}{.}8) \\
& + 11\overset{\mathrm{d}}{.}5\sin(3\overset{\circ}{.}85\,E + 124\overset{\circ}{.}1) + 17\overset{\mathrm{d}}{.}5\sin(4\overset{\circ}{.}56\,E + 307\overset{\circ}{.}2) \\
& + 12\overset{\mathrm{d}}{.}3\sin(9\overset{\circ}{.}12\,E + 71\overset{\circ}{.}8).
\end{aligned}
$$

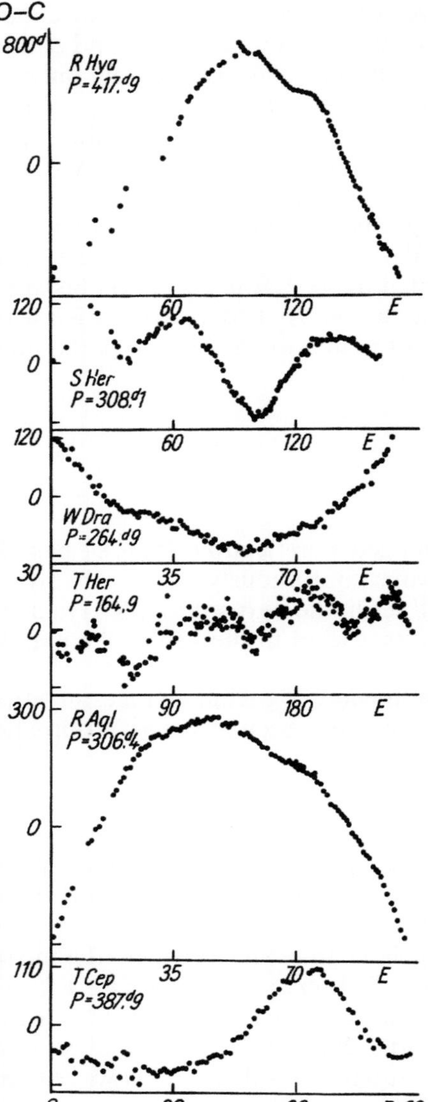

Fig. 26. $O - C$ curves for several Mira stars (after WOOD and ZARRO 1981)

The 4 sine terms in this formula imply periodic changes in the length of the period with cycles of approximately 233, 85, 72, and 36 years. With such ingenious formulae an excellent representation of the given observations can be attained. However, if the basic hypothesis were correct, it would be possible to predict the course of the light-curve for a long time into the future, and it was here that the procedure failed, not only in the case of Mira, but also in many other long-period variables. It is well-known that any arbitrary continuous function – and thus an $O - C$ diagram – can be represented to any desired degree of accuracy by multiple trigonometrical series (Fourier series). This is doubtless the basis for the apparent success of the procedure. However, if we examine a greater number of such diagrams (see Fig. 26), then we realize that in most cases the $O - C$ curves can be reproduced by a succession of linked straight lines. From a physical point of view this would mean that now and then the period values are subject to sudden changes, and that these are at irregular, unpredictable intervals. At Babelsberg, where from 1927 the yearly publication "Katalog und Ephemeriden Veränderlicher Sterne" was prepared, PRAGER and GUTHNICK recognized this state of affairs, and introduced the system of instantaneous elements. In these the epoch and period were taken such as to satisfy the current behaviour of the star, and were changed if observations indicated unequivocal deviations. One can in general reckon on being able to retain any instantaneous formula for 10 years, and in most cases for even longer.

The American theoretician STERNE (1934) brought a surprising new dimension to the discussion. He asserted that the period changes of Mira stars and other variables did not have to be real, as they could be explained by means of an effect which he called cumulative error. He showed that by means of dice, one can draw up $O - C$ curves that are very similar to those of stars.

We will try to explain this briefly in a qualitative manner. If we use two dice, then the smallest throw is 2, the greatest 12, and the mean of all possible values 7. This latter value is the analogue of the mean period of the star. If we were able to throw the mean value on all occasions, then adding the pips, we would obtain the series 0, 7, 14, 21, 28, etc. The figures actually thrown will deviate from this, in other words they will fall around the theoretical value in accordance with the rules of chance. If we assume that at some time a very small value (2 or 3) has been thrown after a previously regular series, then the sum of the pips will be less than expected, so that the $O - C$ is negative. There are now three possibilities for the following throw: it may again be too small, may give a figure close to 7, or may be well above average. In the latter case it will compensate for the deficit incurred by the previous throw. The first possibility increases the deficit, while the second leaves it unchanged. One thus has an approximate 2/3 probability that the following $O - C$ value will again be negative. So there is a certain tendency for any large chance deviation of the sum from the average value to be maintained. With a long series the compensation is such that a series of deviations of opposite sign has the same probability of occurring.

Reverting to the Mira stars, it may be noted that the problem may be reduced to a very simple question. Let M_C be a maximum, the nth in a series calculated with a constant period, and M_{Obs} the observed maximum which occurred markedly earlier or later. The question is: Does the new cycle begin at time M_C or M_{Obs}, and as a result is the next maximum $(n + 1)$ to be expected at time $M_C + P$ or at time $M_{Obs} + P$? The second possibility corresponds to the dice experiment, and it is only under these circumstances that the accumulation comes into effect. But we must seriously consider if this does actually occur under real conditions. For example, if the changes in brightness were controlled by a mechanical process, perhaps a pulsation, then it would suggest that the cause of the deviations should be sought in secondary effects, which do not influence the driving process. This would mean that the $O - C$ diagrams are not produced by cumulative error, but by real changes in the periods.

Objects with sudden, unpredictable, period changes are found not only in the Mira stars and the semiregulars, but also amongst the eclipsing stars (Sect. 4.5).

In complete contrast to the long-period variables, in which stochastic processes obviously play a large part, in the multi-periodic RR Lyrae stars, for example, a large portion of the light-curve may be calculated in advance (Sect. 2.1.2).

Physical Properties

Because of its brightness Mira Ceti is well-known. The *mass* is probably a little greater than one solar mass. Generally speaking one may fix the mass of Mira stars at about one solar mass. The maximum *diameter* of Mira, corresponding to minimum brightness, is between 310 and 540 million km, according to different sources (see also the work by WELTER and WORDEN 1980, which depends upon the determination of stellar diameters by speckle interferometry).

If Mira were located at our Sun's position, the Earth's orbit would lie under the surface, well inside the body of the star. From this we can easily gather how extremely low the mean density of these stars must be. The diameter of Mira stars determined from radiation measurements is at a minimum at the time of maximum brightness. The amplitude of the variation in diameter has a mean value of 18%, and the phenomena are therefore similar to those of the δ Cephei stars. Figure 27 shows the course of the changes in visual and bolometric magnitudes, temperature, diameter, and radial velocity for the star o Cet. The most significant fact is that the change in brightness of more than 6 magnitudes in the visual range contrasts with just the single magnitude change observed in the total (bolometric) radiation. Particularly involved here is absorption in the bands due to titanium oxide – zirconium oxide in stars of class S, or carbon compounds in other objects. At an effective temperature of 2300 K more than 96% of the total radiation falls in the infrared, $\lambda > 760$ nm, and at 1800 K approximately 99%. This fact will be of critical importance in the definitive explanation of the changes in brightness, which is still outstanding.

In recent years many publications have dealt with the observation of Mira stars and red semiregular, and irregular variables in the *infrared* (680–3400 nm); see EVANS (1976), CATCHPOLE et al. (1979), MENNESIER (1981). In the near infrared the amplitudes amount to a few magnitudes.

DEUTSCH was the first to investigate in detail the fact that red giant and supergiant stars in general, and Mira stars in particular, possess *extended circumstellar shells*, and are subject to great *mass loss* through stellar winds (see REIMERS 1977 for a detailed discussion).

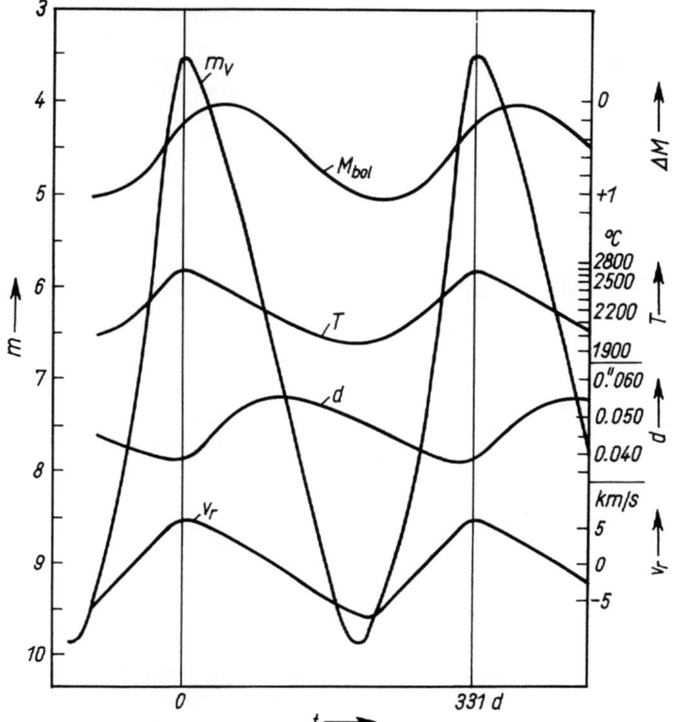

Fig. 27. Changes with time of magnitude, temperature, diameter and radial velocity in Mira Ceti

The existence of considerable mass loss is also suggested by the fact that low mass stars in an advanced stage of evolution (white dwarfs) are found in young clusters (the Hyades and even the Pleiades). In these star clusters only the most massive stars, however, have been able to move away from the main sequence, so the white dwarfs must be the result of very high mass loss.

Circumstellar shells and mass loss in red variables are indicated in many ways:

1) High-resolution spectra show absorption cores, *Doppler-shifted towards the violet*, in low excitation metallic lines. This indicates that above the photosphere a cool gas is expanding at 5–25 km/s.

2) Proof of *dust emission* in the *infrared* spectral region (silicates at 9.7 and 18 μm and silicon carbide at 11.2 μm; carbon dust in R, N, and C stars): The infrared excess in the radiation from Mira stars can be interpreted as the result of thermal emission from heated circumstellar shells.

3) In many Mira stars and class M supergiants expanding circumstellar shells are also indicated by *maser emission* from OH, H_2O and SiO in the *microwave region* (see REIMERS 1977, PERSI and FERRARI TONIOLO 1980 and CLARKE et al. 1981, amongst others). It would lead us too far to explain in full the origin of these spectral lines, which are observed with

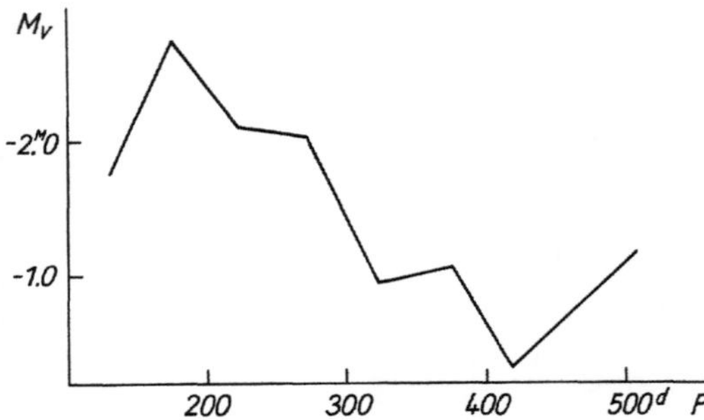

Fig. 28. Period-luminosity relationship for Mira stars of spectral class M, after Osvalds and Risley (1961). M_v is the mean visual absolute magnitude at maximum

Table 23. Period-luminosity relationship for Mira stars

P	M_V	M_J	M_H	M_K	M_L
91^d–149^d	-1.6	-3.4	-4.2	-4.4	-4.8
150–199	-3.0	-5.8	-6.6	-7.0	-7.4
200–249	-1.8	-5.7	-6.5	-6.9	-7.3
250–299	-1.6	-6.0	-6.8	-7.3	-7.7
300–349	-1.3	-5.6	-6.5	-6.9	-7.4
350–399	-0.8	-5.7	-6.6	-7.1	-7.7
400–612	-1.0	-5.7	-6.5	-6.8	-7.3

radio astronomical equipment, but they are particularly suitable for measurement of the motions of the shells, as the Doppler shifts in these lines may be determined very exactly (errors of about ± 1 km/s).

4) Further information about circumstellar shells is obtained through the measurement of *radio* and *EUV emission lines.*

Investigations of the expansion of shells and determinations of the mass loss rates in Mira stars are to be found in the work by Reimers (1977), Dickinson et al. (1978), Kafatos et al. (1977), and Wood (1979) amongst others. A few of these works investigated the links between mass loss and pulsation. Reimers (1975, 1977) found an empirical relationship between mass loss $\mathfrak{\dot{M}}$, mass $\mathfrak{M}$, radius R, and luminosity L for stars of late spectral type:

$$\mathfrak{\dot{M}} = -A\,\frac{LR}{\mathfrak{M}},$$

where $A = 4 \times 10^{-13}\ \mathfrak{M}_\odot$ per year. On the other hand, according to Willson (1981) the mass loss amounts to 10^{-5} to 2×10^{-6} solar masses per year.

Further details of the observation and theory of mass loss from cool stars in general may be found in the comprehensive account by Dupree (1981).

The *absolute magnitudes* of the Mira stars, and of the semiregular and irregular red variables, are in the first place determined from statistical parallaxes, based on objects with known proper motion and radial velocity. The classic work by Osvalds and Risley (1961) and the predecessors whom they cite, is well-known. They found a *period-luminosity relationship*, and this is illustrated in Fig. 28. The visual absolute magnitudes of the Mira stars lie between 0^M and -3^M; for the rarer Mira stars of classes C and Se the values of $M_v = -1.4$ and -1.6 apply. Later work in this field is that by CLAYTON and FEAST (1969), FOY et al. (1975), and CELIS (1981). ROBERTSON and FEAST (1981) analyzed the Period-Luminosity relationship for bolometric and infrared absolute magnitudes. Table 23 gives a summary of their findings. CELIS (1981) gives a three-dimensional correlation between period, spectrum and visual luminosity. On the basis of this P-Sp-M_v relationship it is possible to employ Mira stars as very exact distance indicators in the study of the structure of our Galaxy (Chap. 7).

Causes of the Changes in Brightness

As already intimated, an exhaustive explanation is still missing. It is certain that a *pulsation* is involved. The pulsation constant Q, which was introduced in our discussion of the δ Cephei stars, has a value of $0\overset{d}{.}096$ for the Mira stars. But there is yet another cause involved, which is that of *changes in the transparency* of the outermost layers to visual radiation, due to the formation of *carbon particles*, which are also now considered to play a part in interstellar extinction. As a result the stellar atmospheres would be dimmed by "smoke" or "soot", recurring periodically and controlled by the pulsations. The energy absorbed by this layer would be re-radiated in the longer-wavelength region as heat. This would explain the small bolometric amplitude. As already mentioned, maximum light corresponds to minimum diameter, in other words to the maximum density of the outer layers. However, this makes certain assumptions about the dissipation in such an absorption layer.

One fact plays a large part; supergiants lie close to the natural stability boundary, and under the physical conditions in this region very small variations in the flow of energy coming from the interior of the star may be sufficient to produce large effects in the outer layers. This is particularly so for the stars described in greater detail in Sect. 2.2.2, the semiregular and irregular variable supergiants of the α Ori and α Her type.

Another finding can be mentioned that supports this conclusion. STEBBINS and HUFFER (1930) observed photoelectrically 190 stars not known to be variable, and of spectral classes M0–M6. They found changes of 0.1 mag and greater in a third of them. They believe it possible that no red giant is of really constant brightness. If this opinion is confirmed, then variability would be the normal state in these stars, which immediately suggests that the explanation can be found in the process outlined above. Similar conclusions can be reached from the work by RICHTER, SCHAIFERS and WENZEL (1961).

In addition, a considerable amount of later work is available on the existence of graphite particles in the atmospheres of N stars. This began with research into the origin of interstellar material by HOYLE and WICKRAMASINGHE (1962). Investigations by FRIEDEMANN and SCHMIDT (1967) support the conclusions and also allow the possibility that graphite particles are involved in the changes of absorption occurring in the outer layers of a star, especially as the rise in temperature during the course of a pulsation can lead to re-vaporization of the particles.

Evolutionary State

Model calculations indicate that the Mira stars are objects of about one solar mass, which are located on the so-called *asymptotic giant branch* of the H-R diagram. As already mentioned, their shells are losing appreciable amounts of material.

Some authors have discussed the possibility that on account of this mass loss the Mira stars develop further into symbiotic stars or planetary nebulae; see Wood (1974), Kafatos et al. (1977) and Willson (1980, 1981).

Starting from the observed period distribution of Mira stars in the solar neighbourhood, Wood and Cahn (1977), Cahn and Wyatt (1978) and Willson (1980), try to draw conclusions, by means of theories of pulsation, stellar evolution and mass loss, about the *later evolution* of these objects into white dwarfs or planetary nebulae. They derive a theoretical frequency distribution for the mass of the white dwarfs that remain after the final loss of the circumstellar shells. The authors mentioned have developed a picture somewhat as follows: low-mass main sequence stars (of about one solar mass) develop into Mira stars towards the end of their evolutionary tracks. From these they turn into white dwarfs either directly, or through the intermediate stage of planetary nebulae. Massive stars (of several solar masses), on the other hand, do not pass through a Mira stage according to this theory.

The variables of spectral classes S, R, N, and C present a theoretical problem that is still poorly-understood. According to some authors these stars are the result of a short phase of stellar evolution, in which thorough mixing has taken place, whereby chemical elements, formed in the stellar interior, have been transported to the surface. Alternatively, through substantial mass loss the inner parts of the star have come to the surface. A comprehensive review of our knowledge of variable carbon stars is given by Alksne and Ikaunieks (1971).

2.2.2 Semiregular, Irregular and RV Tauri Stars

Both between the Mira stars and the semiregulars, and between the latter and the irregulars, there are *very ill-defined transitions*, so that classification is not always unequivocal. The physical properties of the various subgroups are so close that these may all be discussed together.

The semiregulars, and particularly the irregulars, are also red giants and supergiants, although occasionally somewhat earlier spectra (F, G, K) are encountered.

Four groups of semiregulars (SRa, b, c and d), 2 groups of irregulars (Lb, c), and 2 groups of RV Tauri stars are recognized, with the following characteristics (Figs. 29–36):

SRa: Giants of spectral classes M, C, and S that are distinguished from the genuine Mira stars by their smaller amplitude. The light-curves are very variable, yet in general the periods are maintained to the same extent as in the Mira stars. Emission lines are less frequent. A typical member of this subgroup is the star Z Aqr.

SRb: Giants of spectral classes K, M, C, and S. A form of period (cycle-length) is evident, but becomes inoperative from time to time. The changes in brightness are then irregular. Later the period returns, but with an altered phase.

SRc: Supergiants of spectral classes G8–M6, having almost completely irregular changes in magnitude with long wavelengths, low amplitudes and occasional standstills. Some well-known bright stars belong to this group, most of which were previously classified as irregulars.

SRd: Yellow giants and supergiants of spectral classes F–K: not a very homogeneous group. It contains objects that are related to the W Virginis stars, but which show irregularities at times, like S Vul, as well as some that were previously classed as RV Tauri-like, and yet others in which 2 periods alternate at irregular intervals.

Lb: Slow, irregular variables of intermediate and late spectral classes (F to M, C, and S), predominantly giants. The changes in brightness are characterized by slow variations without any sign of periodicity, or at the very most, only very slight indications of periodicity. The same applies to the following group.

Lc: Slow, irregular, variable supergiants of late spectral classes.

RV: The variables of the RV Tauri class belong to spectral classes F–K. The typical light-curve is very characteristic, as a form occurs resembling that of a β Lyrae star, but with relatively sharp maxima. After a while the secondary minimum deepens and becomes the primary minimum, so that temporarily δ Cephei-like forms of light-curve may occur. Alongside this type, designated *RVa* (Fig. 35), there exists the *RVb* group (Fig. 36), in which the typical changes in brightness are superimposed on a very long wave of greater amplitude, giving a total range of up to 5 mag. Around the time of maximum the spectra show H emission, and strong expansion shifts. The RV Tauri class exhibits subtypes and borderline cases, but the typical forms are very characteristic. It is therefore justifiable to restrict membership of this class to these more definite objects.

Typical members of these classes are:

SRa	Z Aqr, α Sco
SRb	V UMi, AF Cyg
SRc	α Ori, μ Cep, α Her
SRd	S Vul, UU Her
Lb	CO Cyg, BY Ser
Lc	TZ Cas
RVa	AC Her ($75^{\text{d}}5$), V Vul ($75^{\text{d}}7$), TW Cam ($85^{\text{d}}6$), R Sct ($140^{\text{d}}7$)
RVb	SX Cen ($32^{\text{d}}9$; 600^{d}), DF Cyg ($49^{\text{d}}8$; $780^{\text{d}}2$), AI Sco ($71^{\text{d}}0$; 965^{d}), R Sge ($70^{\text{d}}6$; 1112^{d}), RV Tau ($78^{\text{d}}7$; 1224^{d}), U Mon ($92^{\text{d}}3$; 2320^{d})

The values in parentheses are the period lengths.

The *diameters* of the red semiregulars and irregulars (α Ori, α Sco, α Her, among others), are of the same order of magnitude as those of the Mira stars. (See for example WELTER and WORDEN, 1980.) It is not surprising that

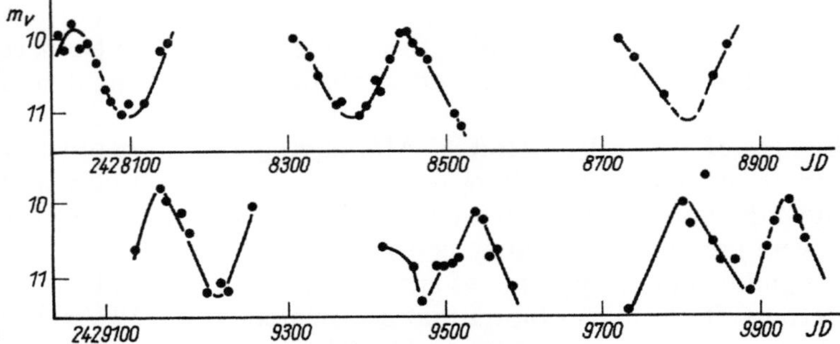

Fig. 29. Light-curve of Z Aqr (after Payne-Gaposchkin); SRa

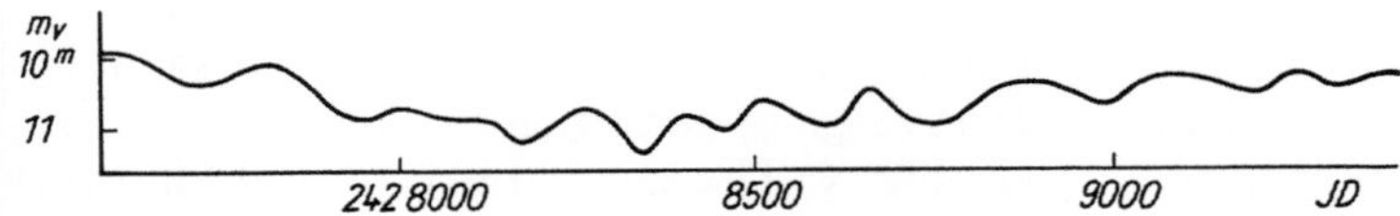

Fig. 30. Light-curve of CQ Cas (after Beyer); SRb

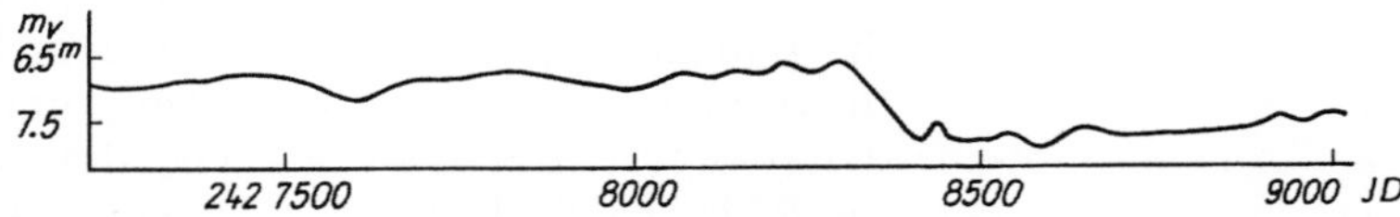

Fig. 31. Light-curve of TV Gem (after Beyer); SRc

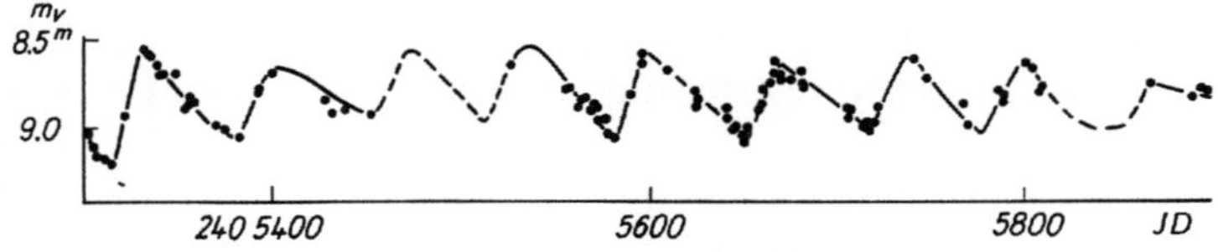

Fig. 32. Light-curve of S Vul (after Schönfeld); SRd

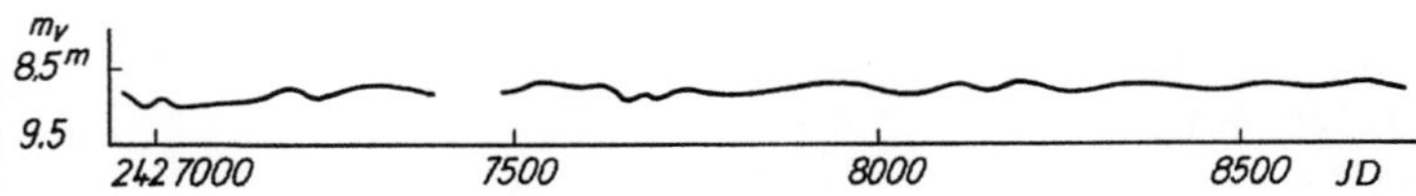

Fig. 33. Light-curve of CO Cyg (after Beyer); Lb

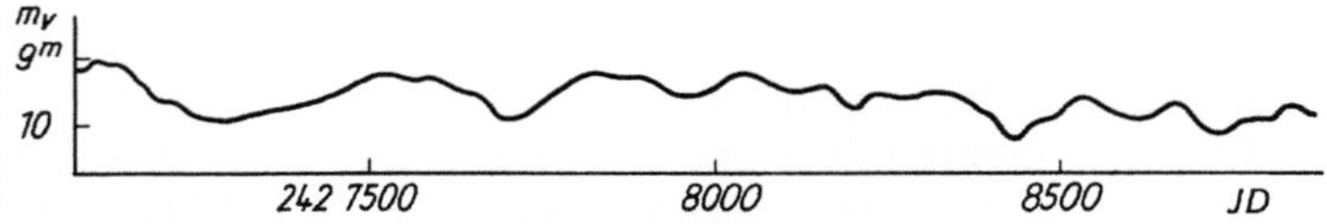

Fig. 34. Light-curve of TZ Cas (after Beyer); Lc

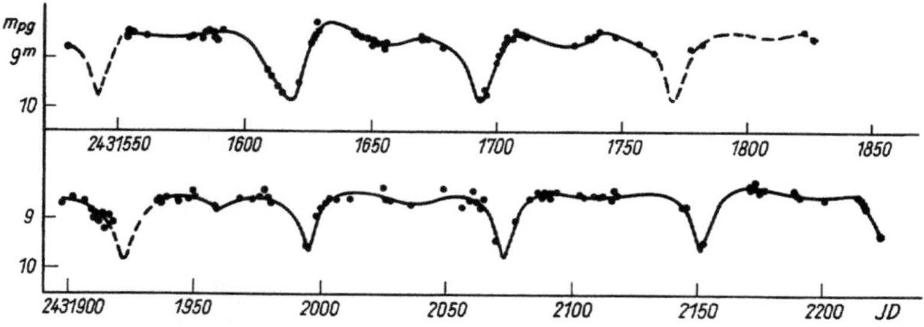

Fig. 35. Light-curve of V Vul (after AHNERT); RVa

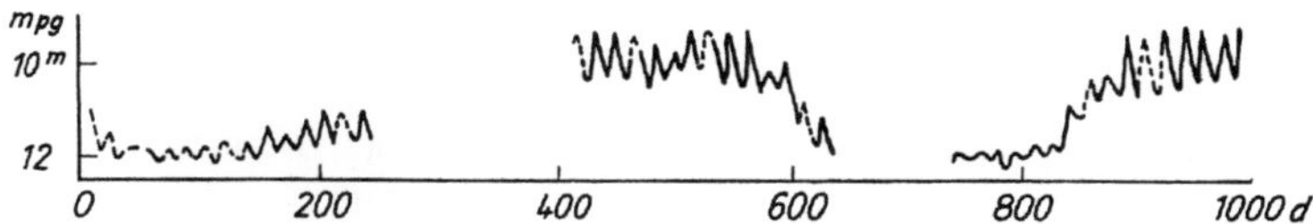

Fig. 36. Light-curve of SX Cen (from Harvard Observatory magnitude estimates); RVb

various methods, such as determinations from the energy radiated, or interferometric measurements, give significantly different values even for individual stars, as it is hardly possible to define a true surface with stars of such low density. With o Cet the uncertainty in the parallax makes it more difficult and increases the mean error in the diameter.

According to JOY (1942) and WILSON (1942), SRb stars of spectral class M have the same mean luminosity as Lb stars of spectral class M. Moreover, the galactic density gradients and the kinematic data for these two classes of variable agree within the limits of error, as do the spectral characteristics. We may therefore suspect that the distinction between these two classes of variables is purely arbitrary, being based upon photometric, rather than physical, differences.

Within the SRb and Lb classes, however, objects of very different characteristics have been lumped together, so that the SRb and Lb stars may be divided into at least *three* physically well-distinct *subclasses* (RICHTER 1967 a, for example):

1) The most numerous are giants of spectral class M (e.g. AF Cyg), and mean periods of 160^d. The spatial distribution within the Galaxy and the relatively high space velocities indicate that the majority of these objects can be classed as being of intermediate Population II.

2) Giants of spectral classes C and S (e.g. UX Cas) have, on average, a period of 280^d, and on the basis of their spatial distribution and mean velocity belong to Population I.

3) The so-called CH stars (e.g. V Ari, TT CVn), have very high space velocities and belong to extreme Population II.

Mean photographic absolute magnitudes:

SRa, SRb and Lb,	spectral class M	0^M
SRc and Lc	spectral class M	-4
SRa, SRb and Lb,	spectral class N	$+1$
SRa, SRb and Lb,	spectral class S	0
SRd		-1

According to FEAST (1980) the SRc stars approximately follow a period/ luminosity relationship:

$$M_{\text{bol}} = -7.20 \log P + 12.8$$

An absolute photographic magnitude of -3^M0 has been ascribed to RV Tauri stars at maximum (JOY 1952, BARNES and DU PUY 1975).

The absolute magnitudes also show that the SR and RV classes are by no means homogeneous. The classification of these stars is a very difficult exercise. The foregoing division, taken from the General Catalogue (GCVS) is not the only one possible. BEYER (1948), one of the most experienced observers of these stars, proposed a classification which differs in many respects, and which forms the basis (with some changes) for that used by SCHNELLER (1952). The difficulties mentioned are due to the fact that any division should not be based on the apparent behaviour alone, but should take into account the physical properties. At present, however, the latter are insufficiently known for this to be possible.

In contrast to the Mira, SR, and L stars, RV Tauri objects are not sources of OH radio-frequency radiation (BOWERS and CORNETT 1974).

The RV Tauri phenomenon is not yet physically understood. Model calculations by DEUPREE and HODSON (1976) attempted to explain the cause of the alternating amplitudes as convection, variable with time, and with a high amplitude. Some conceptual models have been discussed by DAWSON (1979).

A comprehensive explanation of the problems posed by slow variables can be found in IKAUNIEKS (1971). Objects of spectral classes R, N, and C are treated by ALKSNE and IKAUNIEKS (1971); see also ALKSNIS and ALKSNE (1977).

Table 21 gives the period distribution, and Table 24 the spectral classes, of semiregulars and irregulars in comparison with the Mira stars.

Table 24. Frequency distribution of spectral classes of Mira stars, semiregulars (SR) and irregulars (L)

Spectrum	Mira	SR	L	%
K	1	59	42	5
M	865	507	402	80
S	48	18	16	4
R, N, C	53	99	108	11
Total	967	683	568	100

2.3 Non-Radial Pulsators

Non-radial pulsations have already been mentioned in Sect. 2.1.4 as a possible cause of part of the variability of δ Scuti stars. Low-amplitude, *transverse* waves are involved, running across the surface of the star, and producing slight changes in brightness, generally only detectable by photoelectric methods. This phenomenon is rather well-understood mathematically, but quite obviously still awaits a final physical explanation in terms of the excitation mechanism and also integration with theories of stellar evolution. A possible cause of non-radial oscillations is tidal action, due to the existence of a companion star in a binary system. For the time being, however, there is no observational evidence that all non-radial pulsators are binary stars.

Something is said in the two following sections about the formal grounds for accepting that such oscillations occur.

2.3.1 β Cephei Stars

Variables of the β Cephei class (also called β Canis Majoris stars) form a sharply-defined group on the Hertzsprung-Russell Diagram, with a spectral class between B0.5 and B2, and luminosity class IV or III. The amplitudes of the changes in magnitude are usually about 0.1 mag in the visual, and periods are between 3 and 7 hours. There is a displacement between the radial velocity and magnitude curves (the periods of both are identical), so that the greatest velocity towards the observer occurs at phase 0.25 in the light-curve. In approximately half of the known β Cephei stars a modulation of the magnitude changes exists. This is seen as interference between two slightly different periods. From a list given by STERKEN and JERZYKIEWICZ (1980), which contains 37 objects, we have chosen, somewhat arbitrarily, the 13 stars no fainter than $m_v = 4\overset{m}{.}0$ given in Table 25.

Table 25. Bright β Cephei stars

Stars	m_v	Spectrum	P
ϑ Oph	$3\overset{m}{.}3$	B2 IV	$0\overset{d}{.}1405$
γ Peg	2.8	B2 IV	0.1518
β Cru	1.3	B0.5 III	0.1605
ε Cen	2.3	B1 III	0.1696
ν Eri	4.0	B2 III	0.1735
α Vir	1.0	B1 IV	0.1738
β Cep	3.2	B1 III	0.1905
$\varkappa$ Sco	2.4	B1.5 III	0.1999
λ Sco	1.6	B1.5 IV	0.2137
σ Sco	2.9	B1 III	0.2468
β CMa	2.0	B1 II-III	0.2513
α Lup	2.3	B1.5 III	0.2599
β Cen	0.6	B1 III	0.30

The shortest period is shown by IS Vel (= HD 68324) ($0\overset{d}{.}108$, STERKEN and JERZYKIEWICZ, loc. cit., p. 106), and the longest ($0\overset{d}{.}30$) by β Cen, included in Table 25 (BALONA 1977). A Period/Luminosity relationship does not seem to exist. This finding also leads to "the idea that the primary photometric periods of β Cephei variables may correspond to a variety of non-radial oscillation modes" (STERKEN and JERZYKIEWICZ, loc. cit., p. 123, see also JERZYKIEWICZ and STERKEN 1979).

A piece of work, which was also very informative as regards method, was published by JERZYKIEWICZ (1978) on the analysis of the star DD Lac (= 12 Lac), where particular attention was given to the existence of the *equidistant frequencies* (= number of magnitude cycles per day = $1/P$).

He pointed to the fact that LEDOUX (1951) had already shown that closely neighbouring frequencies in a non-radially pulsating variable could occur due to the slow rotation of the star. If f is the frequency of one of the stationary non-radial oscillations, which are possible according to model calculations and which act symmetrically about the rotational axis, then observable frequencies for further potential oscillations are obtained from the formula:

$$f_m = f - mk\Omega \quad (m \text{ is an integer})$$

where Ω is the angular velocity of the star, and k is a parameter dependent upon its internal structure (amongst other factors). The number m can take on values within a range that is determined by f. For a fixed fundamental frequency f we always find between any two consecutive secondary frequencies f_m, the same difference $f_m - f_{m+1} = k\Omega$. One speaks of "equidistant" frequencies, and these represent an important characteristic of non-radial pulsations in slowly-rotating stars. If $m < 0$, then the waves described by f_m travel in the direction of the rotation; when $m > 0$ they run in the opposite sense. When $\Omega = 0$ (the star is not rotating), the phenomenon is not present, and it can similarly be shown that fast rotation disturbs the effect.

By the inclusion of crucial spectroscopic findings JERZYKIEWICZ detected in DD Lac (= 12 Lac) the frequencies given in Table 26 (in cycles per day, slightly rounded; P_i are the corresponding periods).

Table 26. Periods in the β Cephei star 12 Lac

i	f_i	P_i
1	5.1793 d^{-1}	0 1931 d
2	5.0665	0.1974
3	5.4901	0.1821
4	5.3347	0.1875
5	10.5140	0.0951
6	4.2405	0.2358

It will be seen that $f_4 - f_1 = f_3 - f_4$ (equidistant triplet) and that the combination frequency $f_1 + f_4 = f_5$ also occurs, as does another frequency f_2, related to an additional fundamental frequency f (see the above formula),

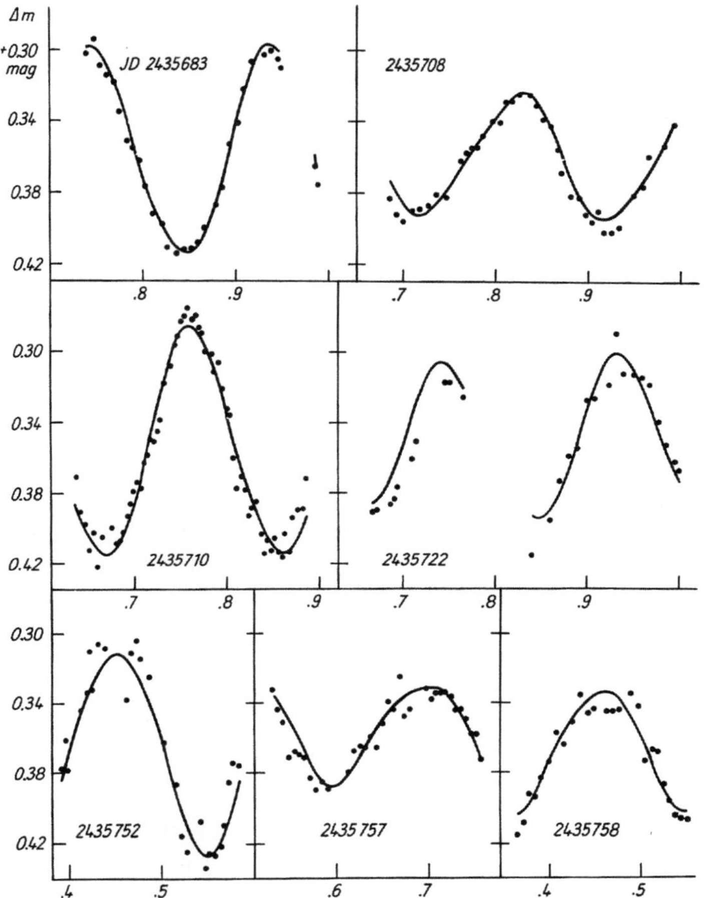

Fig. 37. Light-curves (yellow spectral region) of the β Cephei star DD Lac. The points are photoelectric measurements obtained through an international campaign, and the lines indicate the theoretical curves calculated from the frequencies listed in Table 26 (after JERZYKIEWICZ 1978)

as well as the f_6 term which is, for the time being, indeterminate. This example demonstrates the complexities of the light-curves, as is also observed in other β Cephei stars. Some portions of the light-curve of 12 Lac are shown in Fig. 37.

The cause of this form of variability is still unknown, despite intensive research. In all cases it seems that an ionization zone (such as that of He^+) is out of the question as a driving mechanism of the type found in classical pulsations. This is because at the high temperatures of early B stars these zones lie much too close to the stellar surface to be sufficiently effective.

In connection with the β Cephei stars at least two new classes of B-variables have been discussed recently (LE CONTEL et al. 1981). First there are four variables (amplitudes of about 0.02 mag!) with spectra around B2 IV or

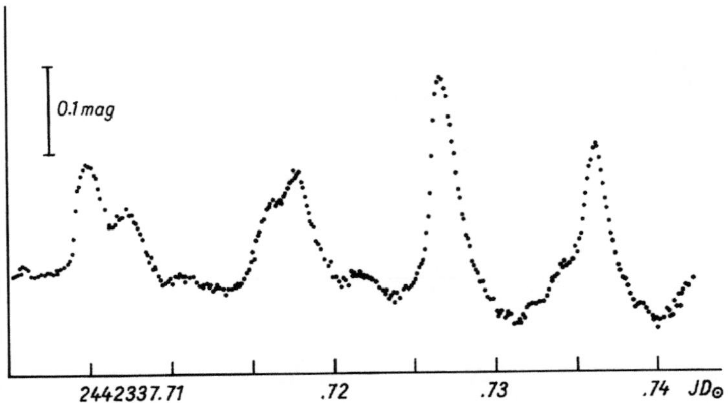

Fig. 38. Light-curve of the ZZ Ceti star ZZ Psc (after PETTERSEN 1980)

V, but with very short periods ($0\overset{d}{.}02 - 0\overset{d}{.}03$, among which is χ Cen) (JAKATE 1979). The second group is not homogeneous and consists of main sequence, or giant, B stars with photometric periods $> 0\overset{d}{.}3$, small amplitudes (a few hundredths of a magnitude) and rather irregular light-curves, a typical member being perhaps 53 Per. We shall not discuss the two groups any further, as their status is still poorly established.

2.3.2 ZZ Ceti Stars

The ZZ Ceti stars are *variable white dwarfs*. Their spectral class is generally DA (D = white dwarf), their surface temperature about 12 000 K, and their density about 1 million $\mathrm{g\,cm^{-3}}$ (see Sect. 1.2). Characteristic cycle-lengths for the variation in brightness are 100–1000 s, and the amplitudes are 0.3 mag at most. The first such variable was discovered accidentally by LANDOLT (1968) in the course of a photometric programme of investigation of the stars in the direction of the dark cloud in Taurus. It had already been included in a list of possible white dwarfs drawn up by HARO and LUYTEN as HL Tau-76, and since 1971 has been called V 411 Tau. Its amplitude (about 0.3 mag) is one of the greatest yet known. On the other hand, ZZ Ceti, the prototype after which these stars were named in 1974, has variations in brightness only slightly greater than 0.01 mag. The light-curves are similar, apart from the scale, to those of classical pulsating stars with multiple periodicities. Compare, for example, the curve of the ZZ Ceti star ZZ Psc given in Fig. 38 with that of AQ Leo in Fig. 13. It may be seen, however, by calculating the pulsational parameter $P\sqrt{\bar{\varrho}/\bar{\varrho}_\odot}$, that normal radial pulsations cannot be present. From the data given above this amounts to about 5^{d}, as against the common value $Q = 0\overset{d}{.}03$. Modern models therefore invoke non-radial pulsations, although many difficulties still complicate the interpretation.

So far as is known at present, all ZZ Ceti stars are *multiperiodic*, and more than 20 periods may occur simultaneously in a single object. The differences

Table 27. ZZ Ceti variables

Star	Designation	Primary periods	Mean amplitude	$\bar{V}$
L 19-2	MY Aps	114; 192 s	0.03 mag	$13^{m}\!\!.75$
R 548	ZZ Cet	213; 274	0.012	14.10
G 117-B 15 A	RY LMi	216; 312	0.05	15.52
BPM 31 594	VY Hor	310; 617	0.21	15,03
GD 385	PT Vul	252; 564	0.03	15.50
GD 99	VW Lyn	260; 590	0.07	14.55
G 207-9	V 470 Lyr	292; 318; 557; 739	0.06	14.64
HL Tau-76	V 411 Tau	494; 625; 746	0.28	14.97
BPM 30 551	AX Phe	298; 823	0.22	15.26
R 808	TY CrB	833	0.15	14.36
G 29-38	ZZ Psc	820; 930; 1020	0.27	13.10
G 38-29	V 468 Per	929; 1020	0.22	15.63
GD 154	BG CVn	780; 1186	0.10	15.33

between the frequencies occasionally show the typical behaviour described for the β Cephei stars. The stability of the periods differs greatly from star to star, and runs the whole gamut from exceptional stability (relative change 10^{-12}) to marked variation within a few hours.

Comprehensive reviews, for example those by HANSEN (1980) and PETTERSEN (1980), give 13 ZZ Ceti variables known up to that date (Table 27). The unusual designations (column 1) derive from catalogues of searches for faint blue stars, objects of large proper motion, etc. As will be seen, but could not otherwise be expected, these variables have very faint apparent magnitudes, and this fact, together with the short periods, sets very exacting requirements for the photoelectric equipment needed for their observation.

Photoelectric measurements by McGRAW (1978) in a special four-colour system (the Strömgren system) have shown in V 411 Tau and ZZ Psc that, within the limits of observational accuracy, the changes in brightness can be explained by variations in the temperature alone; the radius of the objects remains constant. This may be direct evidence that radial pulsations do not occur in the ZZ Ceti stars. It is all the more remarkable that on the Hertzsprung-Russell Diagram these variables lie exactly within an extension of the pulsational strip that applies to the classical pulsating stars. They possess just that range of temperatures, where the evolutionary cooling of white dwarfs gives rise to a hydrogen-ionization zone in an outer shell. ROBINSON and McGRAW (1976) have therefore proposed that in these variables it is again the Kappa mechanism (Sect. 2.1.2) that drives the pulsations.

While in the ZZ Ceti variables of spectral class DA, discussed up to now, it is hydrogen which predominates in the atmosphere (as in other normal stars), it is helium in the rarer, class DB white dwarfs. In the DB stars it appears that non-radial pulsations also exist, as in the case of GD 358, recently discovered by WINGET and VAN HORN (1982). Its amplitude amounts to 0.3 mag, and 26 periods with values between 140 s and 950 s were found. The excitation and physical properties of the pulsation could be analogous to those of the DA stars.

A possible completely new class of variable star was discovered by McGraw et al. (1979) by means of a special photometer on the Multi-Mirror Telescope, which consists of six 1.8 m mirrors, on Mt Hopkins in the U.S.A. In the 14th-magnitude star PG 1159−035 they found a double-period variability of approximately sine-wave nature, with periods of 460 s and 539 s, and with the very small, but definite, amplitude of 0.03 mag. An interesting point is that the spectrum indicates a surface temperature of at least 120 000 K. It shows no signs of hydrogen and consists of a blue continuum and one He II emission line. The temperature of this obviously degenerate star is considerably higher than in a white dwarf. The evolutionary state and causes of the variability are rather uncertain. Some proposed models have been discussed by Starrfield et al. (1983 a, b), but the extremely high temperature and the presence of pulsations (if this is truly what is observed) do not really fall into place.

3. Eruptive Variables

We intend to include under this heading all the stars where eruptive or explosive processes either produce the magnitude changes, or are at least involved.

The variability is frequently marked by rapid, short, irregular changes in magnitude, or by outbursts of great amplitude. This does not prevent a few less striking forms from being included when there is a reason for assuming that they have basically similar causes.

The eruptive variables are divided into various physical groups according to whether the eruptive or explosive activity takes place in a circumstellar shell, in layers close to the surface of the star, or in the stellar interior. Additional factors are whether the object is a single star or a binary system, and the existence of weak or strong magnetic fields.

So much new observational material has become available in recent years concerning the eruptive variables – above all by observations in unconventional spectral regions and through spectroscopic research – that we have made considerable headway in understanding the physical processes at work in the various classes of object. We may mention:

1) The existence of X-ray binaries and of pulsars was completely unknown in the middle sixties.
2) The discovery that the majority of eruptive variables are radiating detectable hard and soft X-radiation, as well as the first measurements in the extreme ultraviolet (only possible in the last few years), have significantly contributed to our physical understanding of these interesting objects. (See CORDOVA et al. 1981 a and 1981 b, and the further references cited there.)
3) The recognition that the U Geminorum stars are not simply an extension of the group formed by the novae, with smaller amplitudes, less energetic outbursts and shorter intervals, but that the two groups are completely different.
4) It has also been shown that it is not always possible, on the basis of the light-curves alone, to identify objects as belonging unequivocally to one or other of the physical groups of eruptive variables.

Three examples may be given:

WZ Sge behaves, from a photometric point of view, like a typical recurrent nova (1913, 1946, 1978) with an amplitude of about 8 mag. On the other hand, the spectroscopic changes during the 1978 outburst are characteristic of U Geminorum stars. MCLAUGHLIN (1945) had already noted that, on the basis of its high proper motion, WZ Sge must be fairly close to us in space and could thus not be a true nova. However, the other recurrent novae observed spectroscopically up to now show typical nova spectra at outburst. In passing it may be mentioned that other observational data for WZ Sge are also atypical for novae.

The second example is *Nova Cyg 1975* (= V 1500 Cyg). On account of its amplitude of > 18 mag, exceptional for a nova, it was initially taken by some authors to be a galactic supernova. However, this is contradicted by the spectroscopic findings.

Thirdly, the star *FU Ori* was originally taken to be a typical slow nova. Only after more precise investigations was it realized that this object should be classified with stars at an extremely early stage of evolution.

3.1 Eruptive Binaries

3.1.1 Cataclysmic Variables – a Summary

If the distance between the two components of a binary system is very small – about as great as the diameter of the larger of the two stars – then mutual gravitational effects produce very strong tidal forces. The high orbital velocities of the two stars also cause considerable centrifugal forces. If the larger of the two stars approaches its stability limit, material streams away from this larger (and less dense) secondary component, under normal circumstances through the L_1 *Lagrangian point*. Exact calculations of the trajectories of individual particles in the stream show that, due to the conservation of angular momentum, a high percentage of them gather either in an *accretion disk* around the denser, "primary" component, or, when controlled by strong magnetic fields, on the primary component itself (Fig. 39). A detailed description of the geometric properties which may arise in binary stars, according to the mass, density and distance of the two components, follows in Chap. 4.

Since 1938 the term "cataclysmic variables" has become accepted for eruptive binaries (see KRAFT 1963 and PAYNE-GAPOSCHKIN 1977 a). This name is derived from the Greek word "kataklysmos" which means "flood, deluge, catastrophe". The expression "cataclysmic variables" may help to illustrate the fact that these objects are, from time to time, flooded with energy and mass, which may be released either slowly or rapidly, and which (in rare cases) may have catastrophic effects upon the object. Unfortunately this class of objects is not uniformly defined in the astronomical literature, and this can lead to confusion in individual cases. In this book we intend to follow the most frequently-used definition, in that we include amongst the cataclysmic variables all those binaries in which a red dwarf (or subdwarf) star and a white dwarf mutually interact (see Sect. 3.1.2–3.1.4). As we shall see, these are the novae, the U Geminorum stars, and the AM Herculis stars. The symbiotic stars and X-ray binaries are not included under the cataclysmic variables, and this is not completely consistent, as in the second type of

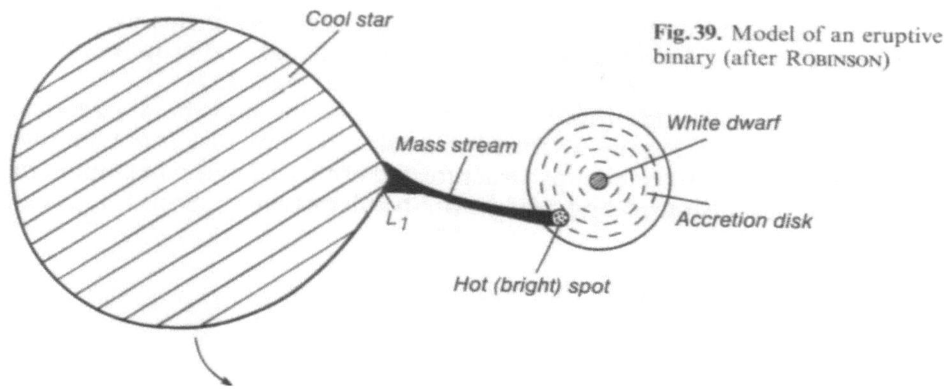

Table 28. Eruptive binaries

Primary component	Secondary component			
	Main sequence (or subgiant)	Giant	White dwarf	Neutron star
Main sequence (or cool subdwarf)	(Classical interacting eclipsing binaries)		–	–
White dwarf — strong magnetic field	Polars (AM Her), Novae (?)	Symbiotic stars (Z And, very slow novae, recurrent novae)	AM CVn stars	–
White dwarf — weak magnetic field	Dwarf novae (U Gem), Novae			
Neutron star — strong magnetic field	Low-mass X-ray pulsars (HZ Her)	Massive X-ray pulsars, Symbiotic X-ray stars (V 2116 Oph)	KZ TrA and 4U 1915−05?	Binary pulsar (PSR 1913 + 16)
Neutron star — weak magnetic field	X-ray bursters			
Massive compact object ($\geqq$ 3 solar masses)	?	V 1357 Cyg (= Cyg X-1)	?	?

object we do indeed find "floods of energy and mass" of a very high order.
Table 28, which anticipates the next sections, indicates (to the best of current
knowledge) the type of stellar components that are involved in the various
forms of eruptive binaries. Here, and in the following sections, the most
compact object is described as the primary. Occasionally in the literature
different definitions are found, particularly in regard to symbiotic and X-ray
stars.

The classical interacting eclipsing binaries, included in this list, are not reckoned to be eruptive binaries. Their mass-exchange and eruptive activity are relatively small. These stars are discussed in Chap. 4.

Quasars, which according to recent suggestions may be a large-scale version of the cataclysmic variables (where the accretion disk is not homogeneous but has a marked clumpy structure, fed not by a single companion, but by numerous disintegrating stars) will be discussed in Sect. 5.3 (Active Galaxies).

3.1.2 Novae

Classification

The appearance of a "new star" means that, completely unexpectedly, a star has appeared where previously no star was visible – in most cases not even with a telescope. Normally the very bright phase lasts a few days, and after a few weeks – depending upon the maximum brightness it attains and differing in individual cases – the star disappears again from naked-eye visibility. Such a sight may make a great impression on a layman who tends to regard the heavens as being unchanging, and accordingly finds it difficult to conceive that a body like the Sun can suddenly appear and then vanish again. Novae of great apparent magnitude are very rare indeed. Between 1900 and 1980 only three really outstanding cases were observed: Nova GK Per in 1901 (0^m2), Nova V 603 Aql in 1918 ($- 1^m1$), and Nova CP Pup in 1942 (0^m5). It will be noted that Nova V 603 Aql was only slightly fainter than Sirius, and that both of the others reached about the brightness of Vega.

From the scientific point of view, the phenomenon naturally appears rather different. A faint blue star, the *prenova*, is found on old photographic plates at the position of the nova. After the outburst a stage is reached, often only several years later, which is known as the *postnova*, and which is recognizable by spectral peculiarities.

The course of the light-curves serves to differentiate the various classes. The amplitudes mostly lie between 7 and 16 magnitudes, but the rise and fall take place at very different rates, and the shapes of the light-curves differ substantially. Four classes have been defined:

Na: Fast novae. The rise is very steep and it lasts one or, at the most, just a few days; the decline is such that the brightness reaches 3 magnitudes below maximum within 110 days at the latest, and generally much sooner. Examples are Nova GK Per 1901, Nova V 603 Aql 1918 and, the extreme case, Nova V 1500 Cyg 1975 (Fig. 40).

Nb: Slow novae. The 3-magnitude decline takes more than 100 days. Many of these novae show a deep, wide minimum about 4–5 months after maximum. This is followed by recovery to a magnitude approximately equal to that expected in an undisturbed decline. Examples are T Aur 1891, DQ Her 1934 (Fig. 41), V 732 Sgr 1936, V 450 Cyg 1942.

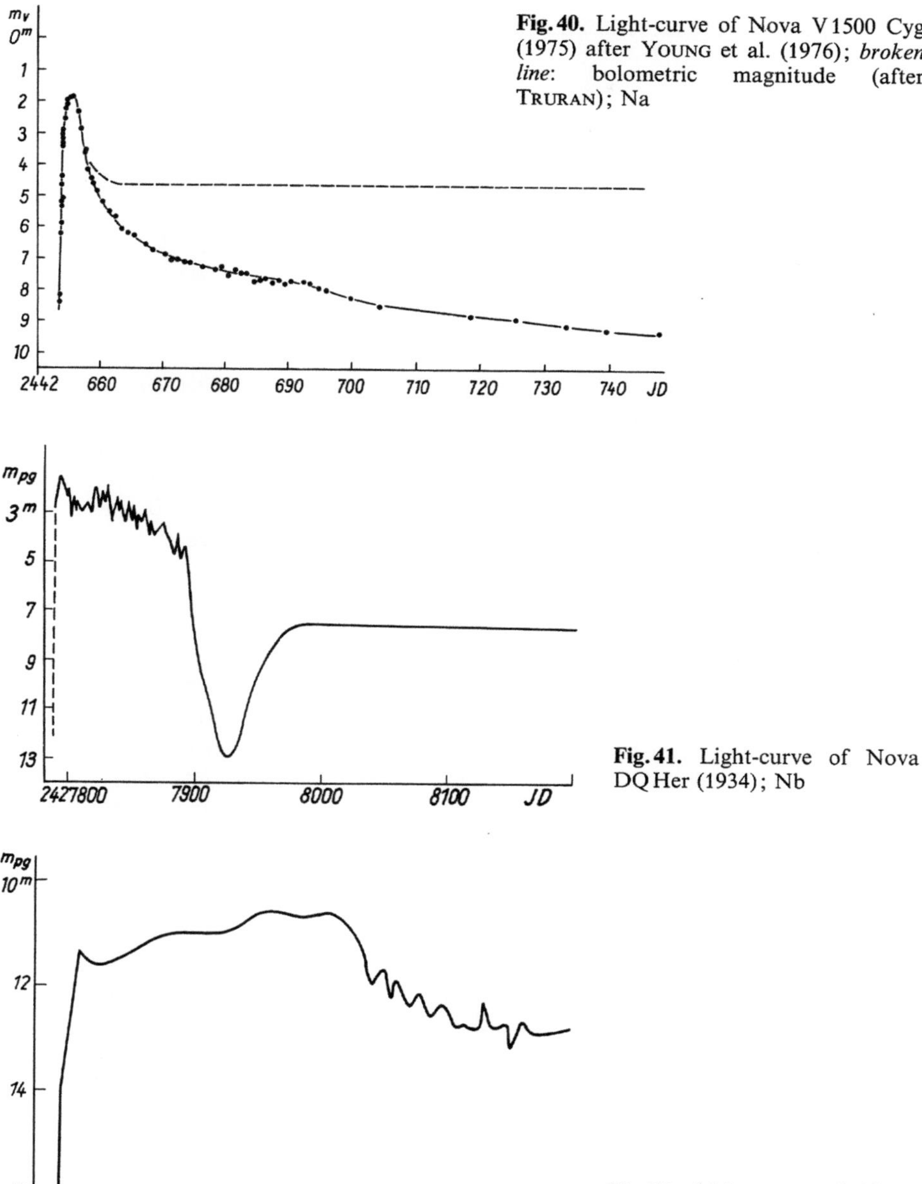

Fig. 40. Light-curve of Nova V 1500 Cyg (1975) after YOUNG et al. (1976); *broken line*: bolometric magnitude (after TRURAN); Na

Fig. 41. Light-curve of Nova DQ Her (1934); Nb

Fig. 42. Light-curve of Nova RT Ser (1909); Nc

Nc: Very slow novae. In 1915 the prototype, RT Ser, rose slowly to 10^m5, remained at this level for almost 10 years, and then began to fade very slowly, reaching 14^m in 1942. The characteristics expected in a nova were shown, quite unequivocally, in the spectrum (Fig. 42).

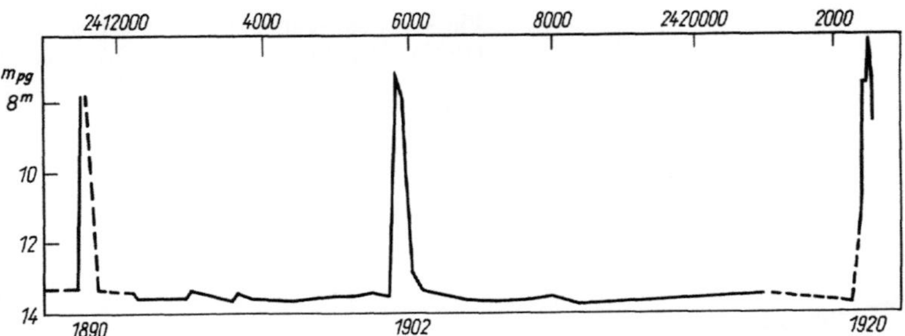

Fig. 43. Light-curve of Nova T Pyx from Harvard Observatory estimates; Nr

Nr: Recurrent novae (Fig. 43). These are novae where more than one outburst has been observed. In the final analysis probably all novae are recurrent, and the fact that as yet only one outburst has been observed in most novae is presumably only because the intervals are generally very long. A summary of the known recurrent novae is given in Table 30.

In the literature another class is occasionally encountered:

Nl: Nova-like variables. This is a very varied, heterogeneous class of stars, where, owing to a lack of suitable photometric and spectroscopic investigations, insufficient research has yet been carried out for them to be classified as belonging to any other, more definite, group of objects. Most must be novae at minimum, where no outbursts have been recorded in historical times, and some may be polars or unrecognized symbiotic stars (see Sect. 3.1.6) or else photometrically atypical U Geminorum stars (UX Ursae Majoris stars, Sect. 3.1.3).

The concept of a "dwarf nova" also exists. This term is used for the U Geminorum stars on the basis of the behaviour discussed at the beginning of Sect. 3.1.3. They are dealt with there in greater detail.

Nomenclature

The designation of novae is by no means fully consistent. When only a few cases were known, they were named according to the constellation and the year, e.g. Nova Per 1901. If several novae had been observed in one constellation, they were numbered N_1, N_2, etc. At first the idea behind this was that novae were something special, and not simply a sub-group of variable stars. Since then, however, advances in astrophysics have brought them into this general category, and they have been appropriately, and retrospectively, named. So Nova Per 1901 is now GK Per; Nova Her 1934, DQ Her; Nova Cyg 1975, V 1500 Cyg. Both designations are used in this book.

Behaviour

The *course of a nova outburst* at visual wavelengths is roughly as follows: the prenova is a blue (hot) object, which in many cases shows slight, irregular changes in brightness. (Only in very rare cases, e.g. V 446 Her, are these as great as 2 mag.) A catalogue of all known light-curves of novae prior to outburst is given by ROBINSON (1975). The rise is generally very steep with increases in brightness of 7 10 mag in 24 hours. Shortly before maximum a short standstill or a slight decline is observed, after which the last rise of about 2 magnitudes takes place. The maximum is generally sharp, except, of course, in the very slow novae of the RT Serpentis type. The decline in magnitude after maximum is smooth initially, down to about 3.5 magnitudes below maximum. Rapid, quasi-periodic variations then appear having amplitudes of about 1 mag. The wavelengths are mostly between 5 and 10 days, but cases of irregular changes in magnitude also occur. In the slow novae this phase ends with a sharp decline to the transitional minimum. This is followed by a second steep rise; after this the decline continues with lesser fluctuations. In normal fast novae the quasi-periodic oscillations die away at this phase. In both forms the transition to the slow decline occurs at about 6 mag below maximum magnitude. The star then gradually approaches the quasi-stable postnova state (also known as the exnova), which approximately corresponds to the prenova, and often shows rapid, low-amplitude changes in magnitude.

One source of error in classification is worth mentioning. When such a variable is discovered there is a danger that it may be included amongst the T Tauri and similar stars, as the rapid, irregular changes approximately correspond to those of this class. Only proper spectroscopic examination (or a fortuitous outburst!) can show a clear distinction. Examples of old novae which were previously classified incorrectly are EM Cyg and VSge.

Statistics

According to PAYNE-GAPOSCHKIN (1957, 1977 b) 177 novae are known, of which only 61 were observed at minimum. Table 29 gives the amplitude distribution of these 61 novae.

Table 29. Amplitude distribution of novae

Amplitude interval	Classical novae	Recurrent novae
6– 7 mag	2	0
7– 8	2	3
8– 9	6	1
9–10	8	0
10–11	11	1
11–12	12	0
12–13	4	0
13–14	6	0
14–15	3	0
15–16	1	0

Extreme ranges were shown by Nova Pup 1942 (more than 16.6 mag) and Nova Cyg 1975 (more than 18.8 mag). Nova Car 1970 had an extremely small amplitude of 3.0 mag, but it is very probable that this object is erroneously classified. The most frequent amplitude in the classical novae lies between 11 and 12 mag; in recurrent novae the amplitudes are below 9 mag. The data are, however, biased towards too low a value for the amplitudes, as two-thirds of all novae are not observed at minimum.

The tables given by PAYNE-GAPOSCHKIN (1957, 1977 b) show the date of outburst, the apparent magnitudes at maximum and minimum, and the galactic co-ordinates. The true maximum is only rarely properly established from observations owing to its short duration. It is possible, however, to deduce the maximum brightness to a high degree of accuracy from the very regular course of the initial decline. Another list published by the same writer (1958) is limited to 81 relatively well-observed cases.

In general, the mean *absolute magnitude* of novae *at maximum* may be taken as $- 7\overset{M}{.}6$; it naturally depends upon the reliability of the distance determinations. Two methods have become very important for the study of distances of novae. One relies upon the fact that the expansion velocity of the ejected nebular shell can be estimated both from the radial velocity in km/s, and from the increase in the angular diameter (in seconds of arc) of the nebular disk. The other possibility derives from the fact that novae, because of their high absolute luminosities, can be observed in other stellar systems of known distances, for example in the Andromeda Galaxy, M31, the spiral galaxies M33 and M81, the Magellanic Clouds and other systems. The individual values for mean absolute magnitude obtained by these various methods range between $- 6\overset{M}{.}7$ and $- 8\overset{M}{.}2$; here and there they may still be distorted by selection effects.

DUERBECK (1981) gives a catalogue of light-curves, absolute magnitudes and distances for galactic novae. According to his re-evaluation of the data, the mean absolute magnitude at maximum is $- 6\overset{M}{.}4$ in the slow novae, and $- 9\overset{M}{.}4$ in the fast novae.

A difference in the amplitudes of the fast and slow novae has not been statistically confirmed.

A result, surprising at first sight, is that (according to both BERTAUD and PAYNE-GAPOSCHKIN) the greatest amplitudes are those of objects with the greatest apparent brightness, and that with lesser amplitude, the apparent brightness at maximum also becomes fainter. PAYNE-GAPOSCHKIN considers three possible explanations, but it must be a question of a trivial selection effect, in that large amplitudes cannot be determined for novae with faint apparent magnitudes, as minimum would be below the limiting magnitude, even with large instruments. If we take a nova with an apparent magnitude of 8^m at maximum and an amplitude of 15 mag, then minimum would be 23^m.

A review of photometric observations of novae in various spectral regions is given by ARKHIPOVA and MUSTEL (1975).

Recurrent Novae

The recurrent novae (Nr) have already been defined at the beginning of this chapter, but there is good reason for these objects to be considered in greater detail. According to recent research (cf. WARNER 1976), most of the *recurrent novae*, in contrast to the "classical" novae, appear to be related to the *symbiotic stars*, which have a *giant star* as the secondary component (see Sect. 3.1.6). According to BRUCH et al. (1981) the recurrent novae are not a homogeneous group of objects.

It is generally suspected that in fact the *classical* novae, in which the secondary is a *dwarf star*, are recurrent, but with very long intervals, probably orders of magnitude longer, so that over historical periods of time a second outburst cannot be expected. (BATH and SHAVIV, 1978, estimate that in classical novae the interval between two outbursts is of the order of 10 000 years.) However, as the physical mechanism for the outbursts of classical and recurrent novae is presumably the same (the photometric and spectroscopic findings suggest this, see below), we have included the recurrent novae in this section.

Table 30 lists the data for known recurrent novae, as given by WARNER (1976). WEBBINK (1978) catalogues all the magnitude measurements of recurrent novae made hitherto in U, B, V, R, I, J, H, K, L, M, and N bands. A discussion of all known spectra of recurrent novae is in BARLOW et al. (1981).

Table 30. Recurrent novae

Designation	Maximum magnitude	Amplitude A	Year of outburst	Interval (mean)	Type	Spectrum of secondary
T CrB	$2^{\mathrm{m}}0$	8.6 mag	1866, 1946	80 years	fast	g M3
RS Oph	4.3	7.2	1898, 1933 1958, 1967	23	fast	g M6 (?)
T Pyx	7.0	7.1	1890, 1902 1920, 1944 1966	19	slow	Main-sequence? (ROBINSON 1976b)
V 1017 Sgr	7.2	7.1	1901, 1919 1973	36	slow	G 5 III
U Sco	8.7	10.6	1863, 1906 1936, 1979	39	fast	d K (?)

The three objects that follow are not included in this list. WZ Sge ($m_{\mathrm{max}} = 7.2$, $A = 9$ mag, outburst dates 1913, 1946, 1978) indeed resembles a recurrent nova photometrically, but is shown to be a U Geminorum star by spectroscopic findings (Sect. 3.1.3). VY Aqr ($m_{\mathrm{max}} = 8.0$, $A = 8$ mag) erupted in 1907, 1929, 1934, 1941, 1942, 1958, 1962, and 1973, according to recent research by MCNAUGHT, WENZEL, RICHTER, and LILLER. A further outburst occurred in 1983. On the basis of the short intervals this is probably also a U Geminorum star. Unfortunately spectroscopic investigations have not been carried out. V 616 Mon ($m_{\mathrm{max}} = 11.3$, $A = 8.7$ mag, eruptions in 1917 and 1975) is an X-ray nova (Sect. 3.1.7).

Where more than 2 maxima have been observed for each object, the following intervals apply (in years): RS Oph 35, 25, 9; T Pyx 12, 18, 24, 22; V 1017 Sgr 18, 54; U Sco 43, 30, 43. It will be seen that the observed intervals are not constant in the individual stars, but may vary in a ratio as great as 4 : 1. (Of course, it is conceivable that one or more of the outbursts have been missed, if the object was in the daytime sky at the time of the eruption.) There appears to be a definite relationship between the mean interval and the amplitude. For two stars with $A \approx 8-10$ mag, it is 59 years, and for 3 stars with $A \approx 7$ mag, 26 years. A similar relationship applies to the shorter intervals found in the U Geminorum stars, and this will be discussed in the appropriate place.

Both fast and slow recurrent novae are included in Table 30. The preceding discussion shows that it can be of great value for old novae to be *continuously monitored*, and this applies especially to objects with small amplitudes (see end of Sect. 3.1.2).

The smaller mean amplitudes of the recurrent novae as against those of the "classical" novae are presumably not due to less energetic eruptions, but rather to the fact that recurrent novae mostly have a giant star as the secondary component. At minimum this radiates considerably more energy than the white dwarf, so that when the latter erupts the amplitude of the whole system is correspondingly smaller.

In 1979, an outburst of a recurrent nova (U Sco) was successfully followed for the first time with photometric and spectroscopic observations in both the visible and ultraviolet regions (see BARLOW et al. 1981).

Spectrum at Minimum Light

Numerous investigations of the spectral behaviour of novae during and after eruption are available. There are two main reasons for this: first, it is only the spectrum that enables us to classify a nova unambiguously, whereas it is easy to make false identifications (U Geminorum stars, symbiotic variables, supernovae, and even Mira stars) on the basis of the light-curve alone. Second, it is the spectrum, above all, that – by means of its continuum and the rapidly varying intensity, width, and Doppler shift of the emission and absorption lines – gives the most important information about the rapidly changing physical processes that occur both during the quiescent phase and throughout the eruption.

When "classical" and recurrent novae return to minimum after an outburst (in the postnova stage), most of the spectra show a *hot continuum*, with distinct *emission lines* (of various widths) of hydrogen, He I, He II, and Ca II. The He I lines are frequently very faint. In good spectrograms, further emission from doubly-ionized nitrogen and carbon is visible (the N III-C III group of lines at $\lambda = 465$ nm). Some novae show a continuum without emission lines, but in these cases only low-dispersion, unreliable spectra are available. Most of the recurrent novae have "symbiotic" spectra (an absorption-line spectrum of a G, K, or M giant is combined with the emission spectrum). In the "classical" nova GK Per there is a superimposed, class K2, main sequence or subgiant spectrum. It may be assumed that all novae have a faint, cool companion, but its absorption lines are generally hidden by the bright continuum.

Detailed descriptions of the spectral behaviour at minimum and during the outbursts, with numerous references to the literature on individual objects, are included in the accounts given by WARNER (1976) and PAYNE-GAPOSCHKIN (1977b).

There are slightly variable objects that have a nova-like spectrum, for example V Sge, whose spectrum greatly resembles that of U Sco (see BARLOW et al. 1981). It is possible that these are novae that erupted in prehistoric times, and which will perhaps undergo a further outburst in the near or distant future. AM CVn (see Sect. 3.1.3) is possibly a completely extinct nova, which, according to WARNER and ROBINSON (1972), has exhausted its supply of fuel for nova outbursts.

So far we have described the spectra of postnovae. As the eruption of a nova is always an unpredictable event, it is not surprising that as yet the spectra of only three classical novae are known prior to outburst: Nova V 603 Aql (1918) – CANNON (1920); Nova V 533 Her (1963) – STEPHENSON and HERR (1963) and GÖTZ (1965); Nova HR Del (1967) – STEPHENSON (1967) and GÖTZ (1968). All these spectra show no emission lines and a more or less blue continuum.

As may be seen, there are considerable differences between the spectra of most postnovae and those of the three prenovae known so far. It must be borne in mind, however, that it is debatable whether these three objects are representative of all prenovae. We must also allow for the fact that these spectra of prenovae were objective-prism ones, where spectral details may be lost due to the small dispersion. Moreover, there are a few spectra of postnovae that show no emission lines, although this may be simply because these were also obtained at low dispersion.

We can only hope that in the near future more spectra of prenovae will be obtained, as these are important for our understanding of the physical processes occurring in novae in the years before they erupt.

Spectral Behaviour During the Eruption

MCLAUGHLIN recognized that all novae follow approximately the same course of spectral development during their eruption. As a result, spectral class Q was then introduced for novae, and the evolutionary stages Q0–Q9 were distinguished. During the decline the spectrum passes through a series of stages, where consecutive systems of emission lines of increasing ionization potential appear. The spectral behaviour is very closely linked to the course of the light-curve (Fig. 44). Nowadays a number of stages are recognized in the development of the spectrum (see Figs. 45 and 46). Each stage begins before the end of the previous one, so that ultimately several spectral states may exist at the same time (PAYNE-GAPOSCHKIN 1957 and MCLAUGHLIN 1965). The stages are as follows:

1) *Pre-maximum spectrum.* From about 2 days before maximum to a few days after, the spectrum is similar to that of a B, A or F star. The absorption lines are nearly always wide, diffuse, and blue-shifted. This blue-shift can be explained as a Doppler effect. A translucent nebular shell, which surrounds the nova itself, is expanding at high velocity. Viewed from our position, certain parts of the shell appear in front of the the hot stellar surface, which

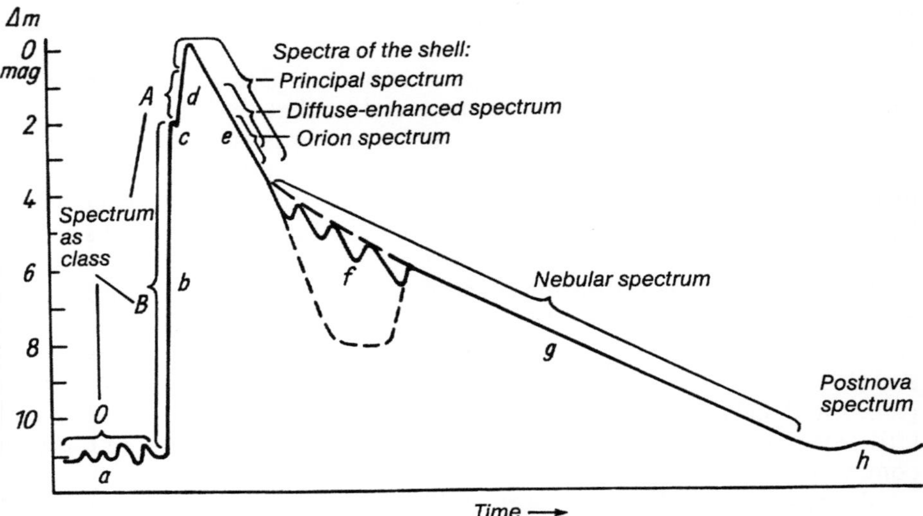

Fig. 44. Schematic light-curve of a nova with an indication of the spectral stages after Pskovski (1978). *a* Prenova; *b* rise; *c* pre-maximum pause; *d* final rise; *e* initial decline; *f* transition stage; *g* final decline; *h* postnova

is radiating a continuum. These portions of the envelope are moving towards us with expansion velocities of some 100–1000 km/s, thus giving rise to spectral lines shifted towards the violet by a corresponding amount.

The pre-maximum spectrum lasts until shortly after peak luminosity, and later fades. The absorption features mostly become stronger and narrower. In some cases the Doppler shift increases, and in others it decreases.

2) *Principal spectrum.* This begins about the time of maximum, after the rapidly expanding nova has reached about 100 solar radii. The spectrum resembles that of supergiants of class A–F. The earliest spectrum that has ever been recorded for a nova at maximum was the B spectrum of Nova V 1500 Cyg (1975). The latest, on the other hand, was the K spectrum of Nova V 1148 Sgr (1943).

New lines appear alongside the absorption lines belonging to the slowly fading, pre-maximum spectrum. Those shifted to the violet (and corresponding to a Doppler shift of − 200 to − 2000 km/s) are sharp absorption lines, while those towards the red (unshifted) are bright, wide emission features, initially of hydrogen, and later of singly-ionized calcium (Ca II) and iron (Fe II). The emission features are wide because they arise from the whole of the still transparent, but expanding shell. Due to the Doppler effect the approaching portion of the shell gives rise to the violet half of the line, and the receding portion to the red half, while those parts that are moving tangentially to us produce the peak emission. Finally, one or two days after maximum, the shell becomes so tenuous as a result of expansion, that "forbidden lines" occur, particularly those of [O I], [N II] and [O III]. These become very intense shortly after their appearance. The duration of the principal spectrum varies greatly from nova to nova.

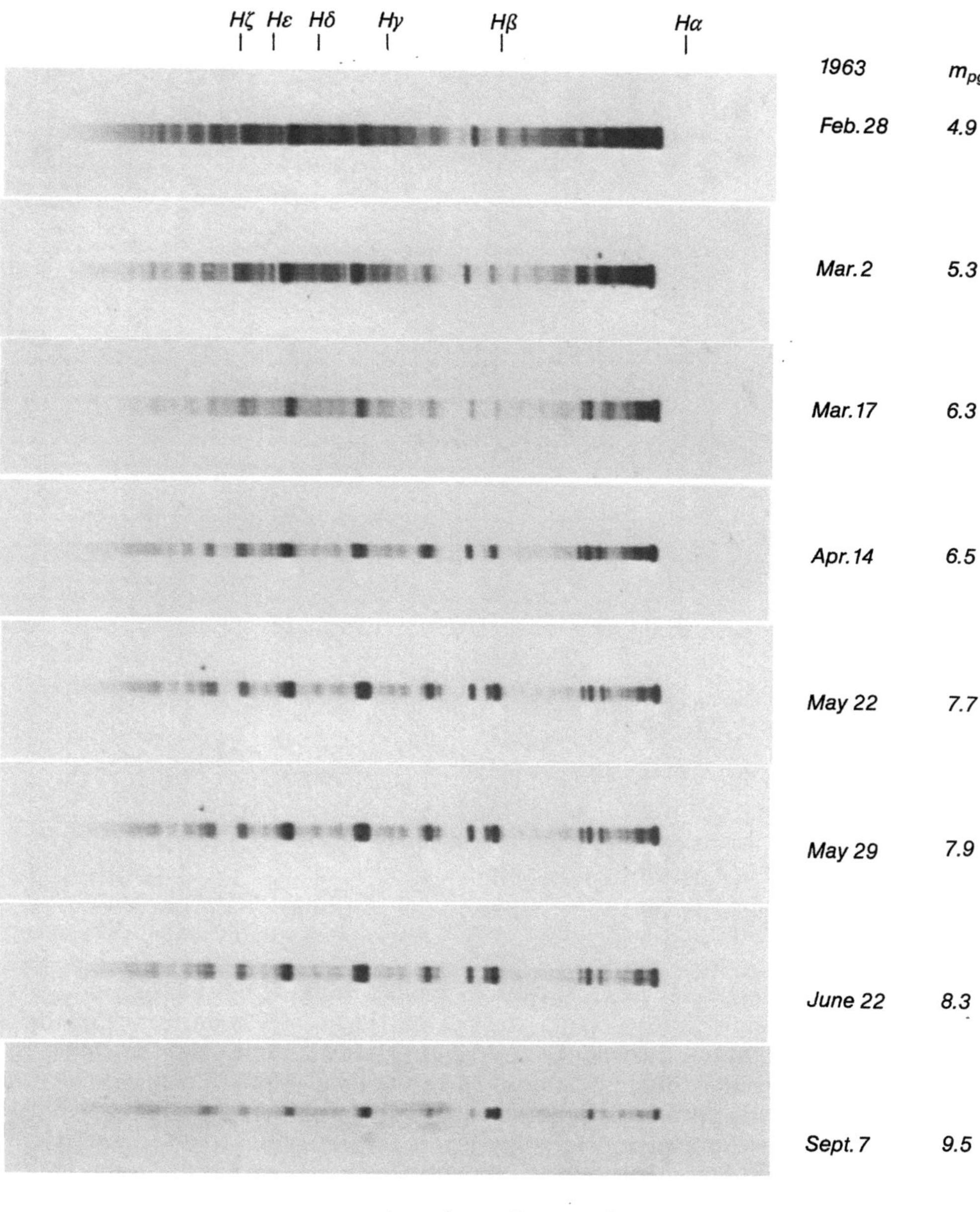

Fig. 45. Spectral development of Nova V 533 Her (1963) after Götz, Sonneberg. Note the complicated profile of the hydrogen lines (absorption and emission) and the forbidden lines of oxygen and nitrogen ions

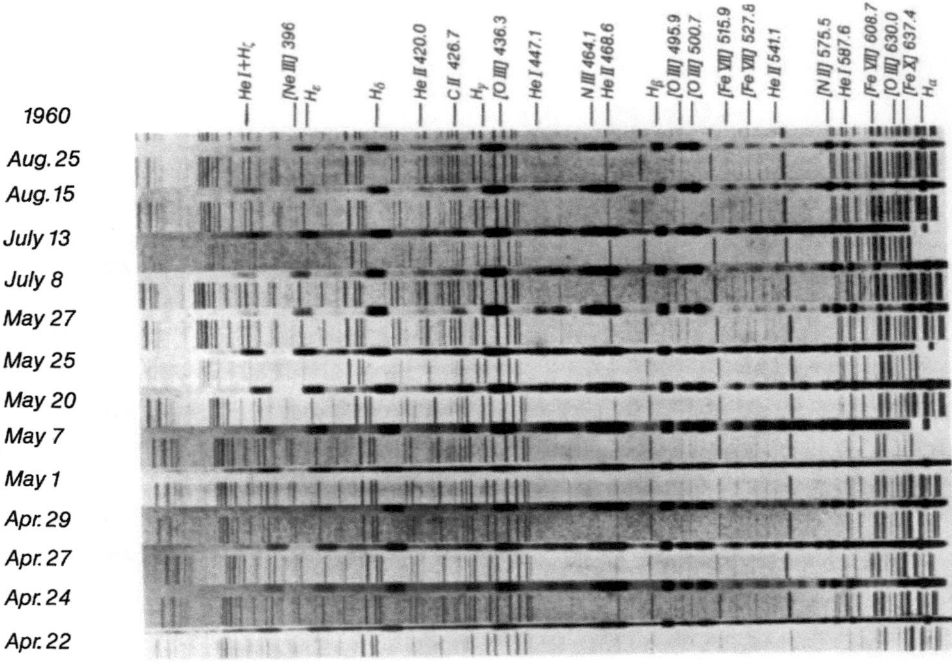

Fig. 46. Spectral development of Nova Her 1960 from 1960 April 22 to 1960 August 5 (after DUFAY et al. 1965)

3) *Diffuse-enhanced spectrum*. The next stage appears even before there has been any marked weakening of the principal spectrum, when the overall brightness has fallen by about 1½ magnitudes. It may last for several days or a few weeks. This third absorption-line system is even more strongly blue-shifted than the previous systems. The lines are very wide and very diffuse, presumably as a result of strong turbulence in the expanding gas cloud.

4) *Orion spectrum*. The Orion spectrum is dominated by absorption lines that are characteristic of "Orion stars"; that is class B stars found in the "Orion Association". These lines are He I, O II, N II, and C II; the Balmer series of hydrogen is missing. In most cases the blue-shift of the lines is even greater than in the diffuse-enhanced spectrum. The strength of the blue-shift often varies in a quasi-periodic manner, which is equivalent to quasi-periodic variations in the expansion velocity. The latter are connected to oscillations in the brightness, the greatest velocity occurring at subsidiary minimum in the light-curve. The Orion spectrum also contains emission lines; these are wide and diffuse. They are most prominent at subsidiary minima in the light-curve.

5) *Nebular spectrum*. When the gaseous shell ejected by the nova has dissipated sufficiently the last absorption lines of the previous stages gradually fade. Now the nova spectrum closely resembles that of a planetary nebula. It consists of bright lines of hydrogen and helium, and of a number

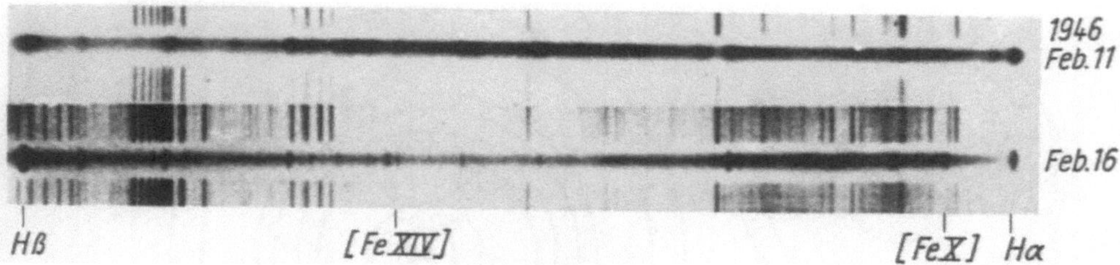

Fig. 47. Spectrograms of the recurrent nova T CrB shortly after maximum (1946 Feb. 9). Note the "coronal lines" of highly ionized iron (after McLaughlin)

of forbidden lines: the "auroral line" at 436.3 nm [O III], the nebular lines at 386.9 nm and 396.8 nm [N III], and forbidden lines of singly- and multiply-ionized iron [Fe II] to [Fe VII] are present. In some objects, in particular in recurrent novae, forbidden lines of iron ionized as many as 13 times [Fe XIV] (Fig. 47) have been found! In a very few novae the expanding *gas shell* has become *visible* during the nebular stage, and the expansion could be followed optically. In these few cases the expansion of the nebula is measurable, as has already been mentioned, not only from the radial velocity (the Doppler shift towards us in km/s), but also from the tangential velocity (in seconds of arc per year). This means that if the expansion velocities are isotropic we have a reliable method of determining the distance of the object concerned. However, it is necessary to be absolutely certain that it is truly the tangential expansion of the nebulosity that is being measured, and not that of the light-pulse into an already existing nebula. But even the tangential expansion of the light (in arc-seconds per year) gives rise to a very exact method of determining the distance, from the fact that one knows that it propagates at 300 000 km/s. Up to now expanding shells have been observed in 15 objects, and light-pulses in 4 cases.

6) *Postnova spectrum.* After the nova reaches its "normal" state at minimum, the nebular stage fades away, and the spectrum becomes that of the postnova stage, described earlier. In a few cases nebular lines (or even the nebula itself), are visible in the postnova stage at least at first, as in the classical novae DQ Her (1934), GK Per (1901) and RW UMi (1956), and the recurrent novae T CrB and RS Oph.

Figure 48 shows a schematic *cross-section through a fast nova*, 3 and 6 days after visual maximum, the individual sections of the shell responsible for the various absorption-line systems being marked. A very good, detailed description of the spectra of a large number of novae is to be found in the compilation "Novae, Supernovae, Novoides" (Centre National de la Recherche Scientifique, Paris 1965).

It is also worth mentioning that the expulsion of shells in a nova does not occur in a spherically-symmetrical manner (see the detailed description by Payne-Gaposchkin 1977a). A *three-dimensional model* of the nova V 603 Aql (1918), about one year after outburst, is given in Fig. 49. Dark areas are regions of high material density, and light ones are "holes".

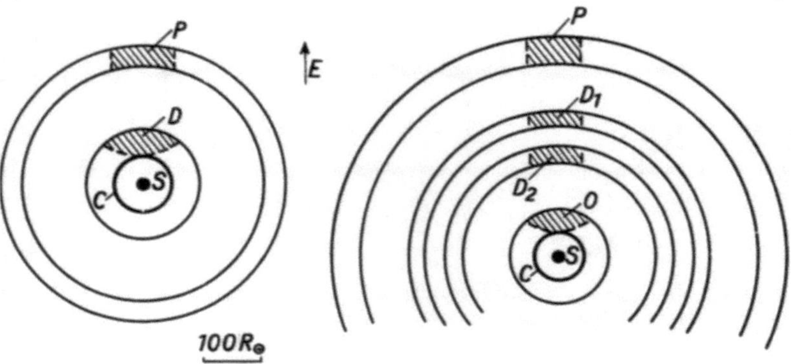

Fig. 48. Schematic section through a fast nova, 3^d (*left*) and 6^d after maximum. The arrow E shows the direction of the Earth. S = the star itself; C = the effective photosphere (i.e. source of the continuum). The "diffuse-enhanced spectrum" arises in the shaded regions D, D_1, D_2; the "Orion spectrum" at O; and the "principal spectrum" at P (after McLaughlin 1965)

Fig. 49. Three-dimensional model of Nova V 603 Aql (1918) (after Weaver 1974)

The spectral development is never exactly as described here, every nova having its own peculiarities.

As we have seen, the changes in the spectrum during the outburst of a nova are very complicated and difficult to determine. However, a few simple *conclusions* may be drawn immediately:

1) The masses of gas ejected from the nova at high velocities indicate that a powerful, *explosive type of process* occurs in the nova.
2) The ejection velocities of a few hundred to a few thousand km/s, measured on the basis of the strong Doppler effect (the blue-shift of the absorption lines), are significantly higher than the escape velocity. In other words the *ejected shell is lost by the nova.*
3) Calculations based on the observed intensities of the spectral lines indicate that in an outburst a nova loses approximately *1/100 000 of its mass,* and *radiates about* 10^{45} erg $= 10^{38}$ J (STRUVE 1962).

It was realized quite early on that the outburst of a nova is caused by an explosion. However, an understanding of the *physical causes* of the outbursts of novae was successfully reached only after the *binary nature* of these objects had been recognized (Sect. 3.1.5).

The Galactic Distribution of Novae

The question of the distribution of novae in the Galaxy cannot be answered simply as despite their large amplitudes they have a low discovery probability (see Sect. 6.3) and as a result they are only incompletely recorded.

The next two figures show the distribution of known novae in the Galaxy. Figure 50 indicates the distribution of novae as projected onto the *galactic plane*. The graduations around the edge indicate galactic longitude in degrees. The used distances from the Sun were determined by means of the known relationship between the maximum brightness of novae at outburst and the rate of decline (see, e.g. DE VAUCOULEURS 1978). The figure shows that there is a strong concentration of novae in the direction of the galactic centre. (The fact that the galactic centre itself and the region beyond it are sparsely populated is primarily a consequence of interstellar extinction,

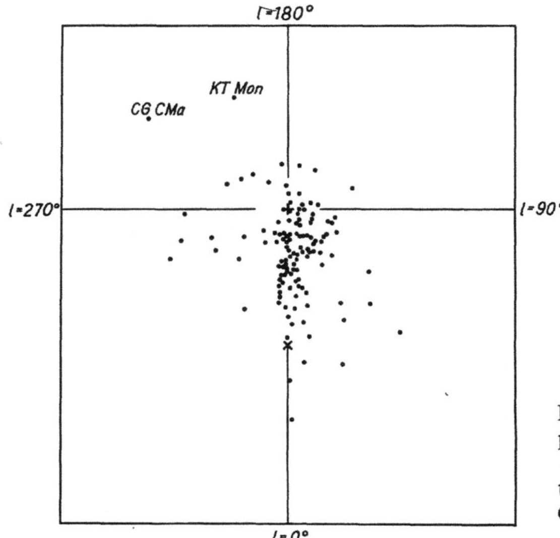

Fig. 50. Distribution of novae as projected on the galactic plane. + position of the Sun, × position of the galactic centre (after PAYNE-GAPOSCHKIN 1977b)

which is strongest in the direction of the galactic centre and absorbs the light of objects far away from the Sun.) Towards the galactic anticentre ($l = 180°$) there are only a few objects. This distribution is typical of old objects (Population II).

Figure 51 shows the number of novae as a function of distance z from the main plane of the Galaxy. The strong concentration of novae close to the galactic plane ($z = 0$) is very striking. The distinct dip between $- 100$ pc and $+ 100$ pc is not real, but a consequence of the interstellar extinction just mentioned, which is strongly concentrated in a very thin layer in the galactic plane. The average of the observed z-distances is 220 pc, and this is typical of relatively young stars (Disk Population). This contradicts the conclusions suggested by Fig. 50. The theory of the structure of our Milky Way system, derived from observations, indicates that there is a relationship between the degree of concentration towards the galactic plane and that around the galactic centre. Concentrations that are high close to the galactic plane and low towards the centre indicate Population I, while the converse is a sign of Population II.

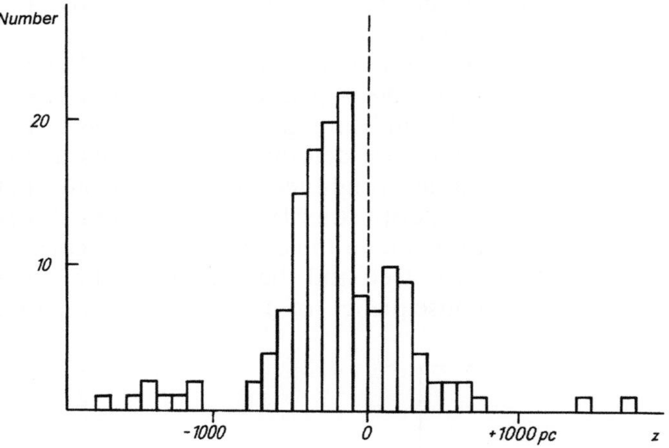

Fig. 51. Frequency distribution of z-distances for galactic novae (after PAYNE-GAPOSCHKIN)

An obvious possible explanation for this finding is that, in both Figs. 50 and 51, we are not seeing all the novae that exist. As already mentioned, the diagrams lack those novae that are invisible due to interstellar extinction and also – to a much greater degree – cannot include the numerous undiscovered objects. In the search for variable stars (discussed in later sections), using blink comparators with photographic plates, regions at low galactic latitudes and in the direction of the galactic centre are given preference, so that the observed distribution is unlikely to agree with the one actually present.

Systematic searches and accidental discoveries at high galactic latitudes show that there are also novae at considerably greater z-distances than can be shown in Fig. 51. The individual objects, with their assumed z-distances in kpc in parentheses, are: BD Pav ($- 3$), RW UMi ($+ 3$), RR Cha ($- 4$),

RT Ser ($+$ 4), X Ser ($+$ 5), U Sco ($+$ 6), V 522 Sgr ($-$ 8), W Ari ($-$ 10), V 351 Car ($-$ 15), V 1548 Oph ($+$ 30). Some of these objects may be falsely identified U Geminorum stars, which are actually fainter and nearer. Only RW UMi, RT Ser, U Sco and RR Cha appear to be completely definite.

A few undoubted novae in M31 and M33 are at extremely great distances from the centres of their parent galaxies (see Sect. 5.2.2). Paradoxically, such remarkable objects are easier to discover in the extreme outer haloes of M31 and M33 than in the halo of our own Milky Way system. This is because a single photographic exposure is sufficient to record the halo of another galaxy. To cover the whole halo of our own galaxy, on the other hand, a very considerable number of plates is required.

To summarize, it can be said that novae apparently occur in all age classes. They cover a whole range from older Population I objects to extreme halo Population II stars.

Parallaxes and proper motions of novae are too small to be measured. Distance determinations must thus be carried out by other methods.

A List of Novae that May Undergo a Further Outburst

The following catalogue of novae (Table 31) was published by Commission 27 of the IAU at the 1967 Assembly, with the recommendation that these objects should be monitored, so that in the event of a further rise it might be recognized and observed as early as possible. The last two objects included come from a list by PAYNE-GAPOSCHKIN (1977 b), which likewise gives a description of potential recurrent novae. The choice was based upon amplitudes, as small amplitudes are statistically linked with relatively short intervals (see above). These are consequently objects that are likely – with a certain degree of probability – to undergo a further outburst in this century, or shortly afterwards. Reference should also be made to the section on recurrent novae. IM Nor has been excluded from the list, as following the recent identification of the prenova, it has been found to have an amplitude of 12 mag.

Table 31. Potential recurrent novae

Nova	Year	Amplitude
X Ser	1903	6.0 mag
V 999 Sgr	1910	8.4
FM Sgr	1926	8.5:
V 1016 Sgr	1899	6.5
V 441 Sgr	1930	7.3
HS Sgr	1900	6.5:
V 1017 Sgr	1919	7.0
FN Sgr	1925	5.0
HR Lyr	1919	8.5
EU Sct	1949	8.4
V 841 Oph	1848	8.9
FS Sct	1952	5.7

In the short accompanying text to the IAU list, RT Ser and FU Ori are also mentioned. RT Ser (1909) is a slow nova, the amplitude of which is not known, and FU Ori is an extreme case of a T-Tauri type (see Sect. 3.3.2), and is therefore a very young star. Both objects should be monitored, even if for other reasons.

A few notes on the individual objects follow:

X Ser: At minimum the brightness fluctuates by up to 2 mag; its mean magnitude is 15^m0. The light-curve resembles that of Z And. RR Tel showed similar behaviour before its eruption. X Ser thus warrants considerable attention.

V 999 Sgr: 26 years after the outburst in 1910 the star had a magnitude of 16^m6. The maximum was about 8^m2.

FM Sgr: The assumed maximum magnitude of 8^m0 is extrapolated from the light-curve. The prenova might be a star of magnitude 16 or 17; the amplitude is thus uncertain.

V 1016 Sgr: It is not certain that the star was observed at maximum. With the extrapolated maximum magnitude of 7^m0 the amplitude amounts to 7.9 mag.

V 441 Sgr: The amplitude given in the table has been calculated from the maximum magnitude of 8^m7, but the maximum magnitude may well have been brighter, leading to an amplitude of about 8.0 mag.

HS Sgr: WOODS gives the magnitude of the prenova as 16^m5; according to MCLAUGHLIN the extrapolated maximum was approximately 10^m0.

V 1017 Sgr: In the interim the star has proved to be a recurrent nova; the predicted eruption occurred in 1973 (see also Table 30). Many authors (see PAYNE-GAPOSCHKIN 1977b) include the object among the Z Andromedae stars.

FN Sgr: Apart from the outburst in 1925, another has been confirmed in 1937. The spectrum does not contradict the possibility that this is a Z Andromedae star.

HR Lyr: The star is very active, especially in the photographic region, around 15^m0. According to ROSINO, the brightness varied between 14^m2 and 15^m3 from 1947–1952, without recognizable regularity.

EU Sct: Maximum was at about 8^m6; according to observations of the prenova between 1918 and 1949 (HARWOOD) normal light was about 16^m8.

V 841 Oph: At maximum the object reached magnitude 4^m2.

FS Sct: At maximum the object attained magnitude 10^m9.

3.1.3 U Geminorum Stars

Classification

The photometric behaviour of most U Geminorum stars – also known as dwarf novae – resembles that of recurrent novae on a reduced scale, where the reduction applies to both the intervals and the amplitudes (Fig. 81 top). The question of whether this apparent similarity indicates some interrelationship will be discussed later. Photometrically the U Geminorum stars may be divided into three main groups:

1) *U Geminorum stars* in the narrower sense (or *SS Cygni stars*), with fairly long, but varying, intervals of quiescence at minimum. The rise of several (2–8) magnitudes lasts 1–2 days, and the decline several days or several weeks. The mean interval between two outbursts, according to the star, is about $10-10^4$ days.

A subgroup of the SS Cygni stars is formed by the *SU Ursae Majoris stars* (cf. the comprehensive discussion in the work by VOGT 1981 and the references given there). Apart from the normal maxima, these objects undergo a *supermaximum* every 3–10 cycles, which may be distinguished from an ordinary maximum by its longer duration and greater brightness. During the supermaxima these objects exhibit periodic peaks – the so-called "super-humps" – the period of which in most of the known cases differs by only a few percent from the binary's orbital period, and with which it is obviously connected. All known SU Ursae Majoris stars belong to the *ultra-short period* cataclysmic binaries (see Sect. 3.1.5 and Table 32): 70% of all ultra-short periodic binaries ($P < 2^h$) are SU Ursae Majoris stars! The following bright members are known: V 436 Cen, YZ Cnc, Z Cha, VW Hyi, WX Hyi, EX Hya (?), AY Lyr, WZ Sge (?), EK TrA, SU UMa, CU Vel. In addition, SS Cyg itself shows several forms of outbursts (as do other objects). It is not, however, included in the SU Ursae Majoris stars. A theoretical explanation of the supermaximum phenomenon is still outstanding, although there have been some initial results (see VOGT 1981).

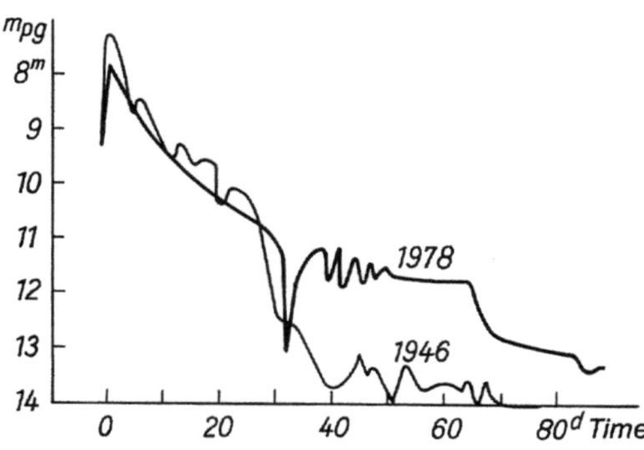

Fig. 52. Schematic light-curve of WZ Sge in 1946 and 1978

The light-curves of U Geminorum stars with very long cycle-lengths, the *WZ Sagittae stars* (BAILEY 1979), can hardly be distinguished from those of the recurrent novae. They can in fact only be recognized as different by spectroscopic features (see Fig. 52). The U Geminorum star with the longest known cycle-length C is WZ Sge ($C = 11\,900^d$, $A = 9$ mag), and the recurrent nova with the shortest cycle-length is T Pyx ($C = 6\,900^d$, $A = 7.1$ mag).

2) *Z Camelopardalis stars*. The intervals at minimum are so short that an almost continuous variation in luminosity takes place. The eruptive cycle occasionally alternates with other, longer intervals at an approximately con-

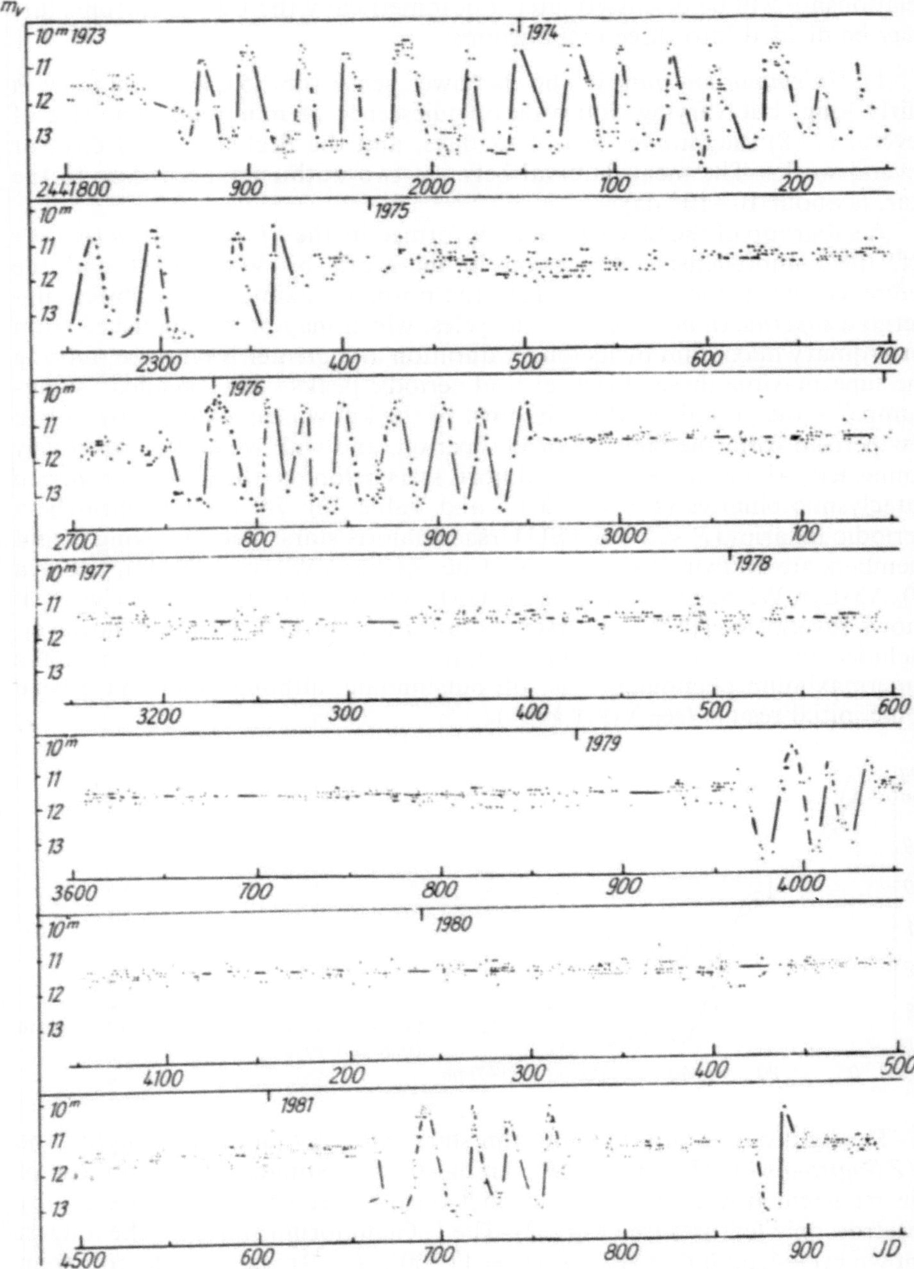

Fig. 53. Light-curve of Z Cam (compiled from GUNTHER and SCHWEITZER 1982)

stant, intermediate magnitude (Fig. 53). These standstills generally begin on the decline, and almost always end by completing the decline. The amplitudes are 2–5 magnitudes and the mean cycle-length C amounts to about 9–40 days. The shortest cycle-length known is that of AM Cas ($C = 9^d$, $A = 2.9$ mag).

Spectroscopically (see below), the U Geminorum stars and the Z Camelopardalis stars cannot be distinguished, so from a physical point of view it appears to be justifiable to regard them as being an essentially homogeneous group.

3) *UX Ursae Majoris stars*, occasionally also known as nova-like objects. These show rapid, low-amplitude fluctuations in luminosity, which in some cases are superimposed on an eclipsing light-curve. The spectra of some of these objects show wide, flat, absorption lines of hydrogen and neutral helium, similar, as we shall shortly see, to those of U Geminorum and Z Camelopardalis stars during outburst. It is possible that these are Z Camelopardalis stars permanently at standstill. One of these objects, TT Ari, appears to confirm the suggestion: after at least 80 years at standstill, it went through a stage at minimum in 1980, when its spectrum was typical of a U Geminorum star. Other objects that have nova-like spectra, however, are probably novae for which there are no observed, historical outbursts.

The best-observed U Geminorum stars are SS Cyg and Z Cam. This is largely thanks to the activities of amateur observers.

VOGT and BATESON (1981) give an atlas of southern and equatorial dwarf novae.

Statistics

KUKARKIN and PARENAGO (1934) were the first to note a linear relationship between the logarithm of the mean *cycle-length* and the *amplitude*. They included two recurrent novae in order to show that there was no fundamental difference between the two types of variables. However, nowadays it is believed that the outburst mechanism in the U Geminorum stars is different from that in the recurrent novae (see below), so that the coincidence of the two amplitude/cycle-length relationships is purely accidental.

Spectrum

A comprehensive description of the known spectra of dwarf novae at minimum and during eruption (with numerous references) can be found in WARNER (1976).

The vast majority of U Geminorum stars show (at minimum) strong *Balmer emission lines* of hydrogen (frequently up to 2 nm wide) on a faint blue continuum (Fig. 54). There are no forbidden lines. The great line-width can be envisaged as Doppler broadening (rotational broadening) in a *gaseous disk* surrounding the white dwarf, and *rotating very rapidly* (500 km/s and more). In those cases where the line width is less we are probably looking at the pole of the objects concerned. Faint to moderately strong emission lines

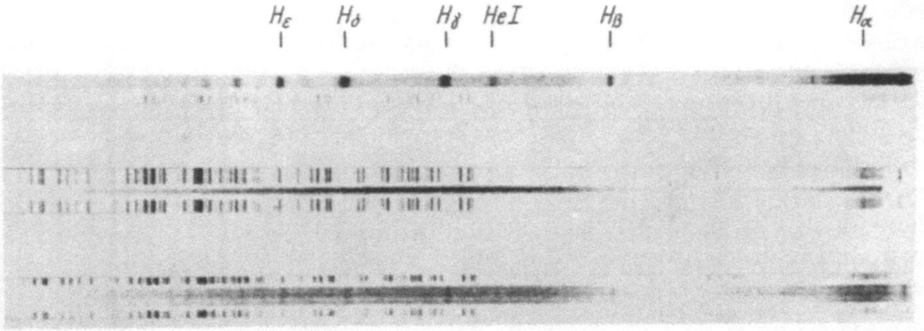

Fig. 54. Three spectrograms of SS Cyg; *above* at minimum light, *centre* at maximum, *below* at intermediate light. Note the emission spectrum at minimum in comparison with the predominant continuum at maximum (after G.S. Mumford 1962)

of neutral helium (He I) are also present, and, in a very few objects, weak emission from ionized helium (He II) is suspected. Lines of doubly ionized nitrogen and carbon (N III and C III) are faintly indicated in only a few, extraordinary cases: BV Pup, WZ Sge (the dwarf nova with the longest cycle-length) and S 10830 Tri (RICHTER et al. 1981). In many objects the emission lines are either permanently or occasionally doubled (Z Cam, Z Cha, BV Cen, EM Cyg, U Gem, EX Hya, VW Hyi, WZ Sge). In a number of cases the absorption spectrum of a G- or K-type main sequence star is visible (the binary's secondary component).

The spectra of U Geminorum stars are thus similar to those of postnovae and recurrent novae at minimum, with the difference that the nova spectra are more highly excited: in contrast to the dwarf novae they have considerable He II emission, and distinct C III and N III emission features. The lower excitation and the lower absolute magnitudes are the only characteristics known at present, apart from the presumed difference in ages, that differentiate the U Geminorum stars from the postnovae. The spectra of U Geminorum stars, like those of the novae, show two forms of change with time:

1) Periodic changes, which are Doppler shifts resulting from orbital motion in a close binary system.

2) Complicated spectral behaviour during outbursts. While the spectral features due to orbital motion in a close binary system are very similar to those of the novae (see Sect. 3.1.5), the spectral changes during eruption are fundamentally different. The spectral characteristics caused by various rapidly expanding shells found in a nova eruption (see above) are lacking in the U Geminorum stars.

As a U Geminorum star brightens, the continuum becomes stronger, but the emission lines grow relatively fainter as their strength does not appear to change significantly. Around maximum, the majority of U Geminorum stars show, in place of the previous emission lines, very wide absorptions with half-widths of up to 5 nm, frequently with central re-emission. The wide absorption lines probably arise from the inner, very rapidly rotating (> 3000 km/s) portions of a disk of material surrounding the primary component (rotational broadening), and the weak re-emission from the outer ("chromospheric"), slower-velocity layers of the differentially rotating disk. Significant differences in spectral development among dwarf novae can be explained, at least in part, by the fact that in many systems our line of sight is approximately parallel to the disk (analogous to the aspect of Saturn's rings from the Earth), while in others it is almost perpendicular to the plane of the disk. With the decline in magnitude the wide absorption lines become invisible, more or less rapidly; the strong continuum gradually fades, and the emission lines regain their prominence.

The Galactic Distribution of U Geminorum Stars

The question of the distribution of the U Geminorum stars in the Galaxy is even harder to answer than in the case of the novae, as, apart from the low probability that they will be discovered (see Sect. 6.3), they have small absolute magnitudes (even at maximum only about $+ 2^M$ to $+ 4^M$). As a result, only the objects in the immediate solar neighbourhood are fairly comprehensively known. The spatial density function in the z-direction (perpendicular to the galactic plane) can therefore only be determined very approximately. So it is not surprising that the relevant data in the literature are contradictory. The spatial density v has a gradient in the z-direction ($- \partial \log v/\partial z$) of 0.17 according to KOPYLOV (1957), corresponding to extreme Population II, and of 3.9, on the other hand, according to RICHTER (1967a), indicating Population I. However, it is significant that searches for variable stars on very deep plates (such as exposures made with the 134/200/400 cm Schmidt telescope at Tautenburg, limiting magnitude approximately 21^m) find considerably fewer U Geminorum stars at very high galactic latitudes than at lower latitudes. Other types of cataclysmic variables (novae, nova-like, and AM Herculis stars) occur at high galactic latitudes at a significantly greater frequency, relative to the U Geminorum stars. This is also confirmed by MEINUNGER (1982). All in all, the U Geminorum stars appear, on average, to belong to a younger stellar population than the novae. This finding still awaits an interpretation in terms of stellar evolution.

A popular scientific description of modern knowledge about dwarf novae has been given by VOGT (1983).

AM Canum Venaticorum Stars

As yet three examples are known of these objects, which are related to the U Geminorum stars, namely *AMCVn*, *GP Com*, and *PG 1346 + 082*, see Table 32. Their form is similar to that of the U Geminorum stars. They differ from these photometrically by a lack of outbursts, and spectroscopically by a deficiency of hydrogen. According to NATHER et al. (1981) the hydrogen abundance is less than that in the Sun by a factor of at least 1500! The spectrum is practically a pure *helium spectrum*. As will be described at the end of Sect.3.1.5, the secondary is a white dwarf. The changes in magnitude have two components:

1) Periodic, *double sine-wave fluctuations in luminosity* (accompanied by radial velocity changes in the spectrum with equivalent period length), due to the *orbital motion of a binary pair*.
Superimposed on this light-curve:
2) *Rapid flickering* with a periodicity of a few minutes. Figure 61 shows the light-curve of AM CVn.

3.1.4 AM Herculis Stars or Polars

The prototype for these objects, *AM Her*, discovered in 1923 by WOLF, was still included as an irregular variable in the 1976 GCVS. In 1976 it was discovered that it was the X-ray source 3U 1809 + 50. The following large-amplitude phenomena contribute to the variations in brightness:

1) *Long-term changes.* These are characterized by the existence of two different states (see Fig. 55), one "active" ("on"), in which the luminosity fluctuates around $13^{m}0$, and the other "inactive" ("off"), where the brightness remains at about $15^{m}0$.
2) *Short-term phenomena.* These can be explained by the orbital motion of a binary system with a period of $3^{h}1$, which is shown by eclipsing light-changes, strongly variable linear and circular polarization (discovered by TAPIA in 1977), and periodic radial velocity changes in the H and He lines (see the report by LILLER 1977). Moreover, in every cycle there is a total eclipse in X-rays that lasts 28 minutes. The position of the primary minimum depends upon the colour! In the blue it is 35 minutes later than in the red, and in the ultraviolet it is almost an hour later. Figure 56 shows schematically the magnitude, radial velocity, and polarization curves.

The high-excitation spectrum corresponds approximately to that of post-novae: among the numerous emission lines those of hydrogen (H) and of ionized helium (He II) at 468.6 nm are the strongest.
The name *polar* was introduced by KRZEMINSKI and SERKOWSKI (1977) for AM Her and related objects, on account of the strong, and variable, linear and circular polarization in the light from these stars.
Apart from AM Her, 10 other objects (Table 32) are now classed as polars (see also WICKRAMASINGHE 1982, LIVIO and SHAVIV 1983, LIEBERT and STOCKMAN 1983).
All the objects have very short orbital periods between a little over 1 hour and a little more than 3 hours (see Table 32). Schematic orbital light-curves, the position of the magnetic poles, and circular and linear polarization curves may be found in CHANMUGAM and DULK (1981). Although they have many features in common, the objects all have their individual peculiarities.
Nearly 130 publications about AM Her alone in 1978–1980 show the increasing interest in these objects. As yet success has by no means been achieved in explaining theoretically all the observed effects (see also ALLEN et al. 1981).

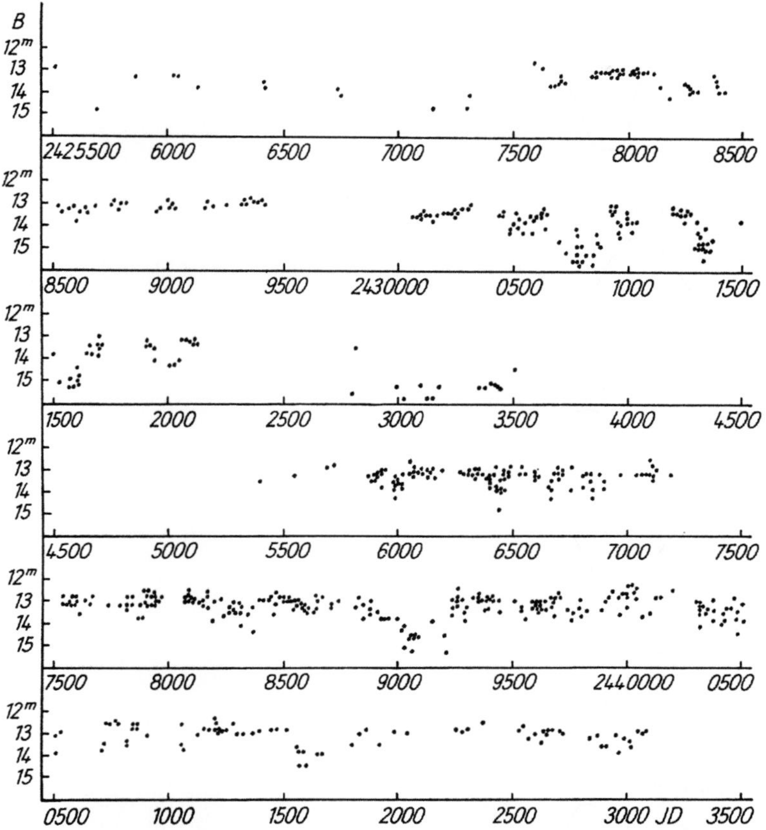

Fig. 55. Magnitude changes in AM Her (after HUDEC and MEINUNGER 1977)

Because of their intense X-ray radiation, which greatly exceeds that of the
novae and U Geminorum stars, the AM Herculis stars are frequently grouped
with the X-ray binaries (to be discussed in Sect. 3.1.7). However, as there is
no gradual transition between the AM Herculis stars and the objects described in Sect. 3.1.7, but because such a gradual transition to the physically
related U Geminorum stars does exist, according to recent research, we have
included them in the cataclysmic objects.

3.1.5 Properties and Models of Cataclysmic Variables

In the preceding three sections the magnitude fluctuations of novae, dwarf
novae (U Geminorum stars) and polars (AM Herculis stars) have been introduced and identified. These differ to such an extent that one would not expect
there to be any close relationship between the three classes of variables.
However, as we shall show, all three types have many similarities, and may
be described by a single basic physical model – the cataclysmic variable
model. We have already established that at minimum light the spectral char-

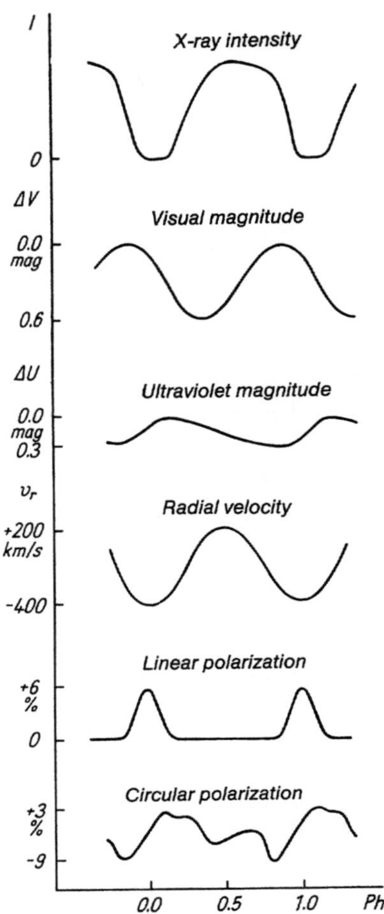

Fig. 56. AM Her: variation of 6 properties with the 3.1 hour cycle (after LILLER 1977)

acteristics are similar, and we shall now show that the same is true of the photometric behaviour.

Luminosity Variations in Novae and Dwarf Novae at Minimum

Only within approximately the last 30 years has it been recognized that novae and U Geminorum stars are variable outside the major eruptions. M.F. WALKER (1954) established that the photometric behaviour is very complex (see also PAYNE-GAPOSCHKIN 1977b):

1) *Irregular fluctuations* in luminosity with amplitudes up to one magnitude or more, occurring on time scales of 10 to several hundred days. An excellent review of earlier observations of this sort is given by ROBINSON (1975).

2) *Rapid, irregular flickering* with a time scale of hours or minutes. SU UMa, for example, shows fluctuations of 0.7 mag in only 5 minutes (Fig. 57).

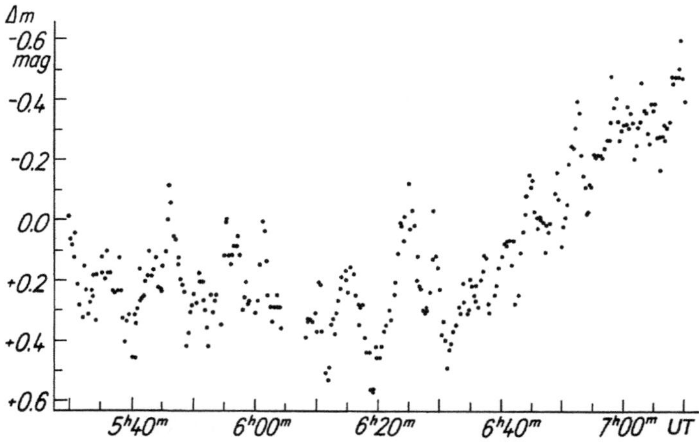

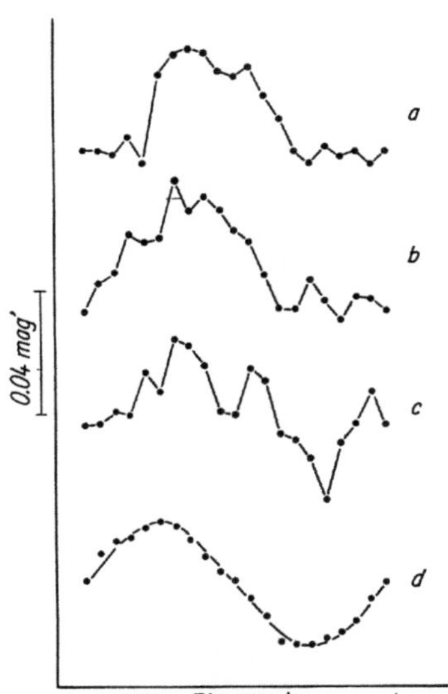

Fig. 57. Rapid fluctuations in the U Geminorum star SU UMa on 1963 Jan. 23, measured at Kitt Peak Observatory (after Mumford 1963)

Fig. 58. 71-second pulsations in Nova DQ Her (1934) (after Nather 1973). *a,b* and *c* are each the sum of 10 pulsational cycles; *d* is the sum of 224 cycles

3) *Rapid, coherent oscillations* on a time scale of a few tens of seconds, with amplitudes of 0.001 to 0.04 mag (see also Chester 1979). For example, Nova V 533 Her (1963) fluctuates with a period of 63.63309 ± 0.00004 seconds (cf. Patterson 1979), and Nova DQ Her (1934) with a period of 71.1 seconds (Fig. 58).

If Pogson's account is to be believed, U Gem showed distinct visual fluctuations on 1856 March 26, with intervals of 6–15 seconds (see Ashbrook 1980). In 1949, Thackeray was apparently able to see visual flickering in VV Pup, with an amplitude of a few tenths of a magnitude, occurring within seconds.

4) The *variability*, already mentioned, which is *synchronized with the orbital motion* of the binary.

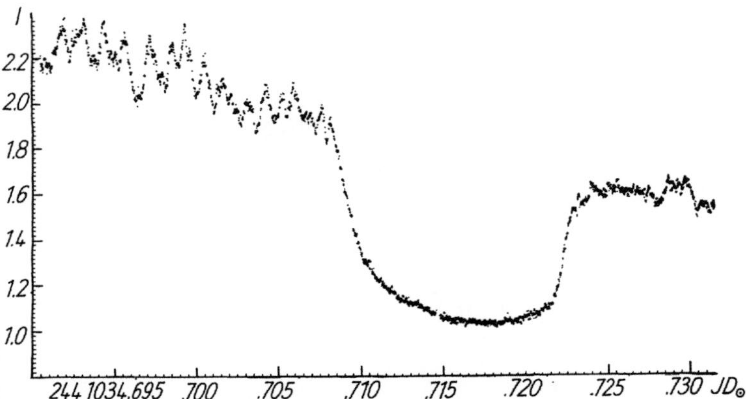

Fig. 59. Light-curve of U Gem; the flickering disappears during the eclipse of the "hot spot" by the secondary (after WARNER 1976). *I* is the intensity in arbitrary units

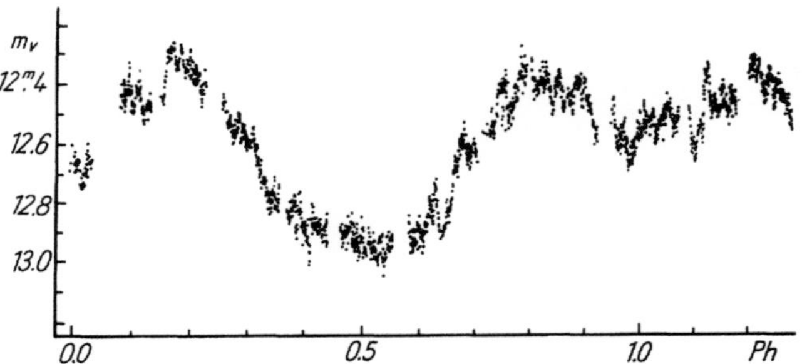

Fig. 60. Eclipse light-curve of AM Her, with superimposed rapid flickering (SZKODY and BROWNLEE, after LILLER 1977)

These 4 components, together with the periodic Doppler shift of the spectral lines, can be explained on the basis of ROBINSON's (1976b) "close binary system" model for the cataclysmic variables, which will be described.

Figure 59 shows the combination of components 2 and 4 (with high time resolution) for the dwarf nova U Gem. In the "true" novae the behaviour is similar.

Polars also exhibit the rapid flickering characteristic of novae and dwarf novae (Fig. 60).

Figure 61 shows a portion of the light-curve of AM CVn. The 18-minute oscillations (orbital period of the two components, $A = 0.04\,\text{mag}$) is combined with rapid, small-amplitude flickering and coherent oscillations (pulsation of the primary, $A = 0.007\,\text{mag}$). The period of the latter (1.98 minutes) could only be obtained by painstaking frequency analysis of the light-curve. AM CVn has the shortest known orbital period of any eclipsing binary.

An explanation of the complicated changes in luminosity will be given in the following section.

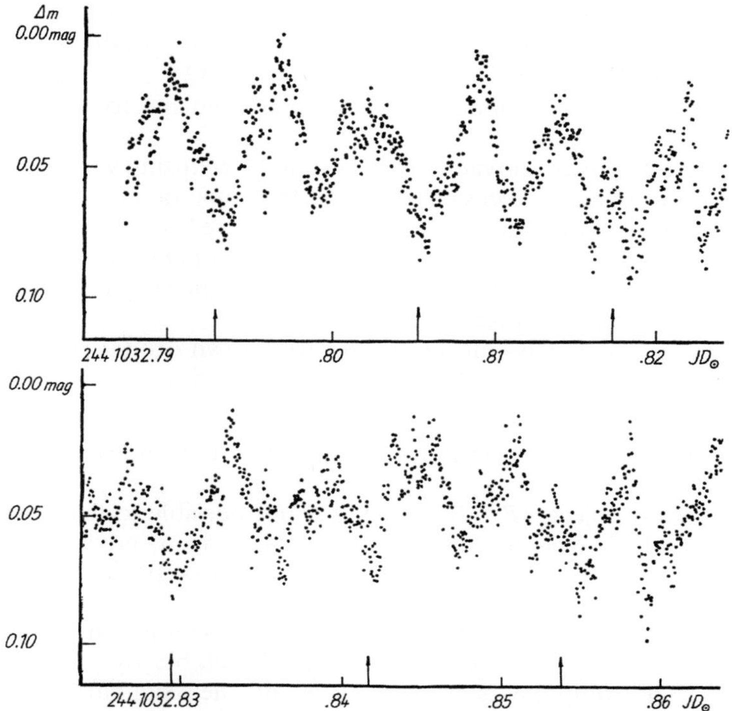

Fig. 61. Light-curve of AM CVn on 1971 Mar. 21 (after WARNER and ROBINSON 1972). The arrows indicate the predicted times of primary minima in the eclipsing light-curve. Rapid flickering is superimposed on the eclipse variations (see text)

Cataclysmic Variables as Close Binary Stars

Even 30 years ago all attempts to provide a reasonable physical theory for the outbursts of novae and dwarf novae came to grief because these objects were regarded as single stars. Only $T\,CrB$ had long been known to have a composite spectrum, the typical, high excitation, nova spectrum being combined with that of a gM3 red giant. To begin with it was considered that a single star might be involved, as it was assumed that the emission features could be produced in the extensive corona of the M star. Then KRAFT (1958) found that the lines from the M star showed a variable radial velocity with a period of $227^{\rm d}6$. Its binary nature was thus established. More detailed studies showed that it is a "semi-detached system" (see Sect. 4.2), in which the red component has filled its Roche lobe. In attempting to expand further it continuously loses material, either to a very extensive atmosphere or, after temporary storage in a disk of material, onto the small, hot companion: a white dwarf. JOY found in 1952 that the dwarf nova $SS\,Cyg$ was a spectroscopic binary, and in 1954 that the same applied to the dwarf nova $AE\,Aqr$.

M. F. WALKER succeeded in making the next notable discovery in 1954, when, by means of photoelectric observations, he recognized that the old nova *DQ Her* (1934) was an eclipsing variable with a period of $0\overset{d}{.}193627$. However, the red component could not be detected spectroscopically. Only recently have infrared observations shown that it is an M3V star. KRAFT (1959) was the first to develop a model for this system from spectrographic observations.

By 1977 5 novae and 6 dwarf novae were known to be eclipsing variables, and 5 novae and 8 dwarf novae recognized as spectroscopic binaries. Since then a number of other novae, dwarf novae and polars have been discovered and shown to be various complicated forms of binary system. Somewhat earlier, M. F. WALKER (1963a, 1963b) had already voiced the suspicion that all novae might be of a binary nature.

Table 32 gives a list of cataclysmic variables with known orbital periods P, principally drawn from WARNER (1976) and RITTER (1983).
Table 32 shows the following points:

1) There are *no* novae among the objects with orbital periods of less than 3 hours.
2) In the longer-period objects ($P > 3$ hours), it is not possible to tell from the orbital period P whether a particular object is a nova or a U Geminorum star! This is an indication of the very close physical relationship between all the cataclysmic binaries.
3) At orbital periods below 6 hours the secondary component is invisible in normal spectra. (Recently it has proved to be detectable by infrared measurements.) The components are very close and the secondary must consequently be very small, as otherwise it would be larger than its Roche lobe.
4) The cycle-length and amplitude of the variations in magnitude are not correlated with the orbital period.
5) The polars (the AM Herculis stars) have orbital periods < 3.1 hours.
6) There is a significant gap in the periods between $P = 2^{\text{h}}$ and $P = 3^{\text{h}}$. At present there are only conjectures as to the cause of this (see the discussion by ROBINSON 1983, for example).

The binary nature of many novae, U Geminorum stars and related objects is not detectable. Despite this, there are good grounds for suspecting that these are close binary systems where our line of sight is approximately perpendicular to the plane of the orbit, so that neither eclipsing behaviour nor the periodic Doppler shift of the spectral lines is visible.

It has already been mentioned that it is believed that the recurrent novae form a group in themselves, the majority of them having a giant star as the secondary component.

From analysis of the exceedingly complicated eclipse light-curves, and of spectral changes in the course of the orbital motion, a *model* for the cataclysmic variables has been established (see for example ROBINSON 1976b, WARNER and NATHER 1972, KRAUTTER 1978 and MARINO 1980). The system contains a *white dwarf* (the primary component, which contributes only slightly to the overall continuum) and a *red companion* (the secondary, generally a main sequence star, which may contribute some absorption lines). This

Table 32. Cataclysmic binaries with known orbital periods

Star	Class	P	Sp	Star	Class	P	Sp
T CrB	Nr	227$\overset{\text{d}}{.}$6	gM3	V 425 Cas	Nl	3$^\text{h}$35$^\text{m}$	
GK Per	N	45$^\text{h}$36$^\text{m}$	K2IVp	RR Pic	N	3 29	
BV Cen	UG	14 38	dG5–8	V 603 Aql	N	3 29	
DI Lac	N	13 3		VZ Scl	Nl	3 28	
V Sge	Nl	12 20	dG	0623 + 71	N?	3 27	
QU Car	Nl	10 54		V 1223 Sgr	Nl	3 23	
V 1668 Cyg	N	10 32		V 442 Oph	Nl	3 22	
AE Aqr	Nl	9 53	K5V	V 1500 Cyg	N	3 21	
SY Cnc	UG	9 7		TT Ari	Nl	3 18	
RU Peg	UG	8 54	K0IVn	PG 1012–029	Nl	3 14	
BT Mon	N	8 1		3A 0729 + 103	Nl	3 14	
Lanning 10	Nl	7 43		MV Lyr	Nl	3 12	M5V
AC Cnc	Nl	7 13	K5V	AM Her	Polar	3 6	M4.5V
EM Cyg	UG	7 0	K5V	TU Men	UG	2 50	
Z Cam	UG	6 56	dK7	AN UMa	Polar	1 55	
V 533 Her	N	6 43		CW 1103 + 254	Polar	1 54	
SS Cyg	UG	6 38	dK5	H 0139–68	Polar	1 54	
RW Sex	Nl	5 56		PG 1550 + 191	Polar	1 54	
RW Tri	Nl	5 34	M0V	UU Aql	UG	1 53	
TV Col	Nl	5 29		TY Psc	UG	1 51	
V 1727 Cyg	Nl	5 14		CU Vel	UG	1 51	
HR Del	N	5 8		WX Hyi	UG	1 48	
RX And	UG	5 5		VW Hyi	UG	1 47	
V 3885 Sgr	Nl	4 57		Z Cha	UG	1 47	
T Aur	N	4 54		HT Cas	UG	1 46	
UX UMa	Nl	4 43	dK8–M6	VW Vul	UG?	1 45	
DQ Her	N	4 39	M3V	AY Lyr	UG	1 45	
SS Aur	UG	4 28		1E 1013–477	Polar	1 43	
TW Vir	UG	4 23		1E 1405–451	Polar	1 42	
BD Pav	N	4 18		VV Pup	Polar	1 40	
U Gem	UG	4 15	M5V	EX Hya	UG	1 39	
WW Cet	UG	4 10		RZ Sge	UG	1 38	
H 2215 – 086	Nl	4 2		EK TrA	UG	1 32	
CN Ori	UG	3 55		OY Car	UG	1 31	
KR Aur	Nl	3 54		V436 Cen	UG	1 30	
PG 1140 + 719	Nl?	3 54		1E 1114 + 18	Polar	1 30	
V 380 Oph	Nl?	3 50		V 2051 Oph	UG	1 30	
CM Del	UG?	3 50		T Leo	UG	1 25	
LX Ser	Nl	3 48		SW UMa	UG	1 22	
3A 0729 + 103	Nl	3 45		WZ Sge	UG	1 22	
E 2003 + 225	Polar	3 43		EF Eri	Polar	1 21	
WY Sge	N	3 41		GP Com	AM CVn	0 46	DB(?)
H 2252 – 035	Nl	3 35		PG 1346 + 082	AM CVn	0 25	
AO Psc	Nl	3 35		AM CVn	AM CVn	0 18	DB(?)

Abbreviations:

Class:
    N = Nova
    Nr = Recurrent nova
    UG = U Geminorum star
    AM CVn = AM Canum Venaticorum star

Polar = AM Herculis star
Nl = Nova-like (i.e. specific classification not yet possible)
P = Orbital period
Sp = Spectrum of the secondary

secondary component fills the largest possible gravitational potential lobe (= Roche lobe, see Sect. 4.2), and loses mass through the system's inner Lagrangian point L_1. Owing to the conservation of angular momentum, this stream of material will, in the absence of any interference from strong magnetic fields, be captured by an accretion ring (or *accretion disk*) of rarified gas, rapidly orbiting the white dwarf. Where the *stream of material* hits the accretion disk a "bright spot" (or "hot spot") is formed (first suggested by SMAK). The *bright spot* can be regarded as the result of a shock front produced by the stream of material, which is subjected to such a high degree of acceleration by the white dwarf's strong gravitation that on impact with the disk a large proportion of the kinetic energy is suddenly transformed into heat and radiation. The accretion disk and the bright spot are the main sources of the continuum and the emission lines. The absolute lack of "forbidden lines", which can only occur in very tenuous gases, gives a lower limit for the density of the material in the disk, while the lack of pressure broadening in the spectral lines sets an upper limit.

Figure 39 shows this model of a cataclysmic variable.

The cause of the *rapid, irregular flickering* mentioned above is not yet entirely clear. It is presumably due to instabilities in either the hot spot or the inner region of the accretion disk (or both). This assumption is supported by the fact that the flickering completely disappears when the hot spot and the inner part of the accretion disk are eclipsed by the secondary (see Fig. 59). As far as the polars are concerned, the rapid flickering presumably arises in an *accretion column* close to the white dwarf (see also the discussion in WARNER and KROPPER 1983). On the basis of the variations with time in the photometric, spectroscopic and polarimetric features, the following model has been drawn up for AM Her (see LILLER 1977).

AM Her consists, like a U Geminorum star, of a cool object (spectrum M4.5V), a compact hot object (a white dwarf), and a stream of material flowing from the cool component towards the hotter one. In contrast to the U Geminorum stars and most novae, however, *there is no ring of material* surrounding the white dwarf. Such a ring cannot be formed, owing to the strong, dipole magnetic field of the white dwarf, which is shown to exist by the polarimetric observations, and which amounts to $\approx 10^8$ Oersted. Instead, under the influence of the magnetic field, the stream of material impacts directly onto the white dwarf close to one or both of the magnetic poles (forming an "accretion column"). (An analogy is offered by the aurorae on Earth, where solar particles enter close to the magnetic poles.)

As a result of the very high gravitational force at the surface of the small, but massive, white dwarf, the material is heated so strongly by the impact that intense X-ray radiation is produced. Because of this the AM Herculis stars are frequently grouped with the X-ray binaries. However, we prefer to count only those objects where the primary is not a white dwarf, but an even more compact object (neutron star or black hole), as being X-ray binaries (Sect. 3.1.7).

As we view the orbital plane of AM Her edge-on, we observe a total eclipse of the white dwarf by the cool star on each orbit, giving rise to an

"X-ray eclipse", which lasts about 28 minutes. In the visual region this eclipse is imperceptible as the collapsed object is so tiny. The optical changes during the period of one orbit consist of several effects:

1) A *double wave* (two maxima and two minima) caused by the rotation of the pear-shaped cool component (= rotational variation, see Sect. 4.3). This variation is observed during the "off" state, when the stream of material and the X-ray radiation it produces are weak.
2) A *simple wave* (one maximum and one minimum), which arises when the side of the cool component turned towards the white dwarf is strongly heated by the X-ray radiation. This bright hemisphere on the red star is periodically invisible once in each orbit, when it is turned away from us.

Finally, the hot stream of material provides an explanation for the complicated nature of the luminosity changes in various colours, by virtue of the way in which its contribution to the system's total light alters with the changing geometry during each orbit.

The *rapid oscillations* mentioned, which are always present in Nova DQ Her, occur in dwarf novae only in eruption. They may be explained either by non-radial pulsations of the white dwarf, similar to those of the ZZ Ceti stars (see Sect. 2.3.2), or through accretion of material from the disk onto a rotating white dwarf having a dipole magnetic field (see KATZ 1975, ROBINSON 1976b and KRAUTTER 1978). Table 33 summarizes (after ROBINSON 1976b) the amplitudes and periods of coherent oscillations in 10 cataclysmic variables (see also Table 8 in PATTERSON 1981).

The *variations that are synchronized with the orbital period* are produced by the red star eclipsing the white dwarf, the accretion disk and the hot spot. There is a phase shift of generally about 0.1–0.2 periods between the eclipse of the white dwarf and that of the hot spot, which arises from the fact that the bright spot does not lie exactly on the line joining the centres of the primary and secondary components.

Table 33. Amplitudes and periods (P) of coherent oscillations in cataclysmic variables

Star	Class	P	Amplitude
DQ Her	N	71^{s}07	0.04 mag
Z Cam	Z	16.0–18.8	0.001
SY Cnc	Z	24.6	0.003
Z Cha	UG	27.7	0.003
AH Her	Z	31.3–32.0	0.003
VW Hyi	UG	28.0–34.0	0.02
CN Ori	Z	24.3–25.0	0.005
KT Per	Z	26.7–26.8	0.006
UX UMa	Nl	28.5–30.0	0.002
V 3885 Sgr	Nl	29.0	0.003

Upper and lower limits show the range of variation in P

The relative contributions of the hot spot and the accretion disk to the total luminosity varies from system to system, leading to considerable diversity in the orbital light-curves:

In *U Gem*, for example, the hot spot is dominant (see Fig. 62). It can be seen as a large hump in the light-curve, which lasts for more than half the orbital period, and which reaches its peak luminosity 0.1–0.2 periods before the eclipse of the white dwarf.

In *UX UMa*, on the other hand, the disk predominates. The light-curve (Fig. 63) hardly shows the "hump" and the rapid, irregular flickering (see above); the eclipse minimum is close to spectroscopic conjunction. The small shoulder on the rise is produced by the eclipse of the hot spot.

Figure 64 shows the total eclipse light-curve in OY Car, according to VOGT et al. (1981). T_1 and T_2 indicate the beginning of the partial and total eclipse of the primary component; T_3 and T_4 mark the beginning of the partial and total eclipse of the hot spot; T_5 and T_6 show the end of the total and partial eclipse of the primary.

From the numerous investigations that have obtained excellent light-curves of cataclysmic variables with high time resolution, we may mention those of MUMFORD (1963) and WARNER and NATHER (1972).

As will be explained in detail in the chapter on eclipsing stars, when both components in a binary eclipse one another, and spectral lines with measurable radial velocity changes are present, the masses of the two objects may be obtained very accurately. However, as the eclipsing light-curve of the *eruptive binaries* is difficult to interpret owing to these systems' complexity, mass determinations are imprecise. Suitable methods may be found in WARNER (1973) and ROBINSON (1976a). The estimates obtained by these methods for about four dozen cataclysmic binaries allow the following statistical conclusions to be drawn:

1) The masses of the primary components lie between about 1/3 solar mass (UX UMa, LX Ser, VZ Scl, AM Her) and 1.4 solar masses (T CrB, BV Cen, RU Peg, EX Hya) and are independent of the orbital period and the type of light-changes.
2) The masses of the secondaries are correlated with the orbital periods. The highest masses $(1 - 3\mathfrak{M}_\odot)$ are those of objects with long orbital periods $(P > 9^h)$, and the lowest masses $(< 0.2\mathfrak{M}_\odot)$ are shown by objects with short orbital periods $(P < 2^h)$, such as the UX Ursae Majoris stars, a few polars and the AM Canum Venaticorum stars (Table 32). WZ Sge and AM CVn have secondaries for which the probable mass is only $0.04\mathfrak{M}_\odot$. The cataclysmic variables with rather longer periods $(P > 3^h)$ show *no* form of relationship between type of variation, orbital period (see Table 32) and the mass of either of the components.

Up to now, as already mentioned in Sect. 3.1.3, only two observable differences are known between dwarf novae and novae at minimum: the mean absolute magnitude of novae $(M_v \approx 4.5$ according to MCLAUGHLIN 1960) is about 3 mag greater than that of dwarf novae $(M_v \approx 7.5$ according

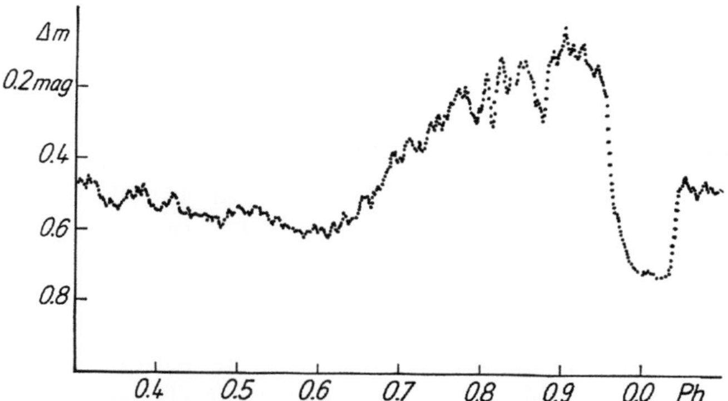

Fig. 62. Eclipse light-curve of U Gem for a little less than one orbital period (after NATHER 1973). Note the lower time resolution when compared with Fig. 59

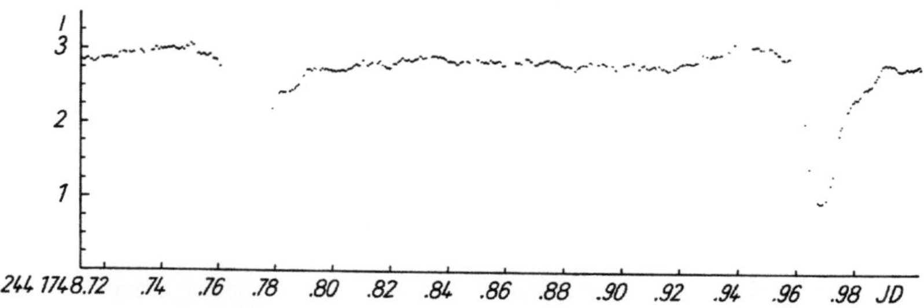

Fig. 63. Eclipse light-curve of UX UMa (after ROBINSON 1975b). I is the intensity in arbitrary units

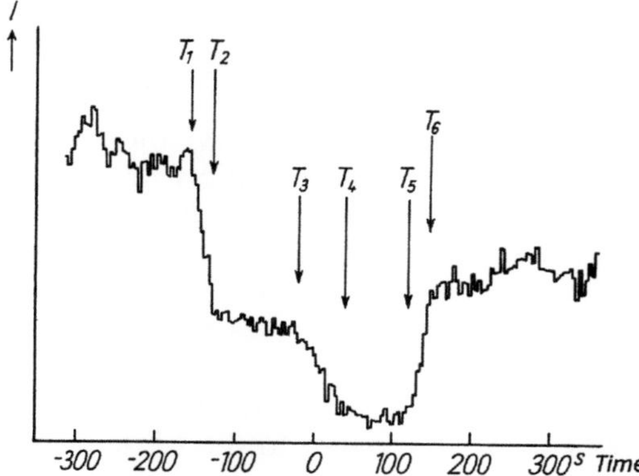

Fig. 64. Eclipse light-curve of OY Car (after VOGT et al. 1981). (For explanation see text)

to KRAFT and LUYTEN 1965); and the emission lines in novae are more strongly excited than those in dwarf novae. Presumably the mass-transfer rate is higher in the novae than in the U Geminorum stars (dwarf novae). The result is that in the novae there is stronger heating and excitation of the "hot spot", and a greater density (and thus a stronger continuum and greater luminosity) in the accretion disk around the white dwarf.

The masses of the (blue) primary components in some novae and dwarf novae (within the limits of error) are "dangerously" close to the upper *Schönberg-Chandrasekhar limit* for white dwarfs, 1.4 solar masses (T CrB, AE Aqr, Z Cam). This means that the mass arriving from the secondary component must be removed from the white dwarf, if it is to remain in that state and not collapse into a neutron star. This naturally raises the question of whether the basic cause of the nova eruption is that the excess mass is violently "shaken off" when it reaches the Schönberg-Chandrasekhar limit. However, this is definitely contradicted by the fact that true nova eruptions are also shown by cataclysmic variables where the white dwarfs lie well below the upper mass-limit (DQ Her, V 603 Aql). Nevertheless nova outbursts probably form a very effective mechanism for preventing the white dwarf from reaching the Schönberg-Chandrasekhar limit, or for postponing that event. However, if for any reason it should not succeed in preventing the mass of the white dwarf from rising above the limit, then the system probably "risks" experiencing a Type I supernova explosion (see Sect. 3.2).

Causes and Physical Development of Nova Eruptions

The first truly useful theories explaining the eruptions of novae were formulated after their development was successfully observed at wavelengths outside the visual range; in the infrared and radio regions and (from satellites) in the ultraviolet. This showed that a gradual decline from peak luminosity only occurs in *visual* light. In the novae FH Ser (1970), V 1229 Aql (1970) and HR Del (1967) the infrared luminosity ($\lambda = 1 - 25 \, \mu m$) continued to rise for 90 days after visual maximum had been reached, before it also began to decline! This finding can be explained by a circumstellar shell of dust, ejected from the star. The shell has a temperature of 900 K (after those 90 days) and a total mass of about 0.0001 solar mass. It becomes transparent, allowing the expanding, and (in many novae) strongly varying, photosphere to be seen. This state has also been observed spectroscopically (see above). The energy radiated at *radio wavelengths* continues to increase even 130–190 days after the eruption. In nova FH Ser 1970 (Fig. 65) the UV-radiation also rose well after the visual maximum, so that the total energy radiated was $\approx 3 \times 10^{45}$ erg, greater than had been previously accepted for novae.

The transfer of mass from the red stellar component to the white dwarf (via the ring of material) obviously plays an important part in initiating a nova explosion. The most direct method of determining the contribution made by this mass loss from the secondary (cf. the discussion in WARNER 1976) derives from the luminosity of the "hot spot" (obtainable from the eclipsing light-curve), on the assumption that most of the kinetic energy in the incoming stream is converted into radiation. Another method derives from the changes in the orbital period produced by the mass transfer.

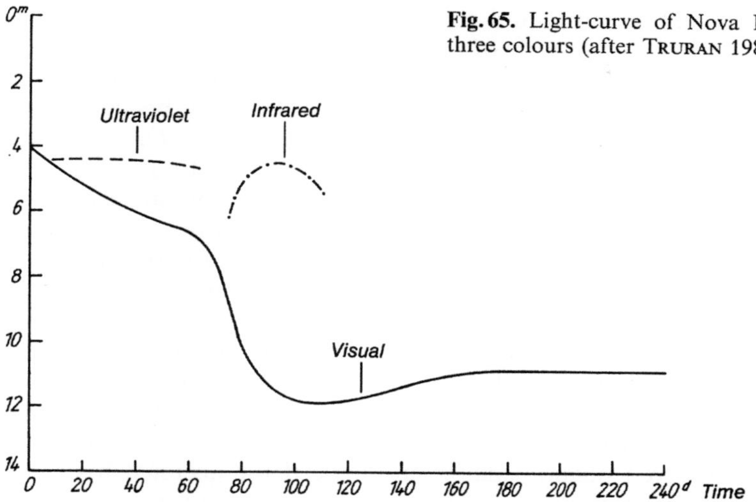

Fig. 65. Light-curve of Nova FH Ser (1970) in three colours (after TRURAN 1980)

Numerical estimates give a mass transfer rate of about 10^{18} g/s ($\approx 1.5 \times 10^{-8}$ solar masses per year) in U Geminorum stars, and about 10 times that amount in novae.

Moreover, it should not be assumed that all the material falling into the accretion disk does finally reach the white dwarf. In some systems there are indications of gas streams escaping from the system (see the discussion in WARNER 1976). It is essential, however, for a considerable fraction of the hydrogen-rich material from the red star to be deposited on the white dwarf. Precise determinations of the mass transfer rate have been discussed by STARRFIELD et al. (1976). TRURAN (1980) assumes an accretion rate of only $\approx 10^{-9}$ solar masses per year in (pre-) novae. In developing a theory for nova outbursts one has to start with the fact that at the surface of a white dwarf, where a whole solar mass is contained in a volume like that of the Earth, the force of gravity is exceptionally high. Consequently, very high energies are required as the material is not only accelerated from the surface to escape velocity (which needs an energy of about 10^{17} erg per gramme), but is also ejected from the system at velocities of as much as 4000 km/s, together with the accretion disk (which lies in the way). Only the release of thermonuclear energy is capable of doing this. The basic idea that nova explosions are due to thermonuclear processes occurring on the surfaces of white dwarfs came from SCHATZMAN (1950, 1951). Comprehensive reviews of modern theories on energy release and subsequent events in thermonuclear explosions as the source of the outbursts of novae, as well as numerous further references, can be found in WARNER (1976), STARRFIELD et al. (1974, 1976) and TRURAN (1980).

Most of the explanations start with the fact that such a white dwarf is the remnant of a star that has exhausted all its sources of energy (through hydrogen and helium burning) and collapsed. It is strongly enriched in the elements C, N and O. As a result of the mass-transfer processes, hydrogen flows from the cool secondary onto the white dwarf, where it forms a layer on the surface. With continuing transfer the base of the layer becomes compressed and heated until finally the critical temperature for *thermonuclear reactions* is reached and, by a runaway process, gives rise to an explosion.

Many models of nova outbursts have been calculated for various masses (0.5–1.25 solar masses) and luminosities of the white dwarf, various accretion rates (10^{-5}–10^{-11} solar masses per year) and various (more or less extreme) degrees of enrichment in the material by the elements C, N and O. (The last point is very important, as these elements play an essential role in the evolution of the nuclear explosion.) The ejecta have actually been found by spectroscopic measurements to be enriched in the elements C, N and O by a factor of 10–100 times when compared with the solar abundance! STARRFIELD et al. (1976) were able to show that a truly effective explosion, explaining the observed phenomena, would arise if the accreted hydrogen atoms did not remain on the surface of the white dwarf but were mixed into deeper carbon-rich layers by convective processes (see LAMB and VAN HORN 1975). It is only then that the special C-N-O cycle can be initiated in a truly explosive manner.

Although an explosion still takes place if the elements C, N and O have only solar abundances, it does not appear to be sufficient to throw off any material.

Theoretical calculations show the following results:

1) It is very difficult to "bring about" a thermonuclear explosion on a low-mass white dwarf of only 0.5 solar masses.
2) The more massive the white dwarf and
3) the higher the initial luminosity (and thus temperature), the less mass has to be accreted before a thermonuclear reaction begins.

For massive white dwarfs, models can be constructed where 10^{-4} solar masses of material are ejected into interstellar space with velocities up to about 100000 km/s! These resemble the properties of supernovae.

Fast and slow novae can be "produced" by varying both the mass of the white dwarf and the enrichment in C, N and O. Models for slow novae are included in PRIALNIK et al. (1978).

During the initial explosive *hydrodynamic ejection phase*, only 10%, at most, of the accreted material is shed. In the following "hydrostatic equilibrium phase" the remaining hydrogen is burnt as a "shell source" – producing an energy of 10^8–10^9 erg per gramme per second in the first few days after the explosion – until the hydrogen is exhausted. During this period of weeks or months, when the nova has an almost constant bolometric luminosity (see Figs. 40 and 65), the effective temperature of the former white dwarf is approximately 10^5 K, and its radius has expanded by a factor of 10–1000, so that it may possibly far exceed the size of the Roche lobe. It then appears (in the Orion-spectrum phase) as a *blue horizontal-branch object*, lying to the left of the RR Lyrae stars on the H-R Diagram, with a close binary system at its core. This configuration is unstable and, as may be observed spectro-scopically, the system loses mass until in the course of its evolution, the nuclear energy sources become exhausted and the radius of the primary component contracts below the Roche-lobe boundary. The primary slowly reverts to a *white dwarf* and the postnova stage is reached. In the "fast novae" the hydrodynamic ejection phase is the most important, and in the "slow novae" the hydrostatic equilibrium phase.

Estimates and calculations of the evolution of the explosions of novae mostly make the simplifying assumption that the hydrogen accretion onto the white dwarf takes place in a spherically symmetrical manner. KIPPENHAHN and THOMAS (1978) established that future theories should start with the more realistic assumption of *non-spherical accretion* of hydrogen around the stellar equator.

We see that current theories of the explosion of novae are able to explain, in a qualitative manner, the observational appearance and the photometric and spectroscopic behaviour, even though all sorts of investigations are still required for a full understanding of novae.

Causes of Dwarf-Nova Explosions

The spectral behaviour of a U Geminorum star at *outburst* shows no blue-shifted, emission-line components in the visible region, unlike classical and recurrent novae. This is a sign that *no* significant amount of *material* is *ejected* from the system, obviously due to the available *energy* in a dwarf-nova outburst ($10^{38}-10^{39}$ erg) being considerably less than that in the explosion of a nova ($10^{44}-10^{45}$ erg).

Theories of dwarf-nova outbursts may be grouped into the following categories (see also ROBINSON 1976 b and BATH 1976):

1) Recurrent, quasi-periodic, dynamic instabilities in the photosphere of the *secondary* (the red star) produce occasional sudden increases in the mass-transfer rate. The outer layers of the red star are stripped away, giving a rise in luminosity (see BATH 1972 and the references cited there). Recent spectro-scopic observations, however, contradict this theory.

2) Intermittent thermonuclear fusion of the hydrogen-rich material ac-creted by the white dwarf *primary*, which is not sufficiently intense to eject material from the system (STARRFIELD et al. 1974). The material thrown off by the white dwarf as the result of such an explosion is slowed down by the disk, which heats up as a result.

In contrast to the novae – violent explosions at long intervals – there are weak outbursts at short intervals as the white dwarf has a different structure from those in the true novae. This reduces the threshold value at which an explosion takes place.

3) Quasi-periodic variations in the mass density and thus the luminosity of the disk, due to *instabilities in the mass transfer* from the red star (BATH et al. 1974, MARINO 1980).

4) The disk of gas absorbs and re-emits *mechanical and optical energy* that originates in the white dwarf (NATHER and ROBINSON 1974).

5) Quasi-periodic instabilities in the *disk* itself, first suggested by SMAK (1971). Since then many publications have appeared to support this hypothesis (SMAK 1982a,b, 1983, CANNIZZO et al. 1982, MEYER and MEYER-HOFMEISTER 1982a,b, 1983). In this theory, cyclic, sudden changes in the viscosity of the outer portions of the disk are responsible for triggering the luminosity outbursts in U Geminorum stars. The mechanism only functions at very low accretion rates ($\leq 10^{-10}\mathfrak{M}_\odot$ per year), as in the U Geminorum stars. At higher mass-transfer rates ($> 10^{-9}\mathfrak{M}_\odot$ per year) like those usually found in the postnovae and UX Ursae Majoris stars, steady accretion takes place. SMAK (1982a) even believes that he may have found a mechanism that permits the existence of two different types of eruptions, as found in the SU Ursae Majoris stars. In objects that are subject to repeated changes between cyclic outbursts and periods of standstill (the Z Camelopardalis stars), the accretion rate is obviously sometimes lower, and sometimes higher than the critical value just mentioned. The interpretation of this oscillation between the two different states in terms of a "relaxation oscillation" is to be found in MEYER and MEYER-HOFMEISTER (1983).

MALLAMA and TRIMBLE (1978) discuss why novae do not show any U Geminorum-type changes, and U Geminorum stars do not undergo nova-like eruptions, although U Geminorum stars and novae are really physically very similar (apart from the fact that the mass-transfer rates in novae are about 10 times as great as those in the U Geminorum stars). We have already seen that at the high accretion rates in the novae, a continuous, stable mass-transfer occurs, only producing small variations at minimum luminosity. On the other hand a moderate mass-accretion rate gives rise to instabilities in the disk and hence the typical U Geminorum variation. The authors mentioned maintain that it might be possible for the U Geminorum stars also to have genuine nova eruptions. But because of the low mass-transfer rates the intervals between eruptions are significantly longer than in the classical novae, so that there is only about a 3% probability that anyone might see one of the approximately 300 known U Geminorum stars explode as a nova during his lifetime. However, as already mentioned, other authors ascribe the difference between novae and U Geminorum stars to the physical state of the primary (especially the chemical composition – the C, N, O content).

Nevertheless, it is worth noting that there is one classical nova in which U Geminorum-like variations began some time after its eruption. This is the nova GK Per (1901) already mentioned (see HUDEC 1981a). The amplitude may amount to 3.5 mag and the cycle-length from 1 to several years. The spectral behaviour during the 1981 outburst, which BIANCHINI (1982) among others was able to follow, was very different from that of a U Geminorum star. Moreover, the spectrum at minimum light resembles that of a nova and not that of a U Geminorum star. This is probably an atypical object.

According to an interesting, but purely speculative hypothesis suggested by VOGT (1983), nova- and U Geminorum-variations are seen as periodically recurring stages of activity in the same binary system. Such a cycle lasts $\approx 10^5$ years and the total lifetime of the cyclic stage $\approx 10^{10}$ years.

Formation and Evolution of Eruptive Binaries

The evolution of normal binary systems by mass exchange between the components will be discussed at the end of Chap. 4.

The experts still disagree over how a "normal" double star may become an eruptive binary through mass exchange. KRAFT surmised that novae and U Geminorum stars are a later evolutionary stage of W Ursae Majoris-type systems. This is suggested by the fact that W UMa systems are also very close binaries with short orbital periods (see WARNER 1974 and discussion in SAHADE 1976). However, many astronomers question whether W Ursae Majoris stars develop into cataclysmic systems (see also DUERBECK 1983).

More recently, attempts have been made to explain the formation of eruptive close binaries from components originally having a greater mass and with a longer orbital period (and greater separation). (See for example PACZYNSKI 1981, RITTER 1976 and 1980, EGGLETON 1976, 1979, MEYER and MEYER-HOFMEISTER 1979 and other references in the works mentioned.) Frictional braking in a resisting medium possibly played a part at times when the two objects had a common shell (and the white dwarf was still a supergiant), as perhaps did tidal braking or magnetohydrodynamic effects, in causing the orbital radius to shrink and thus reducing the period (see also EGGLETON 1983).

KOPAL (1979) suggests that eruptive binaries could be formed by the splitting of certain rapidly rotating stars.

Systems that may be in transition to eruptive binaries may include certain classes of symbiotic stars (Sect. 3.1.6), such as planetary nebulae with central binaries, according to PACZYNSKI (1976); and also, according to RITTER (1980), the two planetaries Abell 46 and Abell 63, whose central stars have properties that are characteristic of cataclysmic binaries (Sect. 3.4.3). According to PACZYNSKI, a further object in the transitional state is possibly V 471 Tau, which will be described in Sect. 4.7.

The later stages of evolution may be as follows:

The main-sequence star, which becomes enriched in He (or possibly C, N and O) in its interior by nuclear fusion processes, continually sheds material from its hydrogen-rich outer layers. It is forced into this mass-loss either because it tries to evolve away from the main sequence and thus tends to expand (see above), or else – as has been recently suggested – because the whole system loses energy (perhaps by gravitational radiation, see FAULKNER 1971 and RAPPAPORT et al. 1982). As a result the main-sequence star spirals closer and closer to the white dwarf. It loses more and more mass; the distance between the two components continually decreases, and the orbital period becomes ever shorter. Finally the main-sequence star loses material strongly enriched in helium (or even C, N or O) – as has been found spectroscopically in WZ Sge and the AM Canum Venaticorum stars, Sect. 3.1.3 – and no further nova eruptions can be initiated (see WARNER and ROBINSON 1972). Finally all that remains is a white dwarf with a rapidly orbiting *planet* that fills its Roche lobe, or else the white dwarf explodes as a supernova if it accretes enough hydrogen-poor material to exceed the *Schönberg-Chandrasekhar* limit (see Sect. 3.2).

Candidates for possible late stages of evolution of cataclysmic binaries are S 10830 (RICHTER et al. 1981), WZ Sge (WARNER and ROBINSON 1972, WALKER and BELL 1980) and, most particularly, the objects already mentioned in Sect. 3.1.3, AM CVn, GP Com and PG 1346 + 082 (WARNER and ROBINSON 1972). NATHER et al. (1981) give the following model for the AM Canum Venaticorum stars:

The objects are very similar to U Geminorum stars (a white-dwarf primary with an accretion disk and a "hot spot", fed by a secondary component). In contrast to the U Geminorum stars, the secondary is a low mass, helium white dwarf, which fills its Roche lobe and loses material to the still smaller, massive white dwarf. The orbital period is correspondingly short ($< 1^h$), and the orbital velocity of the accretion disk around the primary is extremely high (≈ 2000 km/s).

The authors mentioned also discussed the probable past and future evolution of these objects (see also MIYAJI 1983).

ROBINSON (1983) excludes the AM Canum Venaticorum stars from the general category of cataclysmic variables.

Do Single Stars Undergo Nova Explosions?

As already described, a nova outburst probably occurs through the thermonuclear explosion of hydrogen-rich material accreted by a white dwarf. This mechanism functions most effectively in a close binary system, in which a "donor" star is present and contributes most of the hydrogen.

It may be assumed that an isolated white dwarf (a single star) is also able to sweep up hydrogen from space, especially if it moves through a cloud of interstellar material. Calculations have shown, however, that the amount of matter accreted by the white dwarf from gas clouds is relatively low. Nevertheless, the possibility remains that, given a sufficiently long time, one of these objects could also accrete enough hydrogen for it to reach the critical state and produce a nova explosion. There is still disagreement among specialists about this possibility. In any case, the explosion of a single star as a nova, if it does occur, is a very rare event. DURISON and BURNS (1981) point to the relatively great numbers of novae in globular clusters and discuss whether these may be isolated white dwarfs, accreting gas from the cluster.

3.1.6 Symbiotic Stars

Classification

The term "symbiotic star" was first coined by MERRILL, who borrowed the term "symbiosis" from biology, where it means the co-existence of different types of organism to their mutual advantage.

Symbiotic stars, in the widest sense (see BOYARCHUK 1969 and SAHADE and WOOD 1978), are any astronomical objects whose spectra show a combination of the absorption features of a cool star and high-excitation emission lines.

In a narrower sense, symbiotic stars are restricted to objects which meet the following criteria:

1) *Late-type absorption lines* must be detectable (TiO bands, Ca I, Ca II, etc.).
2) *Emission lines* from highly-excited ions must be visible (He II, O III or even more highly ionized states). The Doppler widening of the emission lines should not exceed 100 km/s.
3) *Luminosity changes* with amplitudes of 3 mag (or more) and cycle-lengths ranging up to several years may be present.

In many symbiotic stars no luminosity changes have yet been detected. As far as this book is concerned, we are only interested in those symbiotic stars that are variables. According to current knowledge they are a very heterogeneous group of objects, and they probably represent a number of different evolutionary phases in binary stars. The complicated, irregular changes in magnitude include occasional longer brightenings of greater or lesser duration, so that the description "nova-like stars" has also been employed.

Some symbiotic stars show a very strong *infrared excess,* originating in an extensive *shell of dust* (Z And, V 1016 Cyg and RX Pup, for example). WEBSTER and ALLEN (1975) use this to distinguish between two main types of symbiotic stars: those without dust emission, and those with it, some of the latter having *radio emission* as well. According to SAHADE (1976), MAMMANO and CIATTI (1975) and PACZYNSKI (1976), the symbiotic stars with dust emission (or at least some of them) are precursors for some of the planetary nebulae. The planetary nebulae NGC 1514 and 2346, which both have 2 stellar components (Sect. 3.4.3), may be late-stage symbiotic stars according to MAMMANO and CIATTI (see also the discussion in SAHADE 1976). The fact that some of the symbiotic stars may be the precursors of novae and U Geminorum stars has already been mentioned.

In 1981 four symbiotic stars (all variable) were known to be soft X-ray sources. These were objects frequently also described as "very slow novae": V 1016 Cyg, HM Sge, RR Tel (Fig. 66) – cf. ALLEN (1981) – and AG Dra (L. MEINUNGER 1981). This radiation is presumably released by a flow of material onto a white dwarf. Typical light-curves, spectra and radial velocity curves have been published in the review article by BOYARCHUK (1969): Fig. 67 shows as an example the light-curve of Z And, where quasi-periodic fluctuations in brightness (mean length 714 days, but with considerable variation) are interrupted by outbursts that are up to 4 mag in amplitude. Similar behaviour is shown by BF Cyg, CI Cyg and AX Per, but here the quasi-periodic changes are less distinct than in Z And. The light-curve of AG Peg is completely different: it is remarkable for a great outburst, about 3 mag in amplitude and approximately 100 years in duration, which resembles that of an extremely slow nova (Fig. 68). Outside this outburst changes in luminosity having a periodicity of roughly 800 days predominate. Numerous symbiotic stars resemble very slow novae in their light-curves.

It has already been mentioned (Sect. 3.1.2) that at least some of the recurrent novae are related to the symbiotic stars.

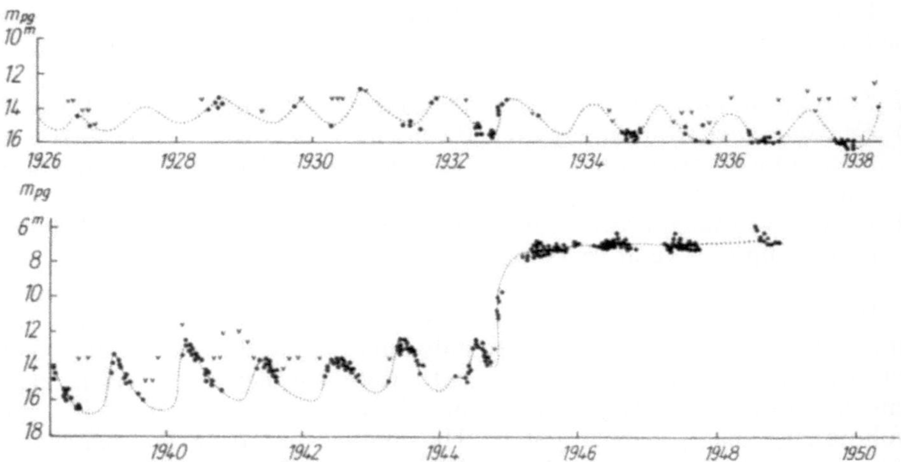

Fig. 66. Light-curve of RR Tel from 1926 to 1949 (after MAYALL)

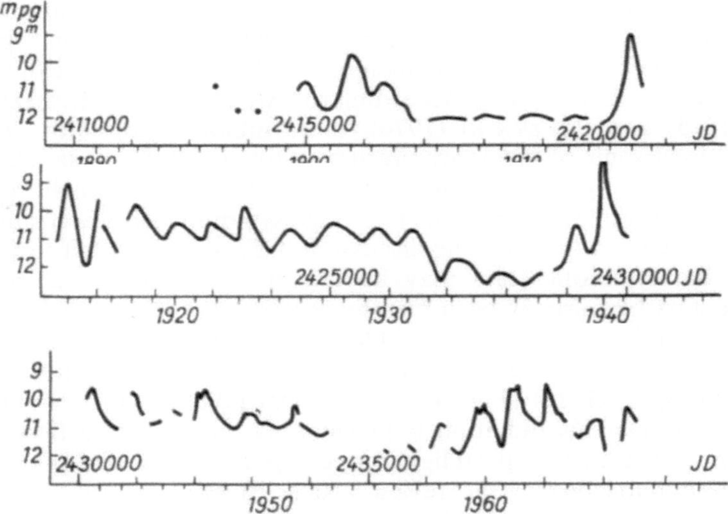

Fig. 67. Light-curve of Z And (after BOYARCHUK 1969)

The changes in luminosity in symbiotic stars are linked with changes in colour. The $B - V$ index increases with a fall in magnitude.

Table 34 – largely from BOYARCHUK (1969) and PAYNE-GAPOSCHKIN (1977 a and 1977 b) with revisions from recent literature – gives a selection of symbiotic stars.

Spectral Changes

When many Z Andromedae stars decrease in brightness they show a strengthening of the absorption-line spectrum and an increase in the degree of excita-

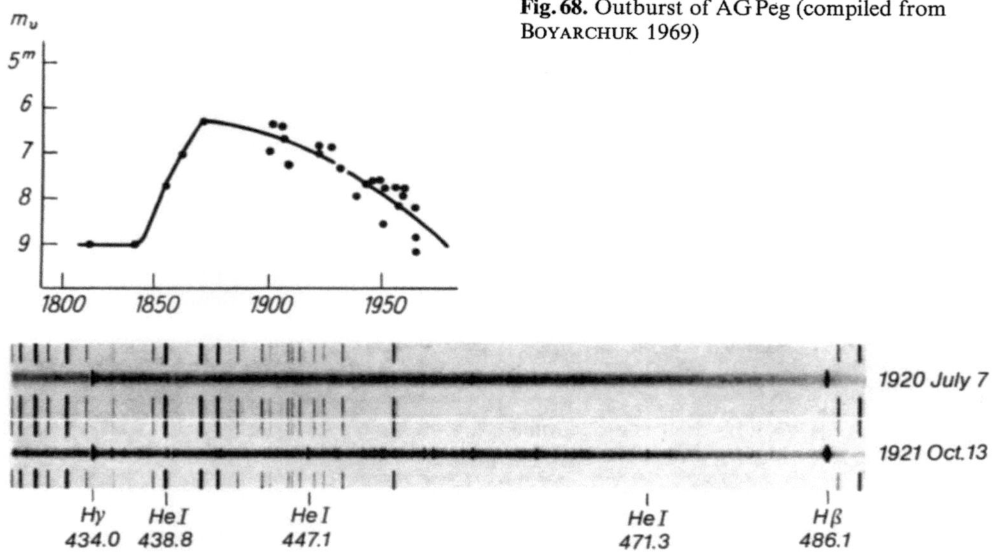

Fig. 68. Outburst of AG Peg (compiled from BOYARCHUK 1969)

Fig. 69. Spectrograms of the symbiotic star AG Peg (see text) (after MERRILL 1959)

tion of the emission lines (see also BOYARCHUK 1969). Other objects, such as AG Peg, occasionally also show strong changes in their spectra at times when there is only a small amount of variability in the total light. Figure 69 shows this in two spectrograms of AG Peg: note the strengthening of the emission lines (H, He, FeII) in the lower spectrum.

Spectral changes during the brighter outbursts may be very complicated (see BOYARCHUK 1969), and, as Fig. 69 shows, P-Cygni features may also occur, indicating the presence of a rapidly-expanding gas shell. Many of the observed phenomena can be explained on the basis of variable transparency of an extended dust shell surrounding the object, and by variations in the conditions controlling the degree of excitation.

Both the Doppler shift of various lines and their changes with time give important information about the physical processes. So too do eclipsing light-curves, which can be observed in some symbiotic stars, and which show that many (indeed probably all) symbiotic stars are binaries. Figure 70 reproduces radial-velocity curves derived from various lines. Open circles indicate FeII, and full circles high-excitation lines.

Absolute Magnitude and Distribution in the Galaxy

According to SAHADE (1976) the absolute visual magnitude is of the order of -3^M to -4^M. The galactic distribution and the space velocities for symbiotic stars correspond to those found for Population II objects.

Table 34. A selection of symbiotic stars

Star	m_{max}	m_{min}	P_{RV}	P_{Ph}	Class	Notes
Z And	8^m0	12^m4		694^d	Z And	
R Aqr	5.8	11.5	9740^d	387	Mira	1
CM Aql	13.2	16.5			Z And	
TX CVn	9.3	11.6			Z And	
RT Car	11.0	11.4			Z And?	
o + VZ Cet	2.0	10.1	$3.6 \cdot 10^4$	332	Mira	
T CrB	2.0	10.8	288		Nr	
BF Cyg	9.3	13.5	750	754	Z And	
CH Cyg	7.4	9.1		97	Z And	2
CI Cyg	9.4	13.7	815	855	Z And + EA	
V 407 Cyg	13.3	[16.5		745	Mira + Nova	
V 1016 Cyg	10.3	17.5	450		x	3, 9
V 1329 Cyg	11.5	18	960		x	4, 8, 9
AG Dra	9.1	11.2		554	Z And	5
YY Her	11,7	[13.2			Z And	
V 443 Her	12.4	12.6			Z And	
RW Hya	10.0	11.2	370	370	Z And	
SS Lep	4.8	5.1	276:		Z And?	
AX Mon	6.6	6.9	232:		x	
SY Mus	11.3	12.3		623	Z And?	
RS Oph	5.2	12.3			Nr	
AR Pav	8.5	13.6	605		Z And + EA	6
AG Peg	6.0	9.4	830	800	Z And	7
AX Per	9.7	13,4	600–880	685:	Z And	
RX Pup	11.1	14.1			Z And	
HM Sge	11	16			x	9
FN Sgr	9	13.9			Z And	
KW Sgr	11.0	13.2		670	Z And (SRc)	
V 1017 Sgr	6.2	14.4			Nr (Z And?)	
V 2416 Sgr	14.4	[17.6			Z And	9
V 2601Sgr	14.0	15.3		850	Z And	
V 2756 Sgr	13.2	15.2		243	Z And?	
V 2905 Sgr	10	14.6			Z And	
FR Sct	11.7	12.5			Z And	
RR Tel	6.5	16.5			Nl	

Abbreviations: (P_{RV}) Radial velocity period, (P_{Ph}) Period of light variation, (Class) Class of variation from the GCVS and its Supplements, (EA) Algol-type eclipsing variations, (x) unique, peculiar objects; the remaining abbreviations are described in the text in this section

Notes:
1) Probably shows eclipsing variations with a period of 44 years.
2) According to Luud, during maximum light of the semi-regular red giant, a temporary disk of material forms around the hot component (a white dwarf). It is the only symbiotic star yet known to show flickering like that found in cataclysmic variables. This was visible during the 1977 outburst.
3) V 1016 Cyg is probably a planetary nebula in a very early evolutionary stage (see the summary in Sahade 1976).
4) "Grygar's Variable". Probably has eclipsing variations ($P = 960^d$).
5) Period after Meinunger (1981).
6) AR Pav is an eclipsing variable with an amplitude of 2 mag ($P = 605^d$). The hot component is physically variable.
7) A model for this interesting and complicated object is described by Cowley and Stencel (1973), see also Sahade (1976), and Ghigo and Cohen (1981).
8) According to Blair et al. (1981) related to V 1016 Cyg and RS Canum Venaticorum stars.
9) In the cases of V 1016 Cyg, V 1329 Cyg, V 2416 Sgr, and HM Sge, opinions are divided whether they should be classed with the Z Andromedae stars or with variable planetary nebulae. They are also covered in Sect. 3.4.3.

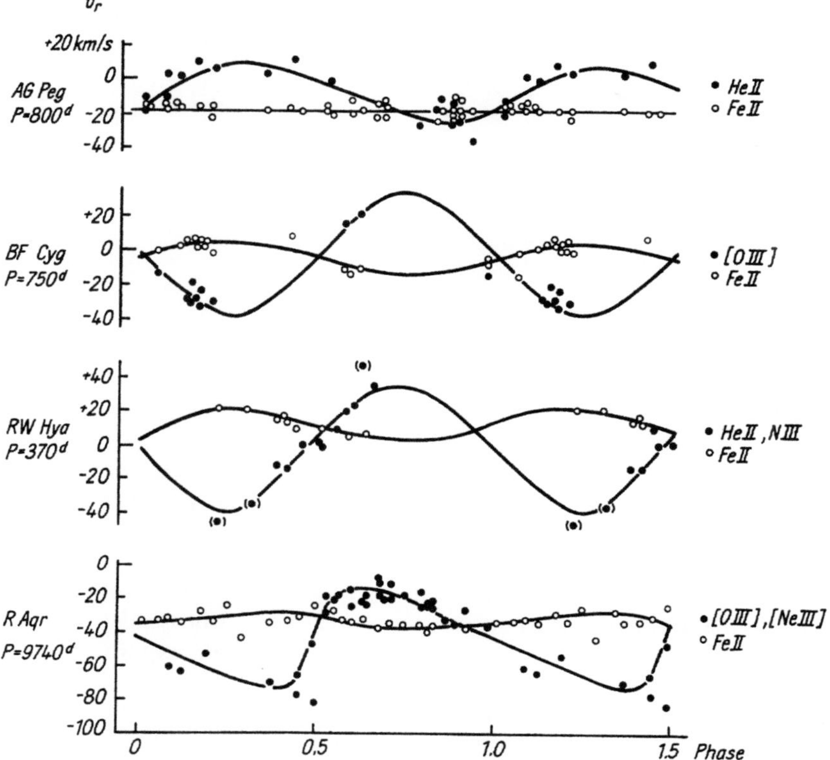

Fig. 70. Evidence for the binary nature of four symbiotic stars based on radial-velocity curves (after BOYARCHUK 1969)

Models of Symbiotic Stars

Although attempts are still occasionally made to explain the spectrum of a symbiotic star by some complicated structure in a single star, the discovery of periodic radial-velocity changes in the spectral lines and of eclipsing light-curves in many symbiotic stars shows that they are binaries.

Nowadays the following general model of a symbiotic object is accepted (Fig. 71, see also COWLEY and STENCEL 1973 and the review article by FEAST 1983): it is a binary with an M (rarely G or K) giant secondary of approximately 100 solar radii, and a hot subdwarf (or occasionally a white dwarf) primary of less than 0.5 solar radii, which has an effective temperature of about 100 000 K. The orbital period is from one to several years, with a mean distance of about 1000 solar radii (5 astronomical units). Both components are surrounded by one or more common shells or disks of gas with radius $\approx$ 50 000 solar radii, electron density $\approx 10^6 - 10^7$ electrons per cm^3, and an electron temperature of about 15 000 – 20 000 K. The source of the gas shell is the red giant, which is losing considerable amounts of mass either through a stellar wind or (as a long-period or Mira star) through pulsation. According

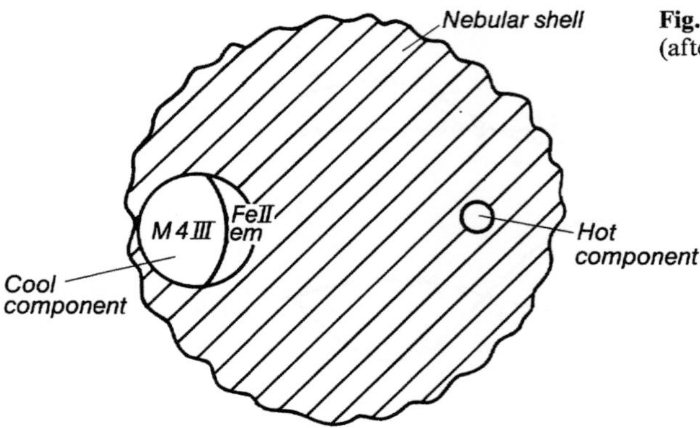

Fig. 71. Model of a symbiotic star (after BOYARCHUK 1969)

to estimates by A. R. WALKER (1976), however, the mean mass-exchange rate in symbiotic stars only amounts to 3×10^{-9} solar masses per year. Only in the case of symbiotic Mira stars is it significantly higher.

The high-excitation emission lines arise in the gas shell by radiation from the hot component, which is accreting mass. As yet it is unknown whether this accretion is spherically symmetrical, equatorially symmetrical or (if the subdwarf has a dipole magnetic field) occurs at the poles. It may differ from object to object.

The lower-excitation lines (Fe II) do not arise in the shell, but in those regions of the cool component's atmosphere that are turned towards the hot star. The radial-velocity curves in Fig. 70 consequently reflect the periodic changes in the velocity of the cool star relative to the portion of the shell that surrounds the hot component. Most symbiotic stars probably do not have the "hot spot" typical of the novae and dwarf novae (see SLOWAK 1980). AR Pav apparently has a ring of material and a hot spot (see SAHADE 1976), like the recurrent novae, as suggested by the flickering that is observed (A. R. WALKER 1976). Flickering is also observed in the case of CH Cyg (MIKOLAJEWSKA and MIKOLAJEWSKI 1980).

In some cases the gas shell is very thin while in others it is very thick. It may then appear like a planetary nebula with two central stars. Some variable planetary nebulae therefore show a close relationship to the symbiotic stars.

All three components in a symbiotic object can contribute to the variability, and the following five effects are found in combination:

1) Variability in the transparency and the excitation of the shell.
2) Variability of the hot (primary) component, probably because of irregularities in the mass-accretion and energy-release rates. In extreme cases (if the primary component is very compact and degenerate) nuclear explosions may occur (recurrent novae like T CrB and extremely slow novae like RT Ser – Fig. 42 – and RR Tel).
3) Long-period or Mira-type changes in the cool (secondary) component.
4) In a few cases (see above), eclipses.

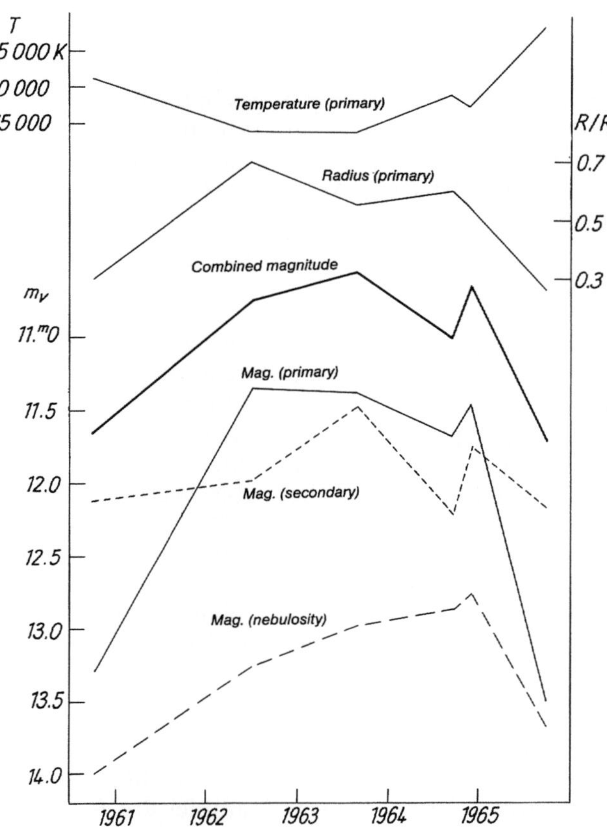

Fig. 72. Z And: respective contributions by three different components to the overall variation in light between 1961 and 1966 (*lower curves*). *Above:* change in temperature and radius of the primary over the same period of time (after BOYARCHUK)

5) Rotational changes in magnitude due to differing surface brightness (arising from the so-called "reflection effect", see Sect. 4.4), which can be observed even when no eclipse is possible.

If the third effect dominates the overall changes in luminosity, then we have a "symbiotic long-period variable", of which the best-known examples (see below) are R Aqr, UV Aur, o Cet + VZ Cet. If the second effect is the most important, then depending upon the properties of the shell and of the hot subdwarf, the object is either a *recurrent nova* (T CrB, RS Oph) or an *extremely slow nova* (RR Tel, see also FEAST et al. 1983). In the *Z Andromedae stars* the first, second and third effects are equally important in controlling the changes in luminosity. Figure 72 shows how these three components contribute to the total variation in Z And.

There appears to be a gradual transition between the Z Andromedae stars and the recurrent novae. For example, CI Cyg had nova-like outbursts in 1911, 1937, 1971 and 1973.

Attempts are nowadays being made to distinguish various types of symbiotic stars, using as a basis the physical conditions assumed to exist (see PACZYNSKI and RUDAK 1980, ALLEN 1980). These are:

Type I Stars

In the hot, primary component energy is released in a stable hydrogen-burning shell. Variations in the accretion rate cause changes in the brightness of the hot component and of the shell (owing to its variation in mass, radius and temperature). Most of the Z Andromedae stars show changes in brightness of Type I. Examples are: Z And, BF Cyg, AG Dra, AG Peg, AX Per. Figure 67 shows the light-curve of *Z And*.

Type II Stars

The hydrogen-burning on the hot component occurs in "flashes", explosions which are similar to those in the classical novae. The difference in behaviour from the novae is probably due to the different composition of the white dwarf (there being no excess C, N or O present), or possibly to the presence of the dense shell, which modifies the course of the outburst.

In the majority of cases the optical behaviour in these objects corresponds to the very slow novae, classified in the GCVS as nova-like "Nl". (However, we prefer to regard the objects defined in Sect. 3.1.2 as being "nova-like".) Some examples are: RX Pup, RR Tel, as well as the objects V 1016 Cyg, V 1329 Cyg and HM Sge, which are also considered to be planetary nebulae. Figure 42 gives the light-curve of *RT Ser* as an example (see also FRIED 1980).

Recurrent Novae

These objects, already discussed in Sect. 3.1.2, have thinner shells than those just discussed. Their changes in luminosity more closely resemble those of the classical novae. They are related to the Type II stars. Examples are *RS Oph* and *T Pyx*.

Symbiotic Stars that Emit Hard X-Rays

As yet only one object of this type is known: V 2116 Oph. In contrast to all the other known symbiotic stars, which, if at all, only show extremely soft X-radiation, this object is identical with a hard *X-ray pulsar, 3U 1728 − 24*. ALLEN (1981) sees this object as a symbiotic star in which there is a neutron star instead of a hot subdwarf or white dwarf (see also WEBBINK et al. 1983). Physically, this object belongs to the X-ray binaries (Sect. 3.1.7).

Symbiotic Long-Period Variables

R Aqr (see WALLERSTEIN and GREENSTEIN 1980, KALER 1981). In this object the M7e spectrum is combined with the spectrum of a variable gaseous nebula, which has high-excitation emission lines. The continuum from a hot companion was also visible from 1922−1933, when it reached magnitude 8. However, the dominant form of variability is a Mira-like variation with a period of $387^{\rm d}$ and a visual range of $5^{\rm m}8$ to $11^{\rm m}5$. An illustration of the nebula is given in WALLERSTEIN and GREENSTEIN (1980).

UV Aur (BOYARCHUK 1969). This red long-period variable (visual range $9\overset{m}{.}8$ to $11\overset{m}{.}1$, *Sp* C8ep) also shows a symbiotic spectrum. The red component is dominant in the changes in brightness.

o + VZ Cet (YAMASHITA et al. 1978). VZ Cet is the physical companion to Mira Ceti (o Cet): *Sp* Beq, maximum distance $0\overset{''}{.}9$, orbital period 100 years. VZ Cet ($9\overset{m}{.}5$ to 12^m) with a cycle-length of about 14 years, is probably not a typical Be star. Its variability is due to interaction with nearby Mira Ceti, a gMe star, which has a 332-day period.

Further information about symbiotic stars may be found in the recent book *The Nature of Symbiotic Stars*, Astrophys. Space Sci. Lib. Vol. 95 (1982), and in *Invited Review at IAU Colloquium 80* by ALLEN (ed. 1983).

Finally it should be mentioned that, despite all the progress, we are still far from a complete explanation of the phenomena shown by symbiotic stars. It is still not clear whether we are dealing with a physically homogeneous class of objects, or whether the stars are of different types but show similar behaviour.

The symbiotic stars should not be confused with the shell stars of the S Doradus (Sect. 3.4.1) or γ Cassiopeiae (Sect. 3.4.2) classes, which are possibly single stars, and from which they may be readily differentiated by spectroscopic methods.

3.1.7 X-Ray Binaries

As has been seen in the previous sections, X-rays are observed from many cataclysmic and symbiotic binaries, and later sections of this book will also discuss other objects that are sources of X-rays. Our Sun, and probably every star, gives off X-rays. How far these may be observed is only a question of the sensitivity of measurement that can be attained.

In the narrowest sense, X-ray binaries are only those objects that *radiate more energy at X-ray wavelengths than in all the other spectral regions combined*. These are objects that could only be discovered and classified following advances in astronomy from Earth satellites, as our atmosphere is not transparent to X-rays.

Although this book deals in particular with stars that vary in the visible region, we need have no hesitation in including variable X-ray sources, as in practically every case precise determination of the position of the source reveals an optical counterpart, which either is a new variable or is already known as a variable object. Thorough photometric investigations in the various spectral regions and spectroscopic work have shown that a large part of the cosmic X-ray sources are binaries, which are more or less related to the novae and U Geminorum stars. The fundamental process for the release of the X-rays and for the optical variation in all these objects is probably the accretion of material onto a compact companion. The smaller and more massive the compact object is, the shorter the wavelength of the X-rays produced. Detailed reviews of the current status of this young field of research have been published by JONES et al. (1974), LIGHTMAN (1976), LILLER

(1977), KUNDT (1981); see also HUTCHINGS (1977) and CULHANE (1977). A catalogue of all X-ray objects known up to the beginning of 1978 had been published by AMNUEL et al. (1979). Discussion of the possible origin of X-ray binaries may be found in FLANNERY and ULRICH 1977.

Nowadays the following types of X-ray binaries are recognized (the AM Herculis stars, which are related to the U Geminorum stars, have already been discussed in Sects. 3.1.4, 5):

HZ Herculis Stars or Low-Mass X-Ray Pulsars

In 1972 LILLER succeeded in identifying the X-ray source *Her X-1*, found by the Uhuru X-ray satellite, with the variable star *HZ Her*, discovered by HOFFMEISTER in 1936. In the 1969 GCVS HZ Her was still identified as "Is", that is a rapidly varying, irregular object, ranging between $13^{m}.0$ and $14^{m}.5$.

HZ Her exhibits rather *complicated changes in luminosity*, but which occur with great regularity.

In *X-rays* the following *effects* are found:

1) Every 1.24 seconds X-ray pulses are observed; hence the name X-ray pulsar (see Fig. 73).

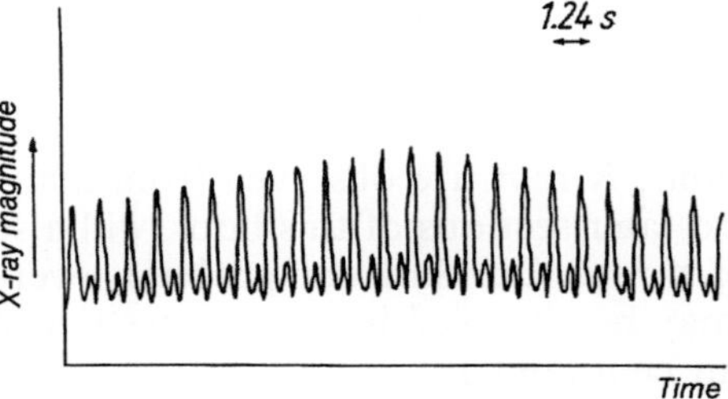

Fig. 73. X-ray pulses in Her X-1 (Uhuru measurements, after KIPPENHAHN 1973)

2) The period of 1.24 seconds is not held rigorously, but shows a frequency modulation with a period of 1.70017 days. The latter may be interpreted as the orbital period of the X-ray object in a binary system: the periodic changes in distance of the X-ray source relative to the Earth cause a periodic increase and decrease in the 1.24-second frequency due to the Doppler effect. Estimates give an orbital velocity of at least 170 km/s and an orbital radius of 8 solar radii.
3) At regular intervals of 1.70017 days the X-ray pulses suddenly disappear for 5 hours, as is shown in the schematic X-ray light-curve, Fig. 74 top. This can be interpreted as a total eclipse, lasting 5 hours, of a very small X-ray source by a normal star (the secondary).
4) The X-ray luminosity also follows a marked 35-day cycle: the source is alternately switched on for 12 days and switched off for 23.

In *visible light* the following changes in brightness are observed:

1) The 1.24-second variations can be detected only occasionally, and with a very small amplitude.

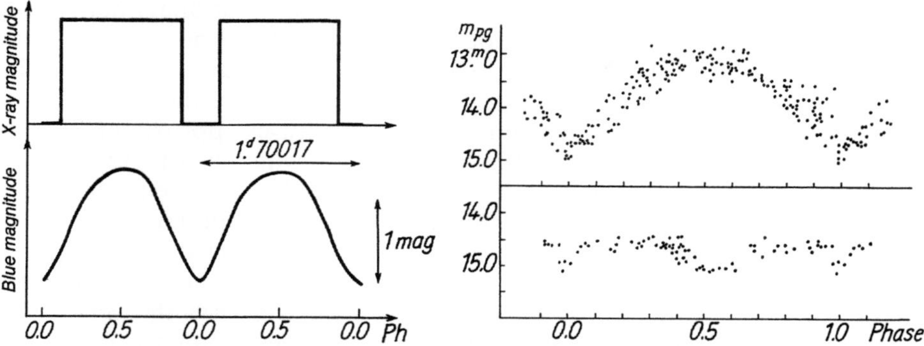

Fig. 74. HZ Her (= Her X-1): schematic light-curves of the 1.7-day period in X-rays and in blue light (after KIPPENHAHN 1973)

Fig. 75. Light-curves of HZ Her ($P = 1\overset{d}{.}7$) during "on" (*above*) and "off" (*below*) states (after HUDEC and WENZEL 1974)

2) Rapid, irregular, low-amplitude flickering on a time scale of several seconds to several minutes, typical of a cataclysmic variable, is found in runs of photoelectric measurements.

3) The principal changes in luminosity come from periodic fluctuations with the 1.70017-day orbital period and an amplitude of approximately 1.5 magnitudes. These are themselves easily observable by amateur astronomers (see Fig. 75 above). Minimum in the visible coincides with the X-ray minimum, but is not caused by the eclipse of the X-ray source. The latter is so faint in the visible that it does not make any significant contribution to the total light of the system. Instead, the optical minimum arises (as in the case of AM Her already mentioned) because one hemisphere of the secondary, strongly heated and thus brightened by the X-rays, is turned away from us at the same time as the X-ray source is eclipsed. The secondary then appears as an F0 star. When we see the heated hemisphere, on the other hand, it resembles a B8 to A0 star.

It is remarkable that the optical, orbital light-curve is only slightly modified by the 35-day, X-ray cycle. During the 23 days when no X-rays are detectable, we still see periodically (every 1.7 days) the secondary's hemisphere that is heated by X-rays. This is an indication that during those 23 days the X-ray radiation is reaching the companion, but not the Earth.

4) HZ Her can be optically inactive for several years, when the brightness drops and exhibits a low-amplitude double wave at about 15^m. This is presumably due to the compact object being inactive in X-rays during this period, the facing hemisphere of the F star being no longer heated (HUDEC and WENZEL 1976). Figure 75 shows the mean optical light-curve ($P = 1.7$ days) during active and inactive states.

Although HZ Her has yet to be fully explained. the current *model* for the object is roughly as follows.

An F0 star, which has already moved slightly away from the main sequence, and which is filling its Roche lobe, is losing mass onto a compact companion of approximately one solar mass. The observational findings can be explained if this companion is not a white dwarf, but a neutron star with a magnetic field of about 5×10^{12} Oersted. The magnetic field causes the flow of material to be accreted onto the magnetic poles of the neutron star. However, as matter falling onto a neutron star releases approximately 1000 times as much energy as it would in falling onto a white dwarf (AM Her for example, see above), the hydrogen is spontaneously converted into heavy elements (principally the iron group), which releases additional energy. Very hard X-ray radiation is therefore produced at the neutron star's (magnetic) poles. The 1.24-second periodicity can be explained, as in the optical pulsars (Sect. 3.6.2), by a lighthouse effect. The neutron star has the very short rotational period of 1.24 seconds. (At this speed a less compact object – a white dwarf – would be torn apart by centrifugal forces.) As the magnetic and rotational poles do not coincide – even on the Earth they do not – the magnetic poles spin around the rotational axis, so that every 1.24 seconds the X-rays from one pole reach the Earth in the form of a short pulse. The 35-day X-ray periodicity is probably due to precession of the rotational axis. This would alter the direction of the beam of X-rays radiated from the magnetic poles of the neutron star in a 35-day cycle. The 23-day interruption in X-ray radiation, mentioned above, can therefore be explained by assuming that for this period the Earth does not intercept the beam of X-rays.

Full descriptions of the interesting system found in HZ Her (Her X-1) have been given by KIPPENHAHN (1973) and JONES et al. (1974). The most recent ideas about this system can be found, in YAHEL (1980), HOWARTH and WILSON (1981), and elsewhere.

Table 35 gives a listing of the pulsational periods of X-ray pulsars known to date, according to DELPINO (1981), who also discusses the possible evolutionary history of HZ Herculis stars.

Table 35. Pulse periods of X-ray pulsars (selection)

X-ray source	Optical object	P
SMC X-1	SK 160	0$.^{\!s}$715
Her X-1	HZ Her	1.24
Cen X-3	V 779 Cen	4.84
4U 1626 − 67	KZ TrA	7.68
A 0535 + 26	V 725 Tau	104
GX 1 + 4	V 2116 Oph	122
Vel X-1	GP Vel	283
4U 1145 − 61	V 801 Cen	292
A 1118 − 61	RS Cen (?)	405
4U 1538 − 52	QV Nor	529
GX 301 − 2	BP Cru	696
3U 0352 + 30	X Per	835

Apart from Her X-1, several other regular X-ray pulsars are known, of which the most important are given in Table 36. This table is compiled from the data in AMNUEL et al. (1979) and RITTER (1982) with some revisions from recent literature.

Table 36. A selection of X-ray binaries with known orbital periods

X-ray source	Optical object	P	m_{max}	m_{min}	Sp_{opt}	$\mathfrak{M}_{opt}$	$\mathfrak{M}_{X\text{-ray}}$	Type
GX 301−2	BP Cru	$41^{d}4$	$10^{m}8$	$10^{m}9$ V	B2Iae	≈30		ecc.
Cir X-1	BR Cir	16.59	13.5	16 r	OBI	≈18	1.5 (?)	ecc.
Cyg X-2	V 1341 Cyg	9.84	14.8	15.4 B	FIII−IV	0.8 ?	1.5 ?	lm.
2S 0921−630	–	8.99	15.3	16.5	GIII:	1:	1.4:	lm.
Vel X-1	GP Vel	8.97	6.7	6.9 V	B0.5Ia	21	1.6	hm.
A 0620−00	V 616 Mon	7.8	12	20 B	K5−7V			N
Cyg X-1	V 1357 Cyg	5.60	8.8	8.9 V	O9.7I	25	≈10	hm.
SMC X-1	Sk 160	3.89	13.3	B	B0I	≈19	2.5 ?	hm.
3U 1700−37	V 884 Sco	3.41	6.5	6.6 V	O6f	27	1.3	hm.
Cen X-3	V 779 Cen	2.09	13.4	B	O6.5V−III	17	0.7	hm.
Her X-1	HZ Her	1.70	12.8	15.1 B	B8−F3V	2.2	1.3	lm.
Aql X-1	V 1333 Aql	1.3:	14.8	19.2 B	G7−K3V			N
Sco X-1	V 818 Sco	0.79	11.1	14.1 B	pec.	≈1	≈ 1	lm.
Cen X-4	V 822 Cen	0.31	12.8	> 19 B	K3−7V			N
2A 1822−371	V 691 CrA	0.23	15.4	16.4		0.2:	1.0:	lm.
4U 2129+47	V 1727 Cyg	0.22	16.9	18.6 B		0.65	1.3:	lm.
Cyg X-3	V 1521 Cyg	0.20						lm.
4U 1915−05	–	0.035		> 22				lm.
4U 1627−67	KZ TrA	0.029	18.2	18.7 B				lm.

Notes:

X-ray source:	for details of the designations see Sect. 6.7
P	= Orbital period
Sp_{opt}	= Spectral class of the optical component
$\mathfrak{M}_{opt}$	= Mass of the optical component in solar masses
$\mathfrak{M}_{X\text{-ray}}$	= Mass of the X-ray component in solar masses
Type: ecc.	= eccentric X-ray pulsar
hm.	= high-mass X-ray pulsar
lm.	= low-mass X-ray pulsar
N	= X-ray nova

Massive and High-Eccentricity X-Ray Pulsars

The massive X-ray pulsars also form binary pairs, which consist of a *neutron star* radiating the X-rays and an *optical companion*. However, in contrast to the low-mass X-ray pulsars, the optical component is not a main sequence star of spectral class F−G and 1 or 2 solar masses, but an *early-type supergiant* of about *20−40 solar masses*, easily recognizable spectroscopically. As well as a greater total mass, the massive X-ray pulsars also have correspondingly longer orbital periods P (see Table 36). As most of the visible light is radiated by the essentially non-variable supergiant, and the fraction of the X-ray radiation converted into visible light remains very small, owing to the greater separation of the components, the *changes in light* only amount to *a few tenths of a magnitude* at the most. This is in contrast to the low-mass X-ray pulsars, which may have amplitudes of several magnitudes.

The principal difference between the two types of X-ray pulsars lies in the mass-exchange mechanism. In the low-mass X-ray pulsars ("semi-detached systems", see Chap. 4) the mass-transfer occurs through overflow at the L_1

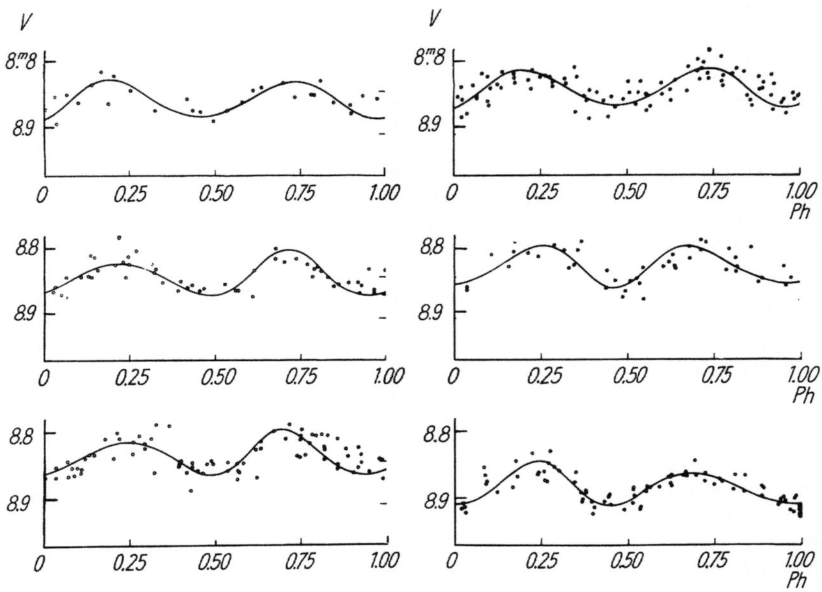

Fig. 76. Light-curves of V 1357 Cyg (at 6 different mean epochs, showing apsidal rotation) (after WILSON and FOX 1981)

Lagrangian point, while in the massive X-ray pulsars ("detached systems") the neutron star accretes material from the supergiant's stellar wind. This interaction with stellar wind material is not particularly effective, as the stellar wind flows out in all directions, and so only a small fraction reaches the compact component. The remainder is lost from the system. However, as the stellar wind is very strong, enough material still arrives on the neutron star to give rise to the intense X-ray radiation. The approximately 20 objects of this sort that are known include GP Vel (Vel X-1), V 861 Sco (HOWARTH and WILSON 1981), V 884 Sco (3U 1700−37) and V 1357 Cyg (Cyg X-1).

The X-ray source *Cyg X-1*, the invisible companion to the optical object V 1357 Cyg, is a *special case*. This is a variable that has been the subject of an exceptionally high number of papers (more than 500), even though the optical variability has only been known since 1973, and the amplitude only amounts to 0.15 mag. The light-curve of V 1357 Cyg resembles that of a β Lyrae star with a period of 5.6 days, secondary minimum being nearly as deep as primary minimum (Fig. 76). An analysis of the light-curve in various colour-systems has been given by BALOG et al. (1981). The *spectrum* is that of a B0 supergiant with emission lines (B0Ibev). The *changes in luminosity* are not due to eclipses, but are probably caused by ellipticity (see Sect. 4.2) in the B star (the result of tidal deformation by the invisible companion). Combined with these ellipsoidal variations are rapid, irregular pulsations with cycle-lengths of 0.3−10 seconds. Figure 77 shows the X-ray light-curve: X-ray minimum occurs at optical secondary minimum (= phase 0.5), when the X-ray source is at superior conjunction, lying on the far side of V 1357 Cyg.

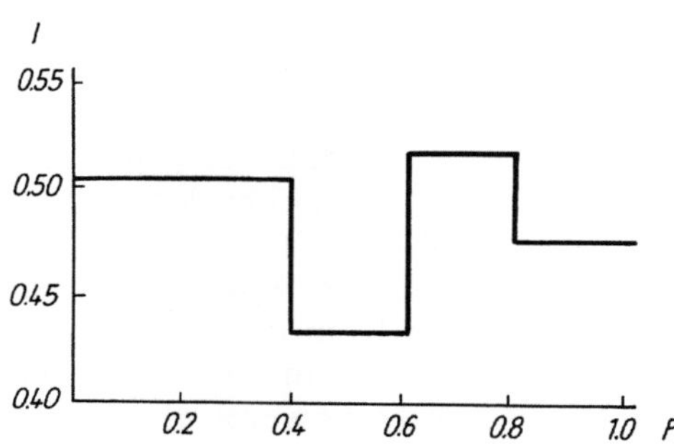

Fig. 77. Schematic X-ray light-curve of Cyg X-1 ($P = 5\overset{d}{.}6$) after HOLT (see LIGHTMAN 1976)

The X-ray radiation is exceptionally hard with photon-energies up to 200 keV, corresponding to a wavelength of 0.006 nm. Recently, γ-rays have also been detected (GILMOZZI et al. 1981). The mean X-ray brightness varies between a low level (90% of the time) and a higher one (10% of the time). The X-ray spectrum and the γ-rays probably result from photon-scattering by extremely hot electrons (temperature $> 10^9$ K!). This scattering process is known as the "inverse Compton effect". The X-ray source is fed ultimately by a flow of material, which probably leaves the optically visible star as a stellar wind.

From the orbital data on the binary the following information can be derived: the *orbital eccentricity* $e = 0.025 - 0.04$, the *mass* of the compact component is at least 5 solar masses. This considerably exceeds the maximum mass, calculated from Einstein's gravitational theory, that is permissible for a neutron star (about 3 solar masses). It would seem that this companion must be a "black hole". However, it must be pointed out that there exist other gravitational theories, not yet refuted, that would permit neutron stars to have masses as high as 30 solar masses. Furthermore, the possibility cannot be completely ruled out that the compact object consists of an ordinary neutron star of perhaps 2 solar masses, surrounded by a very dense and massive disk of material, which under certain circumstances could contain 3 or more solar masses. But the presence of such a stable, massive disk of material causes problems for the theoreticians.

BR Cir (Cir X-1) is an object with a 16.6-day periodicity in the X-ray light-curve, which may be explained as orbital motion in a binary. In addition, irregular X-ray pulsations with a mean period of about 0.5 s are observed, as well as X-ray bursts lasting only 0.01 seconds. Previously, such short outbursts had only been observed in Cyg X-1. The optical companion is an OB supergiant with emission lines.

On the basis of certain similarities in the behaviour of the X-ray radiation with that observed in Cyg X-1, many authors have suggested that Cir X-1 is also a candidate for being a black hole. More precise investigations of the X-ray variation, however, partially support another model. According to this, the BR Cir – Cir X-1 system consists of an OB supergiant of $15 - 20$ solar masses and a neutron star of $1 - 1.5$ solar masses. The two objects move in a *highly elliptical orbit* with an eccentricity of ≈ 0.8. When the stars are far apart (at apastron) the neutron star only intercepts a very small fraction of the stellar-wind material from the OB supergiant, so only weak X-ray radiation is produced, and the side of the supergiant facing the neutron star undergoes only weak heating. But at periastron the neutron star penetrates the supergiant's Roche lobe and almost touches its surface. This leads to an exceptionally high mass exchange (the OB star "boils over") together with a strong X-ray outburst and violent heating of the OB star. Apsidal rotation means that we see the X-ray outburst at periastron from different angles. A comprehensive description of this model is given in GINGOLD and MONAGHAM (1979) and HAYNES et al. (1980). Recently, however, this model has been disputed (ARGUE and SULLIVAN 1982). According to research by SCHLICKEISER (1981), the BR Cir system not only radiates in the X-ray, visible, infrared and radio regions, but in γ-rays as well. The most recent results concerning this very complicated object are to be found in DOWER et al. (1982).

BPCru ($e \approx 0.47$) and 2S0535−668 in the Large Magellanic Cloud ($e \approx 0.7$) are probably related objects (DULDIG et al. 1982, WATSON et al. 1982 and CHARLES 1982).

X-Ray Bursters

This group of stars are low-mass X-ray binaries, which probably consist of a star of < 0.5 solar mass and a neutron star. In contrast to the otherwise very similar HZ Herculis stars (see above), which send out regular pulses of X-rays, these objects show *quasi-periodic bursts* in which the X-ray luminosity increases by a factor of about 10, rising above a more or less stable basic level of activity. The rise in brightness lasts about 1 second and the decline about 5 seconds. The cycle-lengths (the mean interval between 2 bursts) are, according to the object, a few hours to a few days. The bursts have a black-body spectrum of 30 million K!

The first burster was discovered in 1975 in the globular cluster NGC 6624, close to the galactic centre, by the ANS ("Astronomical Netherlands Satellite") X-ray satellite. By the beginning of 1981, 40 X-ray bursters were known, of which 8 were in globular clusters (CHERNYKH 1981). At that time 5 burs-ters had been positively identified optically.

In the summer of 1978, GRINDLAY, McCLINTOCK and CANIZARES were the first to succeed in recording *simultaneous optical and X-ray outbursts* in the object $MXB\,1735-44 = V\,926\,Sco$, see VAN PARADIJS (1981). (MXB = "Massachusetts X-ray Burster") An interesting point is that the optical flash is about 3 seconds later than the X-ray burst. This delayed optical flash can be explained by supposing that part of the X-ray radiation is absorbed by the disk of material, and re-emitted in the visible region (see Fig. 78, where the X-ray and light pulses take 3 seconds to traverse the path NAB, which is about 10^6 km long).

Later, *optical bursts* were successfully recorded in other objects (see LAWRENCE et al. 1983). In the source $MXB\,1636-53 = V\,801\,Ara$ (Fig. 79) the luminosity can rise by a factor of 4 in less than 4 seconds (see PEDERSEN

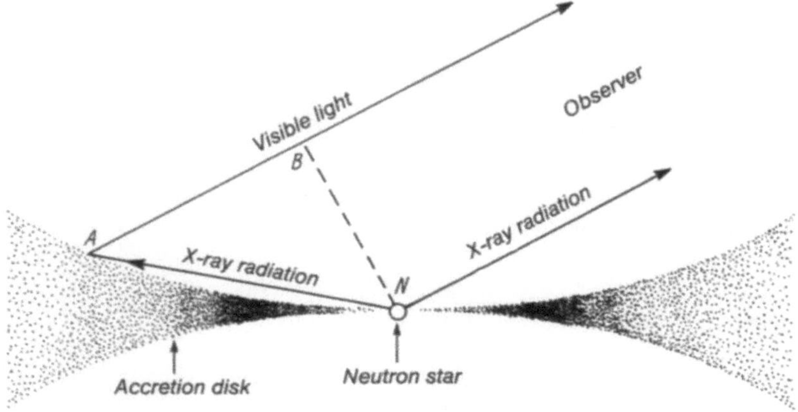

▲ **Fig. 78.** Probable conditions during an optically visible outburst in an X-ray burster (after VAN PARADIJS 1981)

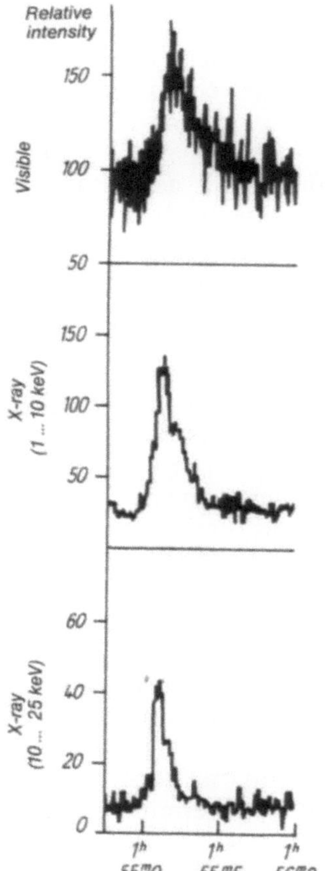

◄ **Fig. 79.** The X-ray burster MXB 1636−53 (= V 801 Ara): simultaneous observations of an outburst in the visible (*above*) and X-ray regions on 1979 June 28 (after VAN PARADIJS 1981)

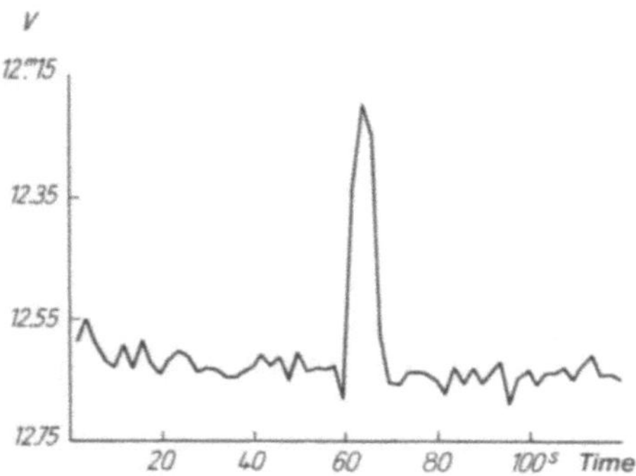

Fig. 80. Optical outburst of V 818 Sco (= Sco X-1) on 1979 Mar. 13, 7ʰ29ᵐ49ˢ UT, observed at La Silla (see MAUDER 1981)

1979, PEDERSEN et al. 1983 a, b)! An exceptional object is *ScoX-1* (=
V818 Sco), see also HUDEC (1981 b), in which only optical, not X-ray, outbursts have been observed to date. Figure 80 shows one of the outbursts of
Sco X-1. It must be an impressive sight to accidentally observe such a flash
through a telescope! Physical models for Sco X-1 are discussed by KAHN et
al. (1981).

In 3 cases (CHERNYKH 1981) the spectral class and mass of the secondary
component, and the orbital period P of the binary, have been successfully
determined. The values lie within the following limits: Sp. = G3−K7; mass
(in solar masses) = 0.5−1; P = 42 minutes−9.8 days.

Some other notable findings about bursters are:

1) The *extremely low optical luminosity*. The X-ray luminosity exceeds the
   optical by a factor of 10000! This is the reason why the few optically
   identified objects are so faint.
2) The frequent occurrence in globular clusters, in the galactic halo, and in
   the direction of the galactic centre, indicates *membership of Population II*,
   and thus a considerable age.
3) The *lack* of X-ray eclipses and other *periodicities*, that indicate a binary
   nature. There are only speculations at present as to the reason for this
   peculiar finding.

At present the most widely accepted model for bursters is as follows.

The bursters are "senile" X-ray binaries in which the neutron star has
only a "weak" magnetic field of at most 10^{10} Oersted. In the X-ray pulsars
the hydrogen that flows down from the accretion disk to the neutron star is
directed onto small areas of the surface by the magnetic field. The compressed jet is so strongly heated that the material is spontaneously converted
into elements as heavy as the iron-group. In the bursters, however, owing to
the much weaker magnetic field, the hydrogen is spread more or less evenly
over the whole surface of the neutron star. The temperature and pressure are
thus much lower, and estimates show that the incoming hydrogen is indeed
spontaneously converted into helium, but not into heavier elements. More
and more helium gathers on the surface of the neutron star, until finally the
temperature and pressure are high enough for the "helium bomb" to be
ignited. This is a similar process to that which we have encountered in the
novae, where the outburst is caused by explosive hydrogen fusion on a white
dwarf.

The most important differences lie in the duration of the outburst, which
in the novae may be months, but in the X-ray bursters only lasts a few
seconds; and in the intervals between outbursts, which are tens or hundreds
of years in the recurrent novae but only hours in the X-ray bursters.

VAN PARADIJS (1981) discusses a possible method of determining the
distances of X-ray bursters.

Excellent reviews of the current state of knowledge about X-ray bursters
are given by LIGHTMAN (1976), LEWIN and VAN PARADIJS (1979), WAMSTEKER
(1979), KUNDT (1981) and VAN PARADIJS (1981).

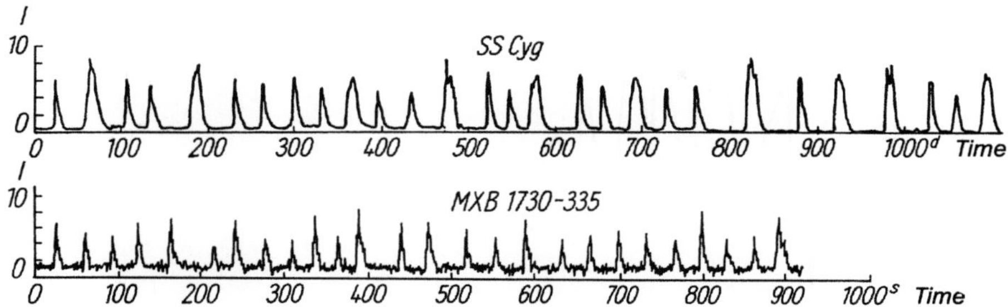

Fig. 81. X-ray light-curve of the rapid burster MXB 1730−335 (SAS 3 measurements), compared with the visual light-curve of the U Geminorum star SS Cyg (AAVSO observations) (after BRECHER et al. 1977). I is the intensity in arbitrary units

A Rapid Burster

Among the X-ray bursters known to date there is one exceptional object, *MXB 1730−335*, that shows, in addition to the normal bursts, as many as 1000 rapid X-ray bursts per day, at intervals ranging from 10 seconds to a few minutes. This rapid burster becomes active about every six months, when it sends out bursts for 2−6 weeks. WAMSTEKER likens the behaviour to X-ray "machine-gun fire" (see Fig. 81).

Apart from X-rays, MXB 1730−335 also radiates in the infrared. The cause of the rapid bursts is probably instabilities in the flow of material, presumably analogous to those in the U Geminorum stars. BRECHER et al. (1977) also compare the X-ray curve of MXB 1730−335 with the optical curve of the U Geminorum star SS Cyg, finding considerable agreement in the pattern of eruptions, except for the time scales being very different (see Fig. 81). MXB 1730−335 also shows fluctuations in the X-ray light-curve with a period-length of 12.2 milliseconds, probably caused by pulsations in the shell of the neutron star. They resemble the "coherent oscillations" that we have encountered in the cataclysmic variables. Four other X-ray binaries similarly show this sort of pulse at intervals of 15−69 milliseconds (LIVIO and BATH 1982).

Another interesting object is *4U 1915−05*, which, according to WHITE and SWANK (1982), may contain a white dwarf that is passing material to a neutron star.

An Exceptional X-Ray Binary: SS 433 = V 1343 Aql

In 1977 STEPHENSON and SANDULEAK published a list of Hα emission-line stars discovered by various workers. A year later, object number 433 on this list caused a great stir, and a veritable flood of publications appeared. In 1978 there were only 5 papers about the object, but in 1979 and 1980, 78 and 65 were published, respectively (according to the Sonneberg card index). Examination of the comprehensive series of plates at Harvard Observatory (LILLER) and at Sonneberg (WENZEL 1980) showed a rapid, irregular fluctuation in magnitude within the range 15−17^m. Photoelectric observations by

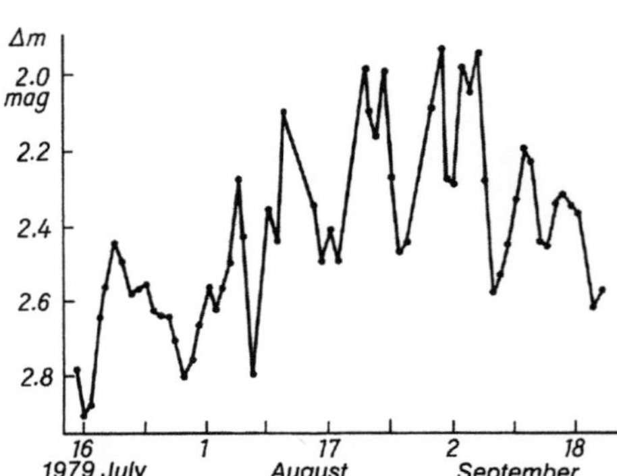

Fig. 82. Luminosity changes in V 1343 Aql (= SS 433) from July to September 1979 (see OVERBYE 1979)

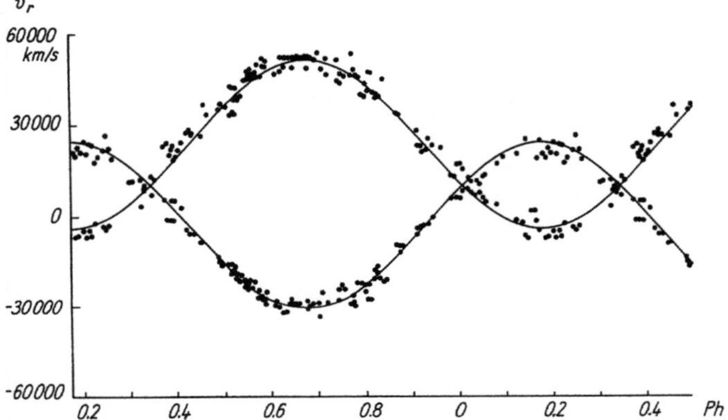

Fig. 83. Radial-velocity changes (Doppler shift of the emission lines) as a function of phase in V 1343 Aql (= SS 433) (after MARGON et al. 1980)

KEMP et al. (see OVERBYE 1979) suggested that a 13-day cycle might be present (Fig. 82).

This object – which is variable at radio, optical and X-ray wavelengths – is particularly interesting because there are enormous, periodic Doppler shifts in the emission lines. Each emission line consists of three components: one (almost) stationary, one red-shifted and another blue-shifted. The almost stationary component shows a slight line-shift with a period of 13.1 days that can be interpreted as orbital motion in a binary. The other two strongly shifted components, however, change their position with a 164-day cycle (Fig. 83). The maximum shift amounts to more than 1/10 of the rest wavelength!

At the moment there are almost as many theories about SS 433 as there are authors. Some papers try to explain the object as a special form of binary, consisting of a normal star and a black hole.

The models most frequently discussed at present are those that postulate a binary having a neutron star (or black hole) of ≥ 1.6 solar masses and an O or WR star of ≥ 2 solar masses, and an orbital period of 13 days. Most of the light comes from a massive accretion disk that surrounds the neutron star. The neutron star's mass-accretion rate probably considerably exceeds the critical value. In other words, an excessive supply of material is producing so much X-radiation that some of the material is being accelerated to relativistic velocities by the enormous radiation pressure. This gives rise to two jets of material at right-angles to the disk, ejected at velocities of 80 000 km/s. The 164-day cycle is presumably the effect of precession in the accretion disk and in the jets perpendicular to it (see the illustration on page 513 in the paper by OVERBYE 1979). Comprehensive reports on current ideas about SS 433 = V 1343 Aql, with further references, can be found in DORSCHNER and GÜRTLER (1980), DORSCHNER, GÜRTLER and FRÖHLICH (1980), OVERBYE (1979), KUNDT (1981) and SCHORN (1981) among others.

Transient X-Ray Sources and X-Ray Novae

In contrast to all the other objects discussed in this book, the transient X-ray sources are not a special group of variable stars, but a phenomenon observed in physically different classes of variables.

In the "persistent" X-ray binaries described in the preceding pages the radiation is visible most of the time, but the transient X-ray sources are usually invisible. Either they show a single outburst (e.g. Nova Oph 1977 = 2107 Oph), or else they flare up to a high X-ray luminosity every few months, years or decades, when they are detectable for a few weeks (recurrent transient X-ray sources such as V 1333 Aql and V 616 Mon).

Only a very small fraction of the numerous transient X-ray sources that are discovered every year – the British satellite "Ariel 5" was very successful in this respect – have been optically identified, and an optical outburst is not always observed at the same time as the X-ray burst. Conversely, only a few of the novae or recurrent novae (Sect. 3.1.2), discovered in the visible, are transient X-ray sources at the same time. (For details of the much fainter X-ray radiation in a "classical" nova outburst see Sect. 3.1.2.)

A few of the objects that have been optically well-observed will be described here: *A 0620–00 = Nova V616 Mon* is frequently mentioned in the literature, including the GCVS (3rd Supplement 1976), as a recurrent nova. This object's X-ray and radio properties, however, differentiate it sharply from the true novae. To date two outbursts have been observed (1917 and 1975), when the blue magnitude rose from $20^m - 11^m3$. Figure 84 gives the light-curve of this object. Optical observations show a 7.8-day, low-amplitude periodicity that can probably be explained as orbital motion in an binary. During its 1975 outburst V 616 Mon does not show the complicated spectral evolution found in typical novae (cf. ILOVAISKY and CHEVALIER 1977 and DUERBECK 1977); the outburst does not seem to be related to an expanding nova shell. It is probably a low-mass binary system with a G or K main sequence star and a compact companion (a white dwarf or neutron star). The X-ray outbursts are probably caused by a large mass of material suddenly falling onto the compact object, while the optical emission arises partly from

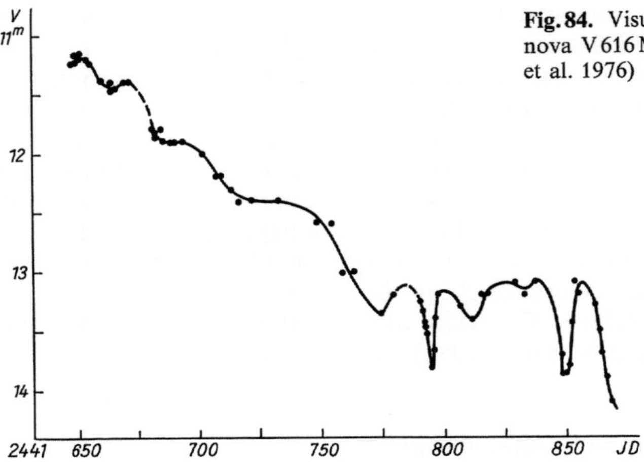

Fig. 84. Visual light-curve of the X-ray nova V 616 Mon (1975) (after ROBERTSON et al. 1976)

the heated disk, and partly from the cool component, where this is heated by the X-rays (cf. CHEVALIER et al. 1980 and the references therein).

Similar optical behaviour was discovered in the 1974 X-ray novae: KY TrA ($=$ A 1524$-$61, brightness $17^m_{.}5 - 22^m$) and V 725 Tau ($=$ A 0535 $+$ 26) – for the latter see RÖSSIGER 1979.

A rather different transient X-ray source is the symbiotic object *V 2116 Oph*, discussed in Sect. 3.1.4.

The transient X-ray source A 0327 $+$ 43 is identical with the classical nova *GK Per (1901)* – see Table 31. The X-ray outburst in 1981 July was accompanied by a small optical outburst (see also HUDEC 1981 a).

Another well-known, but not completely typical, transient X-ray source with the very short cycle-length of only 12–16 months is *Aql X-1 = V 1333 Aql*. Figure 85 shows the simultaneous X-ray and blue light-curves. Here again, the optical and X-ray flares occur simultaneously. Spectra obtained in the quiescent state ($20^m - 21^m$) indicate that a class K companion is present. Spectra at maximum ($\approx 16^m$) show faint emission lines of H, He, C

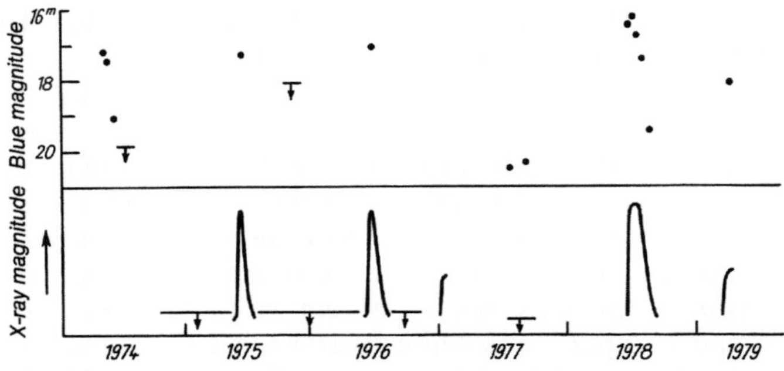

Fig. 85. V 1333 Aql: light-curves (*above*) in blue and (*below*) in X-ray regions (after CHARLES 1980)

and N and resemble those of Sco X-1. Perhaps Aql X-1 is related to this object? Many authors also suggest a relationship with the bursters.

SCHWARTZ et al. (1981) succeeded in identifying the RS Canum Venaticorum star II Peg with a transient X-ray source. GARCIA et al. (1980) estimate that approximately 20% of the transient X-ray sources at high galactic latitudes are *RS Canum Venaticorum stars* (Sect. 3.7.1).

A comprehensive review of transient X-ray sources is given by WILLMORE (1977). CHARLES (1980) discusses some possible models. Reference may be made to DE LOORE (1980) and KUNDT (1981) for details of the possible origins of X-ray binaries.

3.2 Supernovae

As the name suggests, these show an exaggerated form of a nova outburst. The most important difference is the greater absolute magnitude, which is between -16^M and -21^M. That is, on average, more than 10 000 times the luminosity of normal novae!

At present the one factor that makes supernova research more difficult is that as yet *not one* supernova precursor (= presupernova) is known. Investigating the causes of supernova outbursts thus involves dealing with the physics of unknown objects. Nevertheless we need not be completely discouraged. The range of objects that can undergo dramatic explosions can be narrowed down by means of statistical investigations involving known historical supernovae, extragalactic supernovae, supernova remnants and pulsars, in conjunction with stellar evolutionary theories (see OTT 1979 and references therein, and individual review articles in the books on supernovae by COSMOVICI 1974, SHKLOVSKIJ 1976, SCHRAMM 1977 and WHEELER 1980).

Historical Supernovae

In our Galaxy the 5 following cases have been definitely established:

According to Arabic, Chinese, Japanese and southern-European records, there was a supernova in Lupus in *1006*, which rose to nearly the brightness of the quarter-Moon (-9^m to -10^m), and which was visible for more than 2 years. The remnant of this supernova can be identified with the X-ray source 4U 1458−41 and with the radio source PKS 1459−41, which shows the typical ring-like structure found in supernova remnants, and is accompanied optically by filamentary nebulae.

In *1054* a supernova was seen in Taurus, according to Chinese and Japanese sources, which is now known as CM Tau. Its estimated maximum magnitude amounted to -4^m. The nebular remnant is the Crab Nebula M1, the age of which has been determined as about 900 years by various methods. Measurements of the continuing expansion of nebular filaments indicate a date of 1140, but the traditional date of 1054 is doubtless correct, the expansion of the filamentary structure having accelerated (TRIMBLE 1968). The stellar remnant in the centre of the nebula is a neutron star with the mean

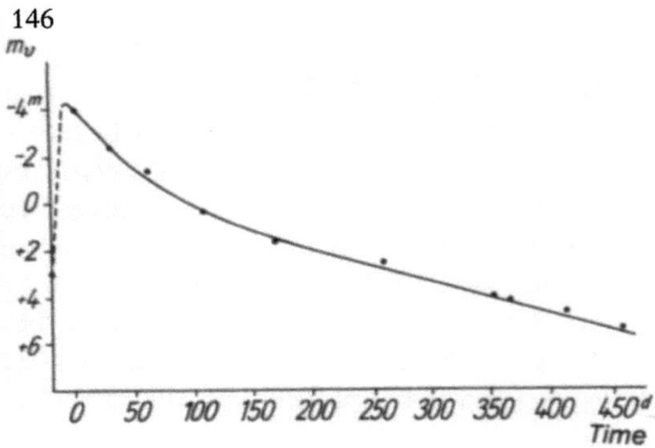

Fig. 86. Light-curve of Tycho's super-nova (1572) (after Clark and Stephenson 1977)

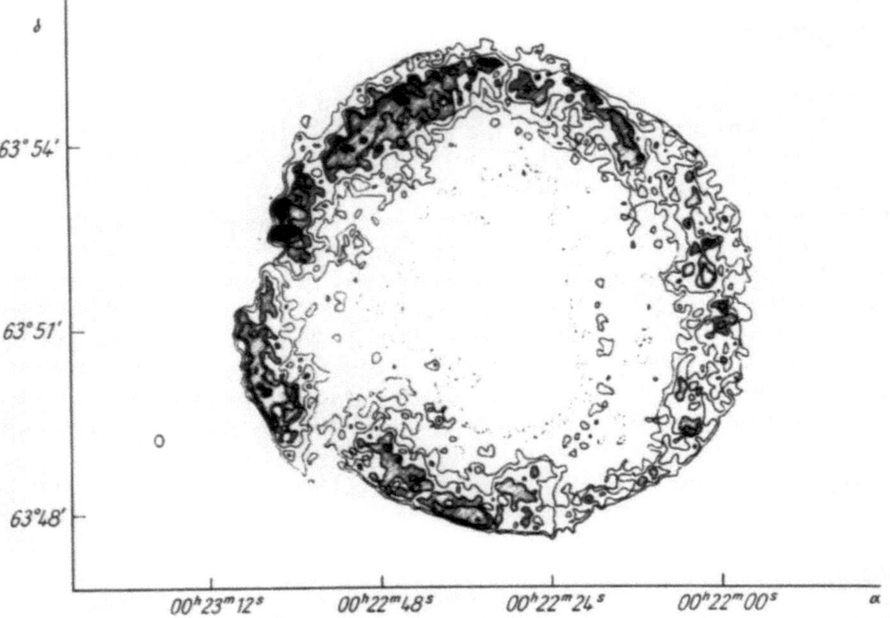

Fig. 87. Radio image at 4995 MHz of the remnant of Tycho's supernova (1572) (see Clark and Stephenson 1977)

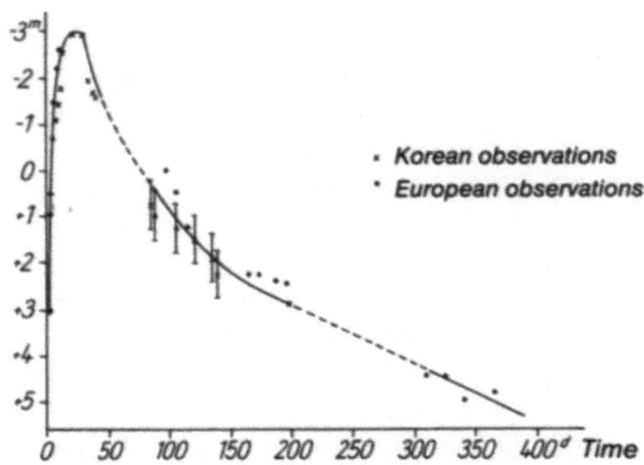

Fig. 88. Light-curve of Kepler's super-nova (1604) (after Clark and Stephenson 1977)

photographic magnitude of $15^m\!.9$. This is the well-known Crab Pulsar, radiating in the radio, optical and X-ray regions with a pulse-period of only $0^s\!.033$ (Sect. 3.6.2). The Crab Nebula itself emits X-ray and radio radiation.

In *1572* a supernova occurred in Cassiopeia (B Cas) with the maximum magnitude of -4^m. This star is usually linked with the name of TYCHO BRAHE who systematically observed it. However, many other astronomers also followed it, so B Cas is the first object for which we have a well-established light-curve (Fig. 86). To the naked eye the star disappeared in the spring of 1574. The amplitude amounted to at least 22 magnitudes. Remnants are found at this supernova's position as well, and consist of a ring-shaped, strongly polarized radio source (5C 10, Fig. 87), optical filaments and an X-ray source (Cep X-1).

In *1604* the supernova now known as V 843 Oph appeared in Ophiuchus, with a maximum magnitude of -3^m and an amplitude of > 21 mag. The star was described by JOHANNES KEPLER and the light-curve is well-known (Fig. 88). This supernova also has a nebular remnant and a radio remnant (3C 358); however, no X-ray source has been observed.

Cas A is a strong radio source (see VAN DEN BERGH 1974 and KAMPER and VAN DEN BERGH 1976), regarded as a supernova remnant on the basis of its morphological and spectral characteristics. At this radio source's position there is a nebulosity, 4 pc in diameter, expanding with a velocity of 7400 km/s. This suggests that the outburst of the supernova that produced this nebula occurred at the beginning of the 18th century. It is rather surprising that the outburst was not observed. The strong interstellar extinction that is present in this area of the sky is probably relevant. It is also possible that when this object was at maximum it was so low on the horizon during the night that it escaped discovery.

There are indications of other historical supernovae that occurred in 185, 386, 393 and 1181, but these cases have not been definitely confirmed.

A comprehensive study of historical supernovae is to be found in the book *The Historical Supernovae* by CLARK and STEPHENSON (1977).

Supernovae in Other Galaxies

Variable stars in other galaxies will be mainly discussed in Sect. 5.2.2. Extragalactic supernovae, however, will be described here. Not one supernova has been observed in our Galaxy since the invention of the telescope. We therefore have to turn to systematic and uninterrupted observations of supernovae in other stellar systems in order to be able to classify these interesting objects into various types, and to carry out statistical research on their frequency. The enormous brightness, which in many cases exceeds the total light from the extragalactic system concerned (Fig. 89), permits supernovae to be observed at very great distances. The number of galaxies is very great, and as supernovae occur in all classes of galaxy (see e.g. OTT 1979), including elliptical and irregular types, their numbers are also very high. Up to 1979 about 380 supernovae had been discovered, around 100 of them by the noted supernova researcher ZWICKY alone.

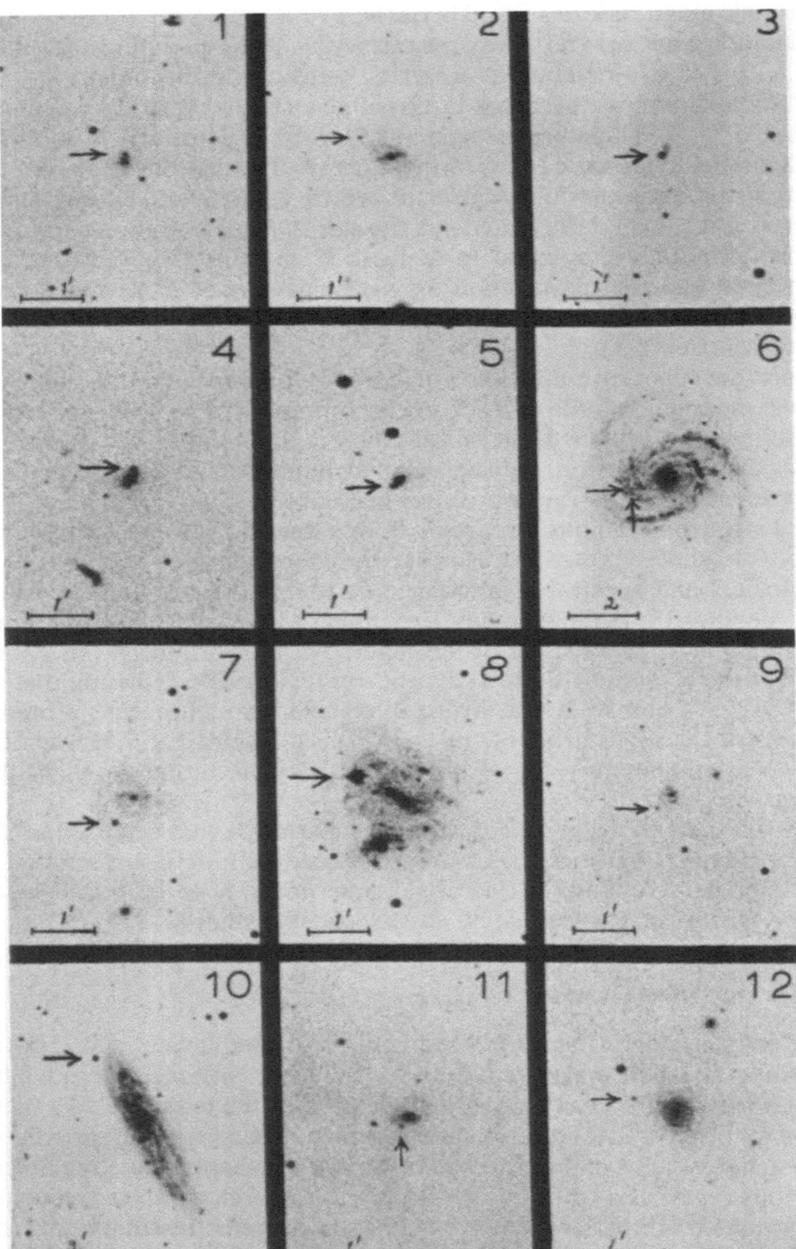

Fig. 89. Supernovae (*arrows*) in extragalactic systems (after HUMASON et al.). Note the brightness of supernova 3 when compared with the system in which it occurred. Negative images

An "intergalactic supernova" (Type I) exploded inside a group of bright galaxies in 1980 (H. A. SMITH 1981).

Classification

Photometric and spectroscopic observations of supernovae in other galaxies show that there are several classes, which differ in their light-curves, absolute magnitudes and spectra.

The *Type I* light-curve resembles that of a fast nova. The decline is comparatively steep, at first about 3 mag in 25–40 days, and then about 1 mag in 60–70 days.

In *Type II* the decline is slower and takes various forms. However, there is a characteristic hump on the decline that begins at about 20 days after maximum.

In both Type I and Type II the rate of rise is approximately the same, but slower than in normal novae. Schematic light-curves are shown in Fig. 90.

The two types differ in their photographic absolute magnitudes at maximum:

Type I: generally between -18^M and -21^M, mean $-19^M_.1$
Type II: generally between $-16^M_.5$ and -18^M, mean $-17^M_.2$

The discovery of Type II supernovae is impaired by their lower absolute magnitude, and that of Type I supernovae, to a lesser extent, by their shorter duration. The majority of Type I supernovae are related to the disk population and Population II stars, although some are possibly members of a younger stellar population (see papers in *Proceedings of the Texas Workshop on Type I Supernovae* 1980). The somewhat rarer Type II supernovae correspond to the spiral-arm Population I, so that their discovery is additionally impeded by interstellar extinction (see also discussion p. 131 ff. in PSKOWSKI 1978).

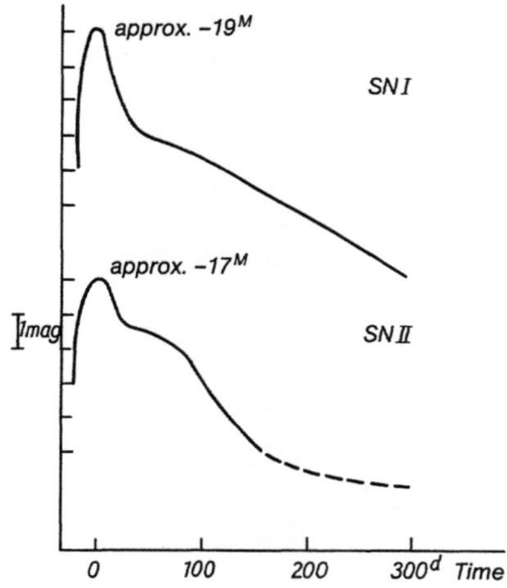

Fig. 90. Schematic blue light-curves of Type I and Type II supernovae (after OTT 1979)

Most of the historical supernovae mentioned above were previously ascribed to Type I. There was some doubt only in the classification of SN 1006, and recently opinions have been increasingly in favour of a Type II classification for SN 1054, which differs significantly from the other objects in its remnant and in other features (amongst them the presence of a pulsar) – see also CHEVALIER 1977.

Full descriptions of the classification of supernovae are given in the contribution by TAMMANN in SCHRAMM's book (1977) and in the paper by DE GROOT (1978), amongst others.

Supernova Remnants

In a supernova explosion a considerable portion of the stellar material is evidently expelled into space as a shell of gas. As mentioned in the description of the 1054 supernova, the best-known example is the *Crab Nebula M1*, one of the strongest X-ray sources, in the centre of which is the famous Crab pulsar. As already mentioned, there are radio sources (generally X-ray sources as well) and expanding nebulosities, visible at optical wavelengths, at the sites of all the other historical supernovae. Yet there are no central pulsars in these objects. This is an indication that the 1054 object was a different sort of supernova, and that pulsars are formed in only some supernova explosions. On the other hand, we cannot be certain when we are dealing with pulsars (Sect. 3.6.2) whether all, or only a fraction of them, are produced by supernovae.

Numerous remnants of *prehistoric supernovae* are known, some being emission nebulae, for instance. The best-known object of this sort is the Veil (or Cirrus) Nebula (Fig. 91), which marks the explosion of a supernova about 50000–100000 years ago.

According to AMNUEL et al. (1979) 13 galactic X-ray sources are recognized as supernova relics.

The remnants observable at radio wavelengths are particularly interesting, because, in contrast to optical and X-ray radiation, radio waves are not subject to extinction by interstellar dust, so that radio remnants can even be discovered on the opposite side of the Galaxy. The Galaxy's "radio spur", a region of stronger radio flux that stretches from about 30° east of the galactic centre towards the galactic pole, was probably caused by a supernova. This supernova must have occurred a long time ago, relatively close to the position of the Solar System, as the radio spur has a very large angular size, and is thus rather close.

Supernova remnants have also been found in nearby external galaxies. The radio emission from very young extragalactic supernova remnants that are only a few years old has been discussed by PACINI and SALVATI (1981).

A summary of all known supernova remnants, including a catalogue of 120 such objects and further references, can be found in Chap. 4 of CLARK and STEPHENSON's book (1977). A generalized description is given by PSKOWSKI (1978).

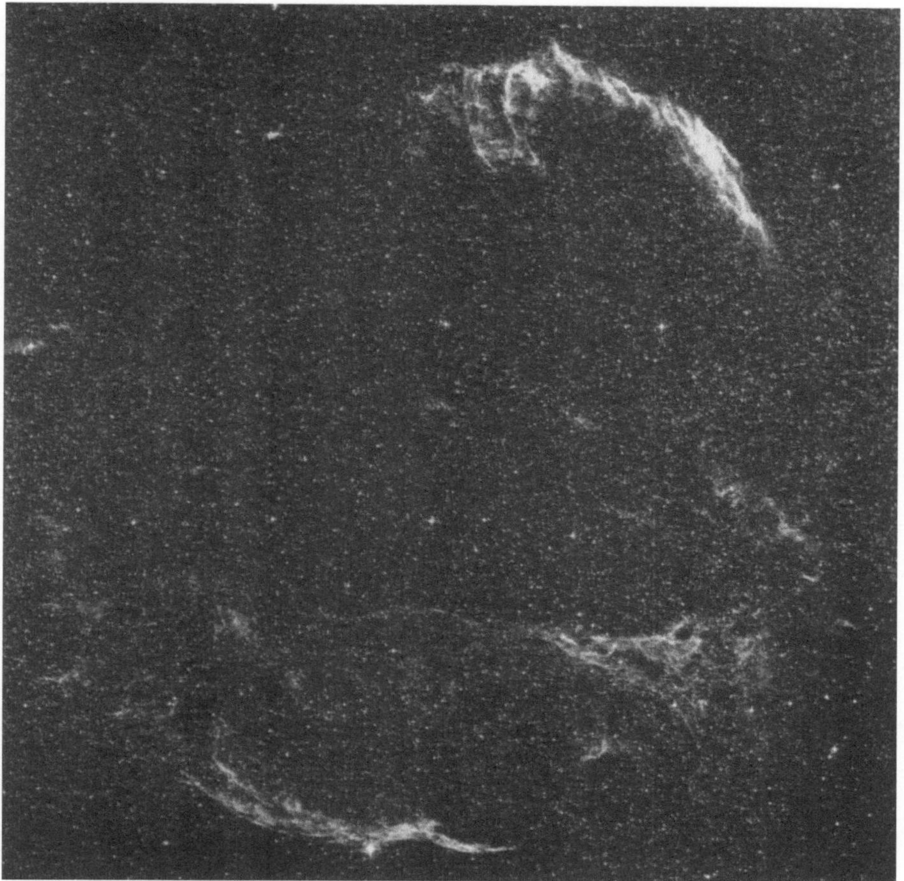

Fig. 91. The Veil Nebula in Cygnus: a supernova remnant. Note the ring-like structure of the gas clouds (photograph by Götz, Sonneberg)

Frequency of Supernova Explosions

The five confirmed supernova outbursts (see above) in the last ten centuries do not reflect the number that occur in our Galaxy every thousand years, as we miss a certain proportion of them due to interstellar extinction. The actual proportion is very difficult to determine. Observation of supernova remnants, estimates of the amount of interstellar extinction, and analogies drawn from the observation of supernovae in other galaxies give values between 30 and 100 events per thousand years; see also the discussions in the books by PSKOWSKI (1978), CLARK and STEPHENSON (1977), SHKLOVSKIJ (1976) and SCHRAMM (1977). According to PSKOWSKI (1978) supernovae of Type II are six times as frequent as those of Type I.

Spectra

The spectra of Type I supernovae at maximum are essentially continuous, without distinct details. After maximum very broad bright and dark features appear that are of low intensity and alternate along the spectrum. Consequently, opinions differ as to whether the spectrum should be regarded as a continuum with bright emission bands, or a continuum with very wide absorptions. Figure 92 shows the spectrum of a Type I supernova: a, 2 days before maximum as well as b, 27 and c, 76 days after maximum. PSKOWSKI (1978) interprets the spectral features as very broad absorption lines, which indicate that the expansion velocity of the shell is very high, amounting to more than 10 000 km/s. There are no hydrogen lines.

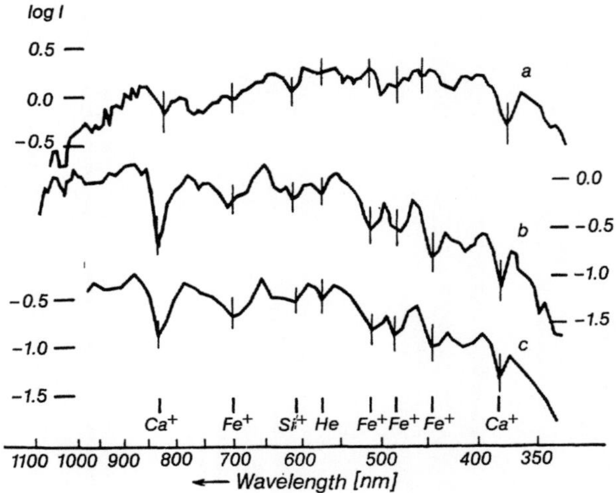

Fig. 92. The densitometry traces of spectrograms of a Type I supernova (1971i), obtained at different times (after PSKOVSKI 1978). I is the intensity in arbitrary units

The spectra of Type II supernovae, which have a certain resemblance to those of ordinary novae, show a continuum after maximum, reaching far into the ultraviolet, and corresponding to a colour temperature of about 40 000 K. In contrast to the Type I supernovae, bright hydrogen lines appear after maximum.

Comprehensive discussion of supernova spectra can be found in KIRSHNER (1974), MUSTEL (1974) and PSKOWSKI (1978).

Table 37 gives the average values for the physical properties of supernova shells, as determined from the spectral evolution, according to PSKOWSKI (1978).

Table 37. Physical properties of supernova shells

	Type I	Type II
Mass of shell in solar units	0.3	1.0 and more
Expansion velocity in km/s	13 500	7000
Shell temperature at maximum light	30 000 K	25 000 K

Physics of Supernova Explosions

As already mentioned, an understanding of the physical processes in a supernova outburst is made significantly more difficult by the fact that no supernova precursor has been seen. We only know the luminosity and spectral evolution of supernova outbursts in other galaxies, and the characteristics of supernova remnants. In addition, it is possible to estimate that the energy released as radiation alone in such an explosion amounts to about $10^{50}-10^{51}$ erg. To this must be added the enormous kinetic energy ($10^{50}-10^{52}$ erg) of the shell, which is expanding at high velocities (see above) against the force of gravity.

Our understanding of supernovae is greatly assisted, however, by the fact that theories of stellar evolution indicate that certain objects experience thermonuclear and gravitational instabilities that may lead to the release of energy on the scale of a supernova explosion.

As we know, in a "normal" star the outward gas and radiation pressures balance the inwardly-directed gravitational forces. The cause of the radiation pressure is the star's internal source of energy, which in a main sequence star is the conversion of hydrogen into helium. If this source of energy fails, perhaps because most of the hydrogen has been consumed, then gas pressure alone cannot compensate for the force of gravity. The interior of the star collapses and is heated in the process, until a new nuclear fuel can be tapped (initially by the conversion of He into O, N and C, and finally into the iron group of elements). In low-mass stars the contraction continues until a new equilibrium is reached at the white dwarf state, where electron degeneracy pressure resists the gravitational forces.

However, electron degeneracy pressure is not able to halt the gravitational contraction in stars above 1.4 solar masses that have exhausted all their sources of nuclear energy. A gravitational collapse (an implosion) ensues, raising the temperature of the centre of the star to more than 5×10^9 K. This causes neutrinos to be released, still further accelerating the collapse, which assumes catastrophic proportions. The *gravitational collapse*, lasting only about *2 seconds* (!), comes to a halt as soon as the star reaches a density of 10^{12} or 10^{15} g/cm^3, having become a neutron star. During the implosion immense quantities of gravitational energy are released, leading to the formation of a violent shock wave, which expels the outer layers of the star at a high velocity. The process is probably assisted by explosive thermonuclear reactions.

It is not yet known whether very massive stars can collapse into black holes, as it is not currently possible to make any exact estimates of the amount of mass-loss that occurs before the start of the gravitational collapse (which in this case probably occurs without a supernova outburst).

If all stars above 1.4 solar masses undergo supernova outbursts then these must be much more frequent than actually observed. Again it is the question of the mass-loss before the star reaches the stage immediately preceding the supernova, and which is so difficult to estimate, particularly for objects on the supergiant branch. It may enable stars that were originally as much as 5 solar masses to end up as white dwarfs of less than 1.4 solar masses. It is probably relatively massive stars, that is Population I stars, in a certain narrow range of masses, that are (Type II) supernova precursors.

The calculation of mass-exchange between stars in close binary systems also provides supernova candidates.

We still lack precursors for Type I supernovae, which are connected with the disk population and Population II objects. As these populations only contain low-mass stars, massive Type I supernova precursors can hardly be involved. On the other hand we can consider burnt-out, low-mass stars (white dwarfs), which, as soon as they accrete sufficient material to exceed the critical mass of 1.4 solar masses, may become supernovae. According to SHKLOVSKIJ (1978) this can occur, for example, in the cataclysmic binaries, if the white dwarf is not able to throw off its accreted material again in the form of nova outbursts. The question of the evolution of nova systems into super-novae is discussed by TRURAN (1980) among others. Some publications (see STARRFIELD et al. 1981) also discuss the possibility of whether some U Geminorum stars may provide suitable supernova candidates. One of the most recent investigations of the physics of exploding white dwarfs is that of CHEVALIER (1981). CHEVALIER suggests that neutron stars are produced by Type II supernovae, but that Type I supernovae leave no compact remnants. Further important investigations of this subject are those by TAAM (1980), references that he mentions there and some of the material in the book *Proceedings of the Texas Workshop on Type I Supernovae* (1980).

It is also conceivable that the white dwarf that will become a supernova is a single star of more than 1.4 solar masses. That is, if it happens to be so hot that the radiation pressure is able to resist the force of gravity for a long period of time, despite the mass being supercritical. When the cooling has advanced so far that finally the radiation pressure can no longer prevent the contraction, it collapses and a supernova explosion results.

We are still far from having a full understanding of the processes occur-ring in a supernova explosion, as the physics of matter under these extreme conditions is still insufficiently known.

Further details of our present ideas about the possible pre-supernova stages, the physics of supernova explosions, and the formation of heavy chemical elements, all of which are outside the scope of this book, can be found in many works. In particular the following may be mentioned: BARBARO et al. (1969), MUSTEL (1974), SHKLOVSKIJ (1976), DE GROOT (1978), PSKOWSKI (1978), TAAM (1980), WHEELER (1980, 1981) MAEDER (1981), REES and STONEHAM (1982), and many individual papers in the book *Supernovae* by SCHRAMM (ed., 1977) and the general compilation by HILLEBRAND (1982).

It is obvious that supernovae, in which elements heavier than iron have been produced by violently occurring processes (the so-called URCA pro-cesses), have been considerably involved in the evolution of our Galaxy.

3.3 Stars at an Extremely Early Stage of Evolution

3.3.1 Historical

Stars at a very early stage of evolution were already known as a group – although their nature was not recognized – in the 1920s, in the form of the irregular variables in the Orion Nebula. The systematic, routine research into this class of variables, begun then and continued for two or three decades – the names of LUDENDORFF (1928), HIMPEL as well as that of HOFFMEISTER (e.g. 1949) should be mentioned here – produced two results: the wide range of spectral classes (from B to M), and the general subgiant, or main-sequence nature of the objects. Specific investigation of the spectra of these variables started in 1945 and is linked (above all) with the names of JOY (1945), O. STRUVE, HERBIG (1962) and HARO. The finding that most of these variables lie in interstellar clouds led first to the attempt to explain their peculiarities by the theory that they were accreting interstellar material (GREENSTEIN, HERBIG, O. STRUVE) and later, as a result of the work by AMBARTSUMYAN (1949) as well as by KHOLOPOV (1951) and PARENAGO from the end of the forties (discovery of the T-Associations), to the suspicion that variables of this class were very young. The fact that these objects are very young was finally confirmed by the comparison of observations with theoretical stellar models, and by detailed astrophysical investigations, which have continued to the present day.

3.3.2 T Tauri Stars and Related Objects

Characteristics, Nomenclature and Sub-Divisions

The T Tauri stars are the best-known of the objects at a very early evolutionary stage. They are named after the star that serves, from a spectroscopic point of view, as a prototype. Their masses lie between about 0.3 and 3 solar masses. Towards higher masses they adjoin the so-called "Ae stars associated with nebulosity" (e indicating the presence of emission lines in the spectra, and, strictly speaking, this type also includes B or early F spectra), and towards lower masses, the flare stars. The latter will be discussed in Sect. 3.3.3. The other two groups have many features in common, so they may both be dealt with here. In addition, according to current knowledge, there are also extremely young stars that have no observed emission lines; these objects, which are shown to belong to these groups by other criteria, likewise belong to the present section.

The photometric characteristics are *irregular luminosity variations*. More can hardly be said, as the range of behaviour is exceptionally great. There are many different types of change within the overall variability. In many stars all of these elements are active, in others there may be only one. Many intermediate forms also exist. We may distinguish the following types of

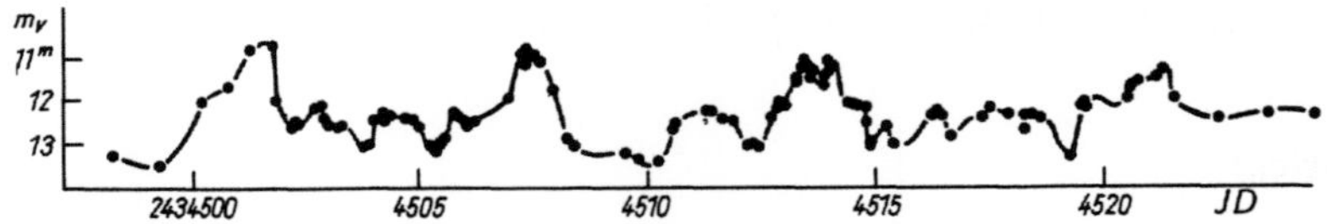

Fig. 93. Visual light-curve of T Cha (after Hoffmeister 1965)

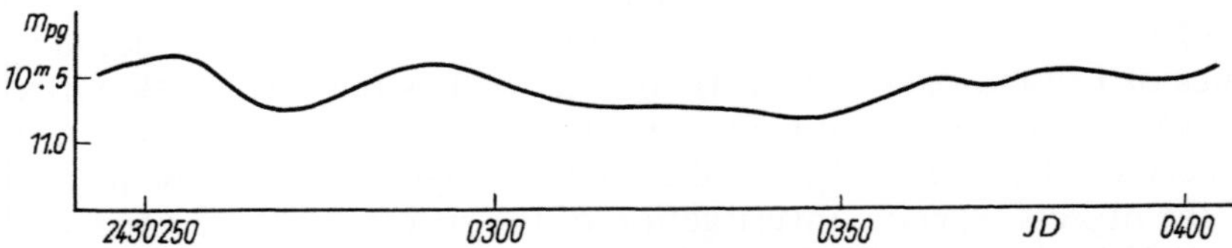

Fig. 94. Slow component in the photographic light-curve of T Tau (after Ahnert, slightly simplified)

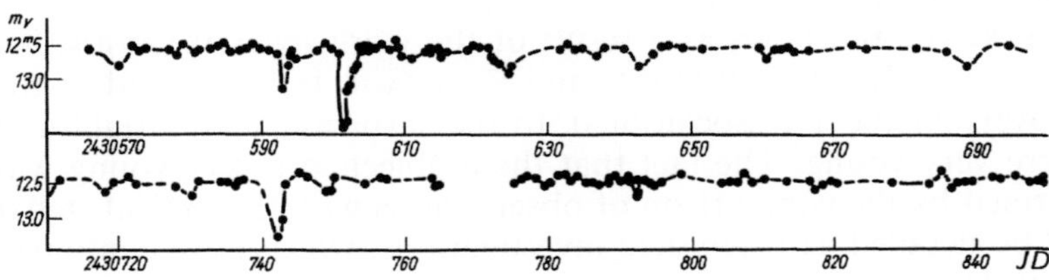

Fig. 95. Visual light-curve of BO Cep (after Hoffmeister 1944)

variation (with typical time-scales, in days, being given in parentheses):

1) Slow fluctuations (100)
2) Minima in brightness (10)
3) Eruptions (0.1−0.01)
4) Variability of the emission lines (1−0.1)
5) Quasi-periodic variations (10)

The name RW Aurigae stars was widely employed and is frequently used, even today, when only the changes in luminosity are being considered. Hoffmeister (1949) distinguished three *subgroups*:

1) Typical RW Aurigae stars. Mostly rapid changes in light, which can be seen over a few hours; the impression of irregularity predominates, the amplitude being 1.5 to about 4 mag. Examples: RW Aur, RR Tau, T Cha (Fig. 93).
2) RW Aurigae-like stars. Characteristics like those of the type stars with the difference that one factor (at least) is weaker. Either relatively slow variations are present, or the amplitude is less than 1.5 mag, or both. Example: T Tau (Fig. 94).
3) Algol-like variants. The irregular minima dominate the picture. Spectral classes mostly F or earlier. Examples: T Ori, BO Cep (Fig. 95), WW Vul (Fig. 96).

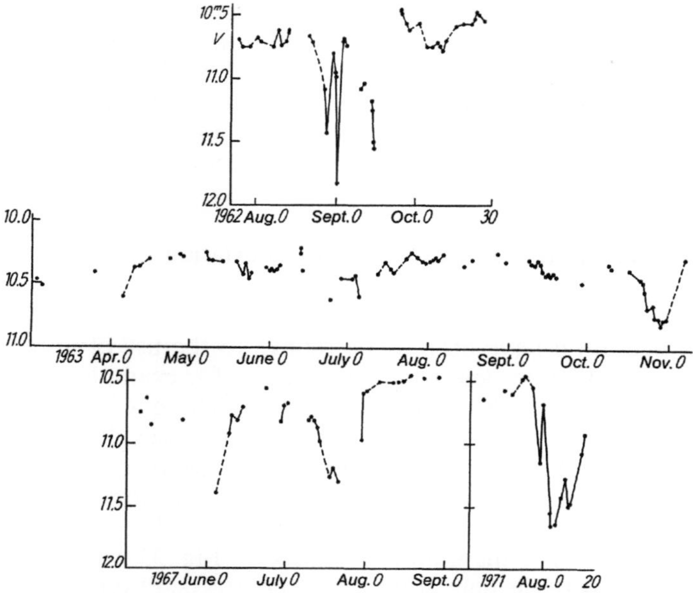

Fig. 96. Photoelectric light-curve of WW Vul (after RÖSSIGER and WENZEL 1972)

The classification scheme proposed by the IAU and employed in the General Catalogue of Variable Stars by KUKARKIN et al. (1969), takes other factors into consideration. Here irregulars in diffuse nebulae and rapidly-varying irregulars are grouped together under the symbol I (with the suffixes "n" for nebular or "s" for fast). A further subdivision takes account of spectral criteria ("a" spectra O–A, "b" intermediate or late spectra, "T" T Tauri spectra). Examples are given in Table 38.

Table 38. Classification of RW-Aurigae stars

Star	Class	Definition
RW Aur	IsT	T Tauri spectrum, rapid, irregular variations
T Tau	InT	T Tauri spectrum, in diffuse nebula
WW Vul	Isa	Rapid irregular, early spectral class
RR Tau	Insa	as WW Vul, in nebulosity
BO Cep	Insb	as RR Tau, but spectral class later than A

Even this classification has weaknesses. For example, the fundamentally different stars BO Cep and T Cha come in the same class (Insb), and the symbol "n" stands both for the strictly limited, surrounding nebulosity, and for the extensive dark clouds in which a particular star may be found.

Naturally there are grounds for confusion due to insufficient information. Among the eruptive binaries (Sect. 3.1), for example, types of variation are sometimes encountered that very closely resemble those in the stars discussed here, and criteria other than just the light-curves are required to differentiate between Ia and γ Cassiopeiae stars (Sect. 3.4.2). In addition it is only too easy for a rapidly changing, short-period star to be concealed by the day-to-day scatter in the magnitudes if insufficient observations are available for any regularity to be apparent. Experienced observers bear in mind the necessary caution in routine searches and in the classification of new variable stars.

Typical Behaviour of Individual Stars, Causes of the Variation

A whole series of recent findings about the *light-curves* of variables in the early stages of their evolution have been made thanks to the use of impersonal photoelectric methods (photoelectric photometry), for example at the observatories in the Crimea and at Sonneberg. Only this technique is capable of investigating colour changes (the alteration in the colour index during the variation in luminosity) and, because of the inherent increase in accuracy, of eliminating spurious variations having their origin in observational errors. Nevertheless, one of the light-curves with the fewest gaps is that of T Cha, obtained by visual methods. The diagram reproduced in Fig. 93 was drawn up by HOFFMEISTER (1965) with the help of the observers JONES, PHILPOTT and BATESON in New Zealand. It was possible to follow the changes in light for 24 hours a day. The first observations took place in 1952 and 1953, and others in 1959. It was found that the general character of the curve remained the same but that the first portion was marked by sharp maxima, and the second by sharp minima. The latter sort of variability was still stronger in the case of RY Lup, which even in 1952–1953 had a tendency towards BO Cephei-type variations, but which showed it far more markedly in 1959. *Histograms* showing the distribution of all the observations in a series (if possible divided into appropriate time intervals) over the whole range in amplitude (perhaps increasing in tenth-magnitude steps) are very informative in this respect. In T Cha one finds there is no preference for any particular magnitude; in RY Lup there is a high frequency at maximum light; in RU Lup the tendency is towards lower magnitudes, and in V 350 Ori there are actually three preferred magnitude levels.

This introduces another typical characteristic of the light-curves: the *rest magnitude*.

One could perhaps simply explain the BO Cephei sub-group as having a strongly marked rest magnitude at maximum. Other objects have their rest magnitudes at other points on the scale. It could well be that these levels are somehow linked to the star's properties; at least they are maintained for relatively long periods of time. Certainly there are also stars in which they are not detectable.

In T Cha one of the characteristics mentioned earlier, the *quasi-periodic variations*, stands out particularly strongly in the form of waves, occasionally superimposed on the irregular changes in brightness. In 1952–1953 the periods $3^{d}.4375$, $4^{d}.1800$ and $3^{d}.2323$ were found successively; observations in 1959 gave a cycle-length very similar to the third value. In RY Lup a similar finding was also indicated. HOFFMEISTER finally published the following typical periods, which nevertheless still required confirmation in RU Lup and AK Sco:

T Cha $3\overset{d}{.}2436$
RY Lup $3\overset{d}{.}7609$
RU Lup $3\overset{d}{.}8375$?
AK Sco $5\overset{d}{.}1480$?

Recently occasional cyclic changes in magnitude have been established for the Insb star SY Cha (SCHAEFFER and MATHIEU 1982): in 1970–1972 the variable showed strongly periodic waves with an amplitude increasing up to 1.6 mag (B) and $P = 6\overset{d}{.}129$. In the years before and after this only weak irregular variations were found. The authors explain the behaviour as the combined effects of the star's rotation and of a bright spot on, or close to, the surface of the star. This spot must certainly have had exceptional properties, although a similar type of explanation had already been suggested by HOFFMEISTER (1970) for the objects just mentioned.

Quasi-periodic variations in the light-curve are also shown by SV Cep, for which a close series of photoelectric U, B and V measurements is available for the period 1962–1966, see Fig. 97 (WENZEL 1969). In this object, relatively sharp minima are observed, which appear not quite irregularly, but recur at an average interval of $15\overset{d}{.}4$. They are of different shape and depth, however, and are accompanied by other elements of the luminosity variations. It is particularly remarkable that there is no alteration in the $B - V$ colour index in these declines (see Fig. 97) – i.e. the B and V amplitudes are the same. In order to explain this element in the variations, WENZEL, DORSCHNER and FRIEDEMANN (1971) proposed the existence of *circumstellar clouds of cm-sized meteoritic bodies* that neutrally weaken the light from the star. The cycle-

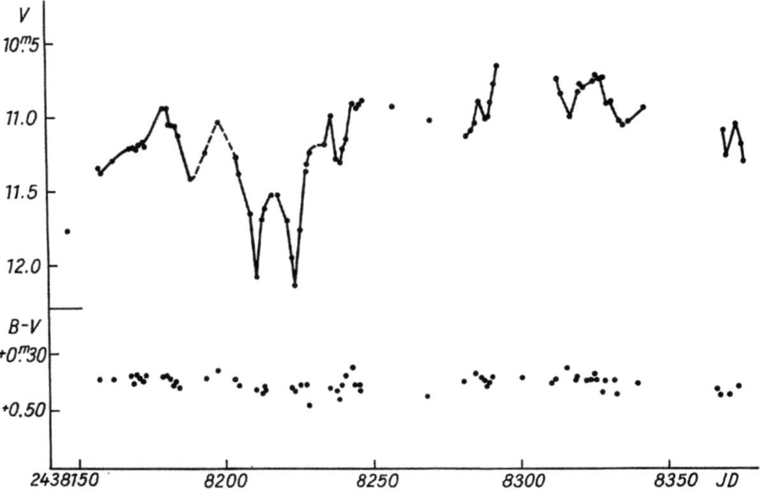

Fig. 97. Photoelectric B and $B - V$ measurements of SV Cep (after WENZEL 1969)

length is statistically determined by the distance of the clouds concerned from the star, and by their number. Estimates showed that some 10^5 clouds, totalling less than 1% of the star's mass, would be required to account for the observations. Only further investigation will show whether this model is any more than a working hypothesis. Eclipse by obscuring clouds was also employed by GAHM et al. (1974) and M. F. WALKER (1978, 1980). On the other hand, from their simultaneous 9-colour measurements, extending to the L band (3.4 µm) in the infrared, RYDGREN and VRBA (1983) concluded that the variability was governed by the alternation of distinctly hotter "plage regions" (corresponding to the faculae on the Sun) and cooler photospheric areas. Stellar rotation (see above) is thus also incorporated in this model. It is probable that a number of different mechanisms are responsible for the complex variations in the magnitude of the variables discussed in this chapter. Only future research will show exactly how these mechanisms occur physically, and the extent of their contribution in any particular star.

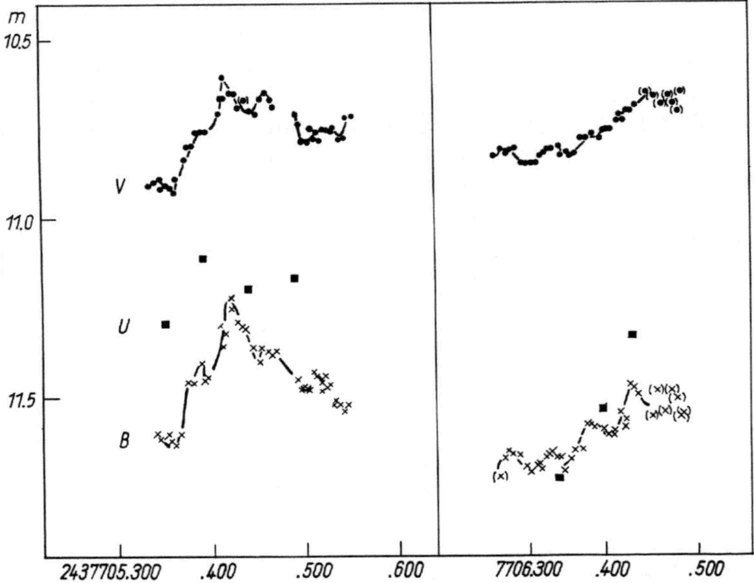

Fig. 98. Photoelectric observations of RW Aur in V, B and U bands (GÖTZ and WENZEL 1967)

GÖTZ and WENZEL (e.g. 1967) obtained fairly close series of photoelectric observations and simultaneous objective-prism spectrograms for a number of T Tauri stars and other sub-classes. The *special status of RW Aur* is particularly striking in this material as, unlike any other star on the programme, it may show variations of about 1 mag from night to night (Fig. 98). It is thus similar in its rapidity of variation to T Cha. The combination of waves with brightenings that have a duration of a few hours and an amplitude of a few tenths of a magnitude, observed visually in the latter star, are photo-

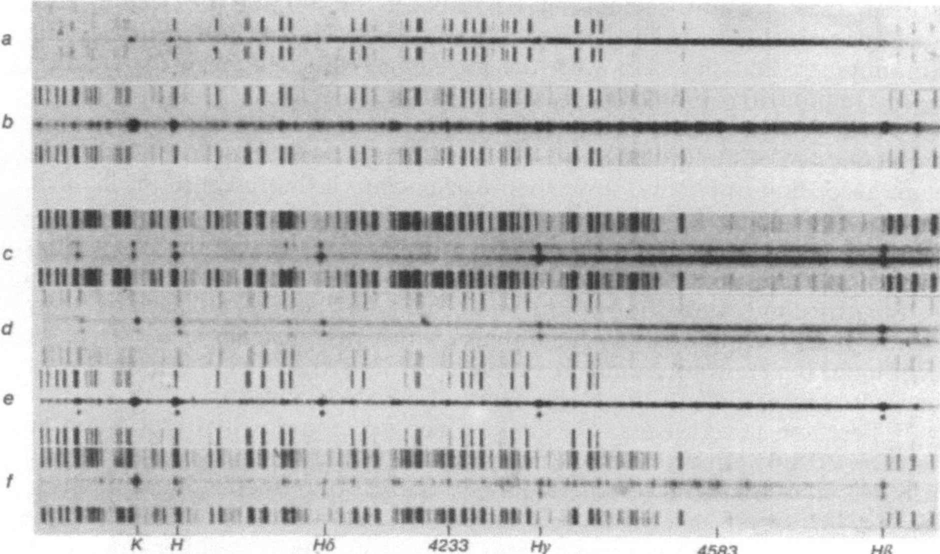

Fig. 99. Spectrograms of extremely young variables at various times (after JOY 1945). RW Aur: *a* 1941 Nov. 2, *b* 1944 Sept. 25; UZ Tau with dMe companion (*below*): *c* 1944 Sept. 23, *d* 1944 Jan. 4, *e* 1945 Jan. 8, *f* 1942 Dec. 28. Note the numerous emission lines; the lines at 423.3 nm and 458.3 nm are those of singly ionized iron. Comparison spectra for wavelength measurements appear above and below each stellar spectrum

electrically confirmed in RW Aur. It is a remarkable fact, also found in other stars on the programme, that changes in the line spectrum do not always occur simultaneously with changes in luminosity. Two asymmetrical, short-duration outbursts, similar to those of flare stars, can probably be ascribed to the red companion RW Aur B. Short-term variability in RW Aur, ranging from 15 minutes to a few days, has, incidentally, also been established spectroscopically by APPENZELLER et al. (1983), from observations made in the long Norwegian night.

Spectrum and Circumstellar Shells

The T Tauri class is defined by spectral characteristics, as mentioned above. Spectral classes are G to M. There are numerous *emission lines with low ionization potentials*, some with complicated line profiles (Hα and other members of the Balmer series, the H and K lines of Ca II and lines of other neutral and singly-ionized metals, e.g. Fe). These indicate the existence of a shell surrounding the star in a state like that of a chromosphere (Fig. 99). The presence of regions, close to the star, with very much higher temperatures, has recently been established from far-UV spectrograms obtained by the IUE (International Ultraviolet Explorer) satellite (e.g. GAHM 1980a). Specific attempts to detect X-rays from T Tauri stars have also been undertaken, for example with the Einstein satellite (e.g. GAHM 1980b). The X-rays appear to arise in a very tenuous corona that lies between the star and the circumstellar shell previously mentioned; they are absorbed by the latter in varying amounts, so that the stars with the most massive shells have the smallest X-ray luminosities (WALTER and KUHI 1981).

Particularly characteristic features in the spectrum are the two iron emission lines of FeI at 406.3 nm and 413.2 nm. These only occur in T Tauri stars and, according to HERBIG, owe their existence to a fluorescence mechanism in which the energy absorbed by FeI at 396.9 nm from the long-wave emission component of the CaI H line (396.8 nm) is re-emitted at the two greater wavelengths just mentioned, among others. The profile of the H line, as well as those of the K line, the Hα line and others, has been established as being due to gases streaming away from the star into circumstellar space. Here we encounter the enormous *mass loss* that, according to KUHI (1964), is the rule, at least temporarily, in the early stages of stellar evolution.

There are, nevertheless, objects in which we temporarily observe not an outflow of gas, but an *inflow*. The spectral lines show the "inverse P-Cygni profile". YY Ori, about which M. F. WALKER has published a series of papers (e.g. 1978), serves as a prototype. M. F. WALKER also sees the cause of the irregular changes in brightness in this star as being due to variable absorption processes within the extended shell. GQ Lup appears to be a brighter example (APPENZELLER et al. 1978, with additional references).

Another important feature found in stars at a very early stage of their evolution is the 670.7 nm absorption line of neutral lithium (Li⁷); analysis of this shows that the *lithium abundance* is about 100 times that found in the Sun. Igneous rocks on the Earth, as well as chondritic and other stony meteorites, show an average over-abundance of lithium of about the same order. These two remnants from the early history of the Sun and its surrounding "solar nebula" have retained the lithium that was once produced in the Sun. In the T Tauri stars the lithium has only "recently" been formed, but in the course of thousands of millions of years it vanishes from stellar atmospheres. The theories describing this phenomenon are still both numerous and contradictory.

The existence of the *gaseous shell* around T Tauri stars is also shown by the ultraviolet, blue and infrared *radiation excesses* found in the continuum. In T Tau, radio-frequency radiation at centimetre wavelengths appears to arise in the surrounding gaseous nebula, which is also visible optically (BERTOUT 1980). The infrared excess can, nevertheless, be explained by thermal emission from dust particles in a *circumstellar dust shell*. There are contradictory opinions as to which processes are dominant. Names such as STROM, COHEN, KUHI and RYDGREN may be mentioned in this context. It is certainly possible that both effects (that of the gas and that of the dust) contribute, and by different amounts in various stars.

A detailed description of the complicated situation prevailing around T Tauri stars has been given by COHEN (1981) under the title "Are we beginning to understand T Tauri stars?". The author does not, in fact, describe the attempts made to account for the changes in brightness, but rather offers a new point of view with his statements about the possible existence of rings of

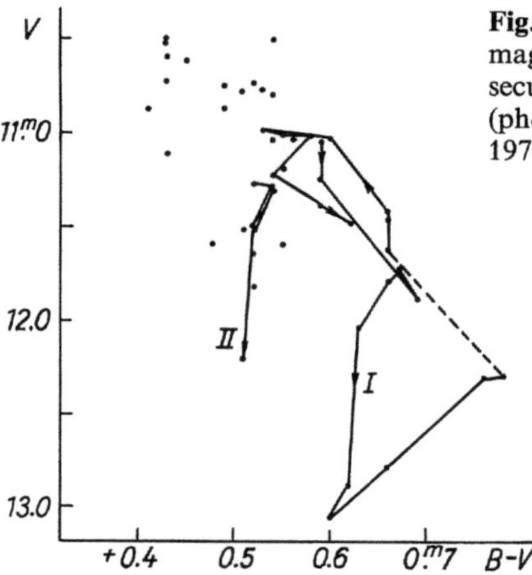

Fig. 100. Motion of RR Tau on a $V/(B-V)$ colour/magnitude diagram during luminosity variations; consecutive points on the decline to two minima are joined (photoelectric observations, after RÖSSIGER and WENZEL 1974)

dust (instead of dust shells) and of strongly directional flows of material (instead of an isotropic stellar wind – see also COHEN 1982).

The most important spectroscopic criterion for *"Ae stars in nebulosity"* seems to be the predominance of one or more normal "shell spectra": emission lines (principally of hydrogen) are particularly distinct when there is a considerable flow of material that is being lost by the star. The spectral lines then show the well-known P-Cygni profile (absorption components on the short-wave side of the emission lines), completely analogous to that in the T Tauri stars. The narrow, deep, hydrogen absorption features that are formed in the shell and overlie the wide stellar lines occasionally resemble those found in the spectra of supergiants.

It is not only mass loss and the formation of a circumstellar shell that are common to the Ae stars and the T Tauri objects (there are definite intermediate spectral types). As mentioned previously, the form of the irregular variability is also of a similar nature. *Minima* are, however, strikingly dominant in the Ae stars, and they take place without any marked changes in the absorption line spectrum (i.e. in the spectral type). In some individual events not even the colour index changes, as sometimes happens in the well-observed object RR Tau, for example (HERBIG 1960, RÖSSIGER and WENZEL 1974) – Fig. 100, and in SV Cep (see above).

A constant spectral class, or spectral behaviour that is not correlated with the changes in brightness, is also found, as indicated earlier, in the true T Tauri stars (DI Cep: GAHM 1979, RW Aur: GÖTZ and WENZEL 1967). This may be taken as strong evidence that no dramatic changes are occurring in the effective temperature of the stellar component in these objects, but that variations in either the gas or the dust shell (or both) are involved. The paper by GAHM (1979) contains many other recent references to the problems posed by the shells in T Tauri stars.

In their "Second Catalog of Emission-Line Stars of the Orion Population" HERBIG and N. K. RAO (1972) compiled a list of all pre-main-sequence stars that have emission lines and for which slit-spectrograms exist. There are

323 objects. As well as spectroscopic and *UBV* photometric data, the catalogue gives details of the respective types of light-curve, the presence or absence of a preferred magnitude serving as a means of classification (PARENAGO 1954, HERBIG 1962). Although the listing is not suitable for any statistical examination of these stars' properties, the fact that only 18% of the objects with known spectral types belong to classes B to F should be mentioned. This is not solely owing to the rarity of massive pre-main-sequence stars. It is rather that conspicuous emission lines are much rarer in early spectral-class variables of the types included here. Variables such as WW Vul, BO Cep, BH Cep or IP Per are not included in the list because of the way in which it was defined, as they are without observed emission features.

Group Properties

The first real possibility of linking observations with theoretical considerations about the pre-main-sequence stage followed from research on *T associations*. An association is defined as a locally concentrated group of stars with specific characteristics, where the total density is insufficient to warrant it being called an open cluster, for example regions with an above average frequency of O and B stars. The concept was first applied to the RW-Aurigae stars by KHOLOPOV (1951). He coined the term "T association" after T Tauri. A famous association of this type is that in the Orion Nebula. It is also obvious that the 7 longest-known RW Aurigae stars are all found in Auriga, Taurus and Orion, and are thus relatively concentrated in a region full of nebulae and dark clouds. RW Aur itself is in a region free from nebulosity and darkening, but the edges of the great Taurus dark cloud are only about 1° away. It is possible that RW Aur once belonged to the cloud and has since moved away from it.

New T associations were principally found from the $H\alpha$ emission of stars in low-dispersion spectroscopic catalogues, and which were usually faint. Pioneering work in this field was carried out in the fifties in Mexico, the USA and the USSR by CHAVIRA, DOLIDZE, IRIARTE, JOY, HARO, HERBIG and MANOVA. The subsequent work by KHOLOPOV (1951), GÖTZ (e.g. 1961) and others established that a high percentage of, if not all, $H\alpha$ stars show RW Aurigae-type variations, but that there are also numbers of such variables in T association regions that do not display any marked $H\alpha$ emission (at least from time to time). GÖTZ (1973, 1980 b) called attention to a number of statistical relationships, between spectroscopic and photometric properties and the evolutionary state, found in stars in T associations and extremely young clusters.

Some important T associations (including extremely young clusters) are given in Table 39; see also Figs. 101–103.

These associations cannot be much older than a few million years as they are strongly subject to dispersion from the effect of differential galactic rotation. In addition their colour-magnitude diagrams show, essentially, the structure that is expected from *modelling* of extremely young and *still-contracting* objects at an early stage of evolution. Moreover, similar colour-

Table 39. Important T Associations

Name	α	δ
IC 348 (Perseus)	3^h38^m	$+32°$
Taurus-Auriga Complex	4^h-5^h	$+16°$ to $+30°$
B 30 (Orion)	5^h25^m	$+12°$
Orion-Nebula	5 30	-6
B 35 (Orion)	5 40	$+9$
IC 446 (Monoceros)	6 25	$+10$
NGC 2264 (Monoceros)	6 36	$+10$
ε Chamaeleontis	11 00	-77
B 228 (Lupus)	15 40	-35
Scorpius-Ophiuchus Complex	16 25	-25
M8, M20 (Sagittarius)	17 56	-24
Corona Austrina	18 55	-37
IC 5070, NGC 7000 (Cygnus)	20 50	$+44$
NGC 7023 (Cepheus)	21 01	$+68$
Cepheus Complex	23 55	$+65$

magnitude diagrams, and corresponding two-colour plots, are also found in individual clusters; these are likewise at the beginning of their evolution.

The position of these extremely young stars in such a colour-magnitude (or Hertzsprung-Russell) diagram is indeed primarily determined by the stage of contraction which they have reached. However, secondary effects cause considerable difficulties in the analysis of such diagrams, for example the radiation excesses mentioned above, the absorption or extinction effects of circumstellar shells of dust particles, and the age range; in a region of space that has suitable physical properties stellar formation can obviously extend over several million years (e.g. GÖTZ 1973).

Overall one can say that T Tauri stars with spectral classes between G and M are, on average, 2.5 mag above the main sequence, and thus have *absolute magnitudes* between $+3^M$ and $+7^M$. The first definite indication of this was probably given by PARENAGO (1953) in his work on the variables in the Orion Nebula. By contrast the "Ae stars in nebulosity" are scarcely 1 mag above the main sequence.

The latter are rare, not only because high-mass stars are less common, but also because their rate of evolution is so high that they spend a *very short time* in the stage that interests us, so that the chances of discovering such a star are very small. Typical time scales for the contraction stage before nuclear energy processes begin (before the main sequence is reached) are, for example:

$$5 \; \mathfrak{M}_\odot \quad 5 \times 10^5 \text{ years}$$
$$1 \; \mathfrak{M}_\odot \quad 10^7 \text{ years}$$
$$0.3 \mathfrak{M}_\odot \quad 10^8 \text{ years}$$

It is for this reason that T associations with several hundred members frequently contain no more than about five massive objects of this sort. They often illuminate nebulosity in their immediate neighbourhood – thus their name Ae stars "in nebulosity".

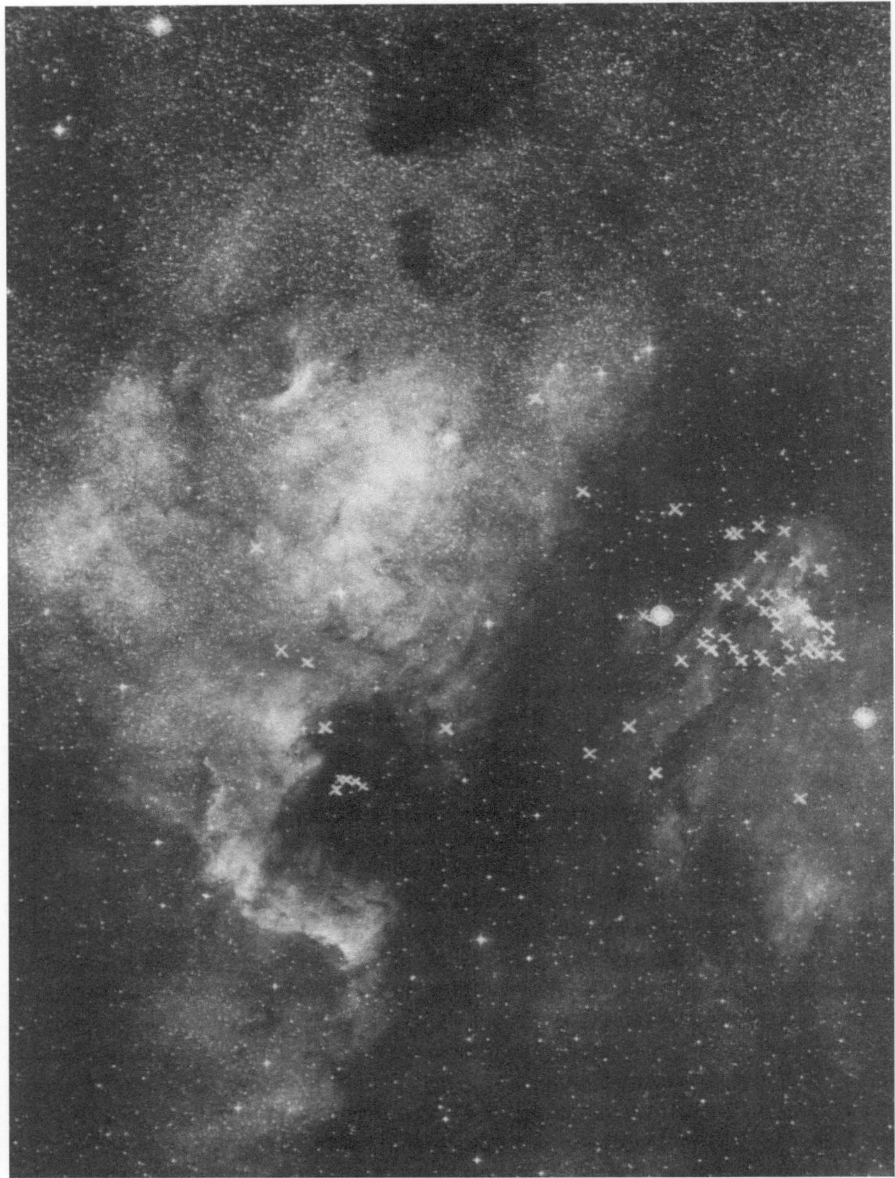

Fig. 101. The North America Nebula (NGC 7000) and the Pelican Nebula (IC 5067). The positions of some very young stars are marked by crosses (Hα stars after HERBIG 1958b, variables after WENZEL 1963 and GIESEKING 1973). (Photograph by GÖTZ, Sonneberg)

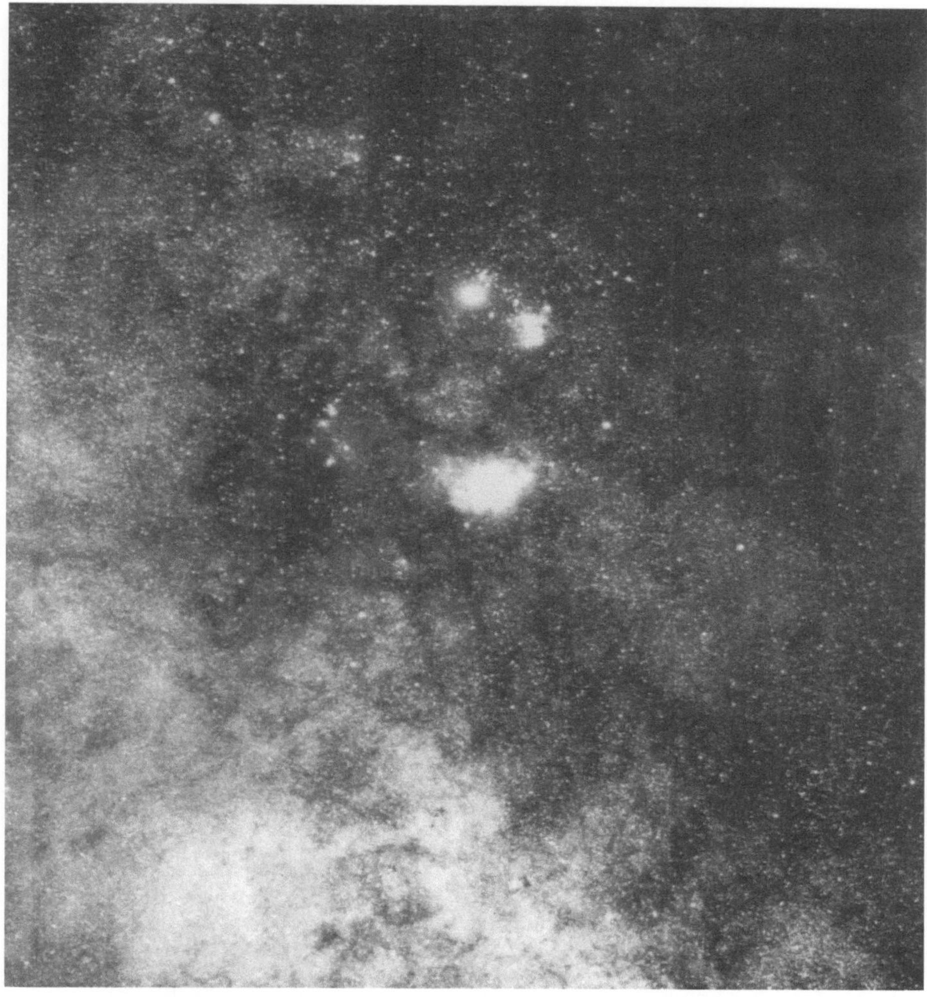

Fig. 102. Gas and dust nebulae in Sagittarius; in the centre is the Lagoon Nebula M8, which contains a T association. Photograph by HOFFMEISTER (Boyden Station)

Other Relationships with Interstellar Matter

The general field is constantly and rapidly enriched by young stars from T associations (clusters disintegrate much more slowly). Whether the fixed stars can be formed individually, or at the most in twos and threes, however, remains open, from an observational point of view. Modern theories of stellar formation from cool interstellar clouds no longer categorically exclude this possibility. T associations are also observed that have very few members, such as the group around BD + 40°4124 (HERBIG 1960; WENZEL 1980a), which contains two variables with different spectral classification and type of

Fig. 103. The "Coalsack" dark nebula in Crux (the Southern Cross); the subject of a number of searches for extremely young stars. Photograph by HOFFMEISTER

variation, LkHα 224 and 225. The first is a typical RW Aurigae star with a probable Ge–Ke spectrum and the second is a Be/Ae star with a predominantly bright normal light and irregular minima.

Both stars, which have a separation of only about 20″, are obviously affected by differing amounts of interstellar extinction, which may indicate the effect of dust in the immediate neighbourhood of LkHα 225.

A similar result had been found much earlier by GÖTZ (1961, p. 136) in the course of an intensive investigation of RW Aurigae stars in T associations by star counts in the areas immediately around the variables. The number of neighbouring stars varied from star to star; this was attributed to *differential extinction* by interstellar dust – in other words, the latter was not of uniform density. It was found to be the rule that the greater the intensity of the Hα emission, the greater the density of the dust cloud. This result can be explained in a quantitative manner from the confirmatory finding that the greater the mass loss (the stellar wind) – KUHI 1964 – the stronger the emission features (WENZEL 1975). A fraction of the gas flow condenses into non-volatile ("refractory") dust particles, so that the stellar neighbourhood (demonstrably out to distances that are perhaps of the order of 1 pc) is enriched with a *new generation of interstellar dust*. The calculated density of this material is such as to agree with the observed increase in interstellar extinction close to T Tauri stars. It is worth noting that some of the brighter RW Aurigae stars that show BO Cephei-like variations in brightness do not appear to exhibit greater extinction than their surroundings (for example IP Per, BO Cep, BH Cep, WW Vul, SV Cep as well as RR Tau). This can be determined from the comparison of the colour-excesses in these stars with those of invariable neighbouring stars of known luminosities and distances (e.g. RÖSSIGER and WENZEL 1973). The procedure cannot be applied to the true T Tauri stars, as in their case it is difficult to separate the circumstellar and interstellar contributions, owing to the abnormalities in the continuum spectrum from the star itself.

There is good agreement in other respects between observation and the theory of stellar formation in cool interstellar clouds (HERBIG 1962, 1977): the *characteristic velocities* of members of T associations are similar to the internal kinematic motions of neutral H I regions and molecular clouds (and not to those of the "hot" H II ionized-hydrogen regions). The existence of such associations is tied up with the presence of additional dense and extensive dark clouds of interstellar dust, which serves to provide the necessary cooling for H I gas. Here one only has to think of the T associations in the constellations of Taurus and Auriga, in Ophiuchus and in the North America Nebula. The conclusion to be drawn is obvious – and has been confirmed by recent velocity measurements – that these young variables are still within their parent cloud (HERBIG 1981). Isolated objects, such as RW Aur, already mentioned, and the special case, TW Hya, which is 13° from the nearest dark cloud (RUCINSKI and KRAUTTER 1983), present unexplained contradictions to this view.

In some clouds of an appropriate type there appears to be a concentration of Hα stars in the neighbourhood of hotter, more massive stars that have greater absolute luminosities. Some examples are the Orion Nebula, NGC 2068 and NGC 2264. It is purely hypothetical whether such positions favour the formation of low-mass stars.

Peculiar Objects

We conclude this section by describing three classes of peculiar variables that are frequently mentioned in modern literature.

FU Orionis stars (occasionally referred to by the abbreviation "Fuors") are T Tauri stars that have great outbursts, which are accompanied by considerable mass ejection and an increase in the surface temperature. The prototype FU Ori has been known since 1937. However, V 1057 Cyg has been far better investigated. Until 1969 this star behaved like a normal T Tauri star with a typical spectrum (HERBIG 1958b) and weak luminosity variations (WENZEL 1963). In the autumn of 1969 there was a sudden outburst of almost 6 magnitudes (from 16^m to 10^m) within about 300 days (Fig. 104). Subse-

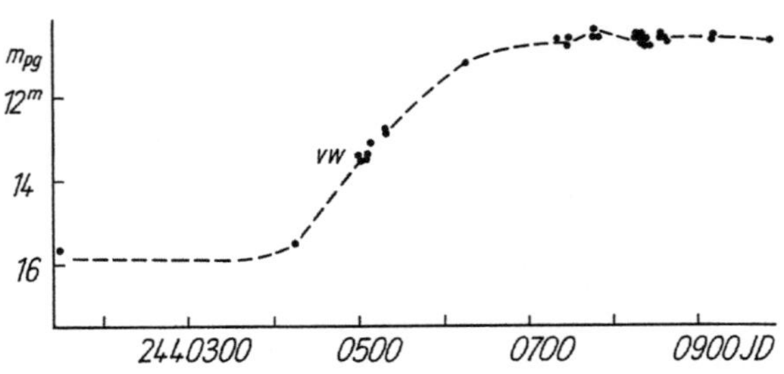

Fig. 104. Outburst of the FU Orionis star V 1057 Cyg (after MEINUNGER and WENZEL 1971)

quently the magnitude fell very gradually by about 2.5 mag (photographic) in 10 years. At maximum the spectrum resembled that of an A-type shell star. V 1515 Cyg, which brightened in 1950, may be a further object of this sort (WENZEL and GESSNER 1975). In a detailed discussion HERBIG (1977) suggests that the brightening in FU Orionis variables is probably a recurrent phenomenon in stars at a young evolutionary stage. This conclusion may be reached from the number of cases known to date (3), together with the period during which the sky has been monitored for eruptive objects (80 years), and the estimated number of potential T Tauri candidates within a reasonable range of distance and luminosity (500). The author, after a thorough discussion of the available material, came to the conclusion that as yet there was no convincing explanation for the FU Orionis phenomenon.

Herbig-Haro objects are small, faintly-luminous regions of nebulosity in dark clouds inside T associations, generally having irregularly variable knots that are almost star-like in appearance (Fig. 105). The infrared sources found in the neighbourhood ("Herbig-Haro stars") may be T Tauri-star precursors, hidden by exceptionally dense circumstellar dust shells. An important property of the H-H objects is the very high negative radial velocities found for the knots' emission lines, relative to the surrounding dark cloud. It still has to be resolved whether these emissions arise in the shell of a H-H star, and are only reflected in our direction by the nebular knot, or whether they are formed directly within the latter. In either case, however, it would still be an indication of high rates of mass loss from the extremely young objects, the ages of which are estimated at 10^5 years (STROM 1977). The discussion by COHEN (1981), already mentioned in connection with the circumstellar effects found in T Tauri stars, also goes into detail about the H-H objects.

Post-T-Tauri stars form a class of objects that are very difficult to observe, and the evolutionary stage of which lies between the T Tauri stars, marked by strong activity, and the stable main-sequence stars. Objects at this evolutionary stage should have T Tauri characteristics (for example H and Ca II emission, Li^7 over-abundance, irregular variability and infrared excess) in a more or less weaker form. HERBIG (1973), who expressed this point of view at the time, saw FK Ser as a possible representative. It is (unfortunately) a visual binary, where both components have K5peV spectra, a strong lithium line at 670.7 nm, and absolute magnitudes M_v about 3.5 mag above the main sequence. The object appears to be associated with a nearby dark cloud.

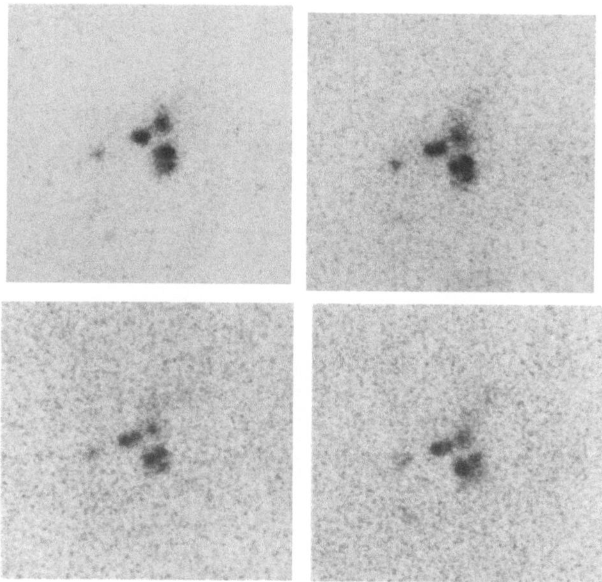

Fig. 105. Variability of the central region of Herbig-Haro object No. 2. The length of each side of the squares is about 2'. Dates are, from top left to bottom right, 1947 Jan. 20, 1954 Dec. 20, 1958 Nov. 9, 1968 Jan. 5 (after HERBIG 1969)

The changes in brightness, however, are BY Draconis-like (CHUGAINOV 1973) possibly combined with flares (Sect. 3.7.1). On the whole the photometric properties of this hypothetical group of variables is still poorly defined. One of the brightest, possible representatives is HD 36705 (a cycle-length of $0\overset{d}{.}514$ is occasionally active: RUCINSKI 1982, with further references about this problem), but which is regarded by other authors as being a BY Draconis or RS Canum Venaticorum star (Sect. 3.7.1), or is seen as being similar to V 471 Tau (Sect. 4.7), which consists of a white and a red dwarf pair. A relationship to the G III star, FK Com, a possible ellipsoidal variable (Sect. 4.3) has also been suggested. This confusion, deliberately described in some detail, should serve as a warning against the inclusion of any variable in this group, especially since the presence of the lithium line alone is a necessary, but not a sufficient, criterion for determining a T Tauri classification.

3.3.3 Flare Stars

Light-Curves and Spectra

The characteristic features of this group are very strong outbursts that often occur in minutes. The prototype is the star UV Cet, a nearby red dwarf (distance 2.7 pc) that has hydrogen emission lines in the spectrum and a visual minimum magnitude of $13\overset{m}{.}0$. Large numbers of flare stars are found in T associations and young clusters; following a suggestion by HARO, who carried out the first fundamental work on these objects (e.g. HARO and MORGAN 1953), these have been called "flash stars". For the sake of brevity we retain this expression here, although it has now gone out of fashion as a

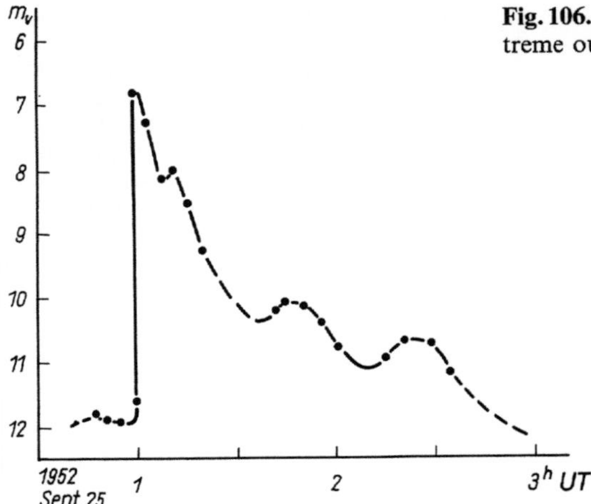

Fig. 106. Visually observed flare of UV Cet (extreme outburst, after OSKANYAN 1964)

rather different meaning has been given to the term "flash" in the context of stellar evolution. Their positions show that flash stars are often associated with bright or dark interstellar clouds; they are then classified as UVn.

Flares and flashes should be differentiated from "flickering" as they are isolated outbursts rather than a continuous or frequent flickering. They recur at irregular intervals, which, as a rule, are much longer than the duration of the flare.

Two principal types of light-curve are distinguished. In Type I the rise to maximum is extremely steep, completed in a few seconds or minutes. The decline may last from 10 minutes to about 2 hours (Fig. 106). In Type II everything takes place about 10 times more slowly. The UV Ceti stars in the solar neighbourhood only show Type I events, but the flash stars, on the other hand, have both forms, even in one and the same object. The rate of rise in a Type I outburst is generally of the order of 0.05−0.1 mag s^{-1}. However, UV Cet, which has been well-observed, showed rates of 0.6 mag s^{-1} in several events, and on one occasion a rate of 2.8 mag s^{-1} when the brightness increased 420-fold in 31 seconds (JARRETT and GIBSON 1975). In this case the *B* amplitude amounted to 6.5 mag. Similar bright outbursts have been repeatedly observed, and it should be noted that the *B* amplitude is always intermediate between the greater *U* and the smaller *V* amplitudes. Random small flares occur, but the degree to which they can be detected naturally depends upon the instrumental accuracy. The frequency that can be determined for the flares depends upon the colour, and also on the absolute magnitude of the star concerned. A rough average is 1 flare per hour ($\geq$ 0.1 mag) in the *B* band. In recent years there has been great progress in the photoelectric observation of flares at high time resolutions (1 s and better) − see for example EVANS 1975, MOFFETT 1974. This work has shown that high-speed flares exist, outbursts with an overall duration of less than 10 s ("spike flares", Fig. 107). These observations have shown that the flare con-

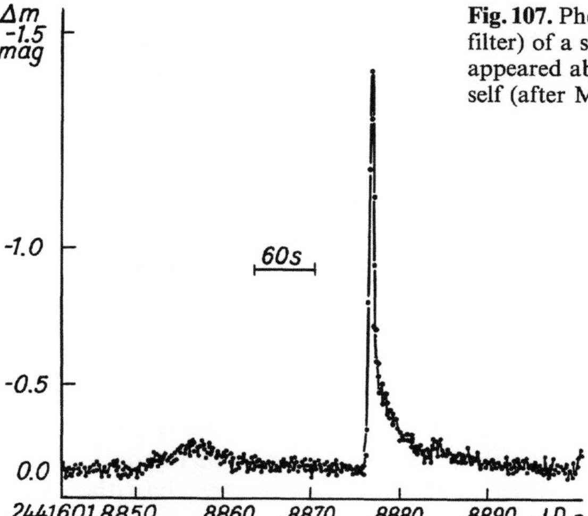

Fig. 107. Photoelectric light-curve (without colour filter) of a spike flare in UV Cet; the "precursor" appeared about 3 minutes before the outburst itself (after Moffet 1974)

ditions are far more complicated than was considered to be the case from photographic (or even visual) evidence.

During the outbursts the emission lines from hydrogen and neutral helium are generally considerably brighter than in the normal state, and the continuum has a *blue* and *UV excess*, which occasionally swamps the absorption spectrum. In many carefully-monitored cases, *radio outbursts* have been observed at wavelengths between 20 cm and 15 m that were simultaneous with the optical flares. These bursts begin approximately at maximum light, or a few minutes later, and last about as long as the flares in the conventional spectral bands.

Gurzadyan's monograph (1980) contains a list of 71 UV Ceti stars in the solar neighbourhood, although it does include a number of doubtful cases and some BY Draconis objects (Sect. 3.7.1). Excluding the latter, the spectral classes are, without exception, dMe, with a distinct increase in frequency towards later spectral sub-classes. All the objects included lie within 20 pc of the Sun. A similar list was given in a review by Kunkel (1975); however he included some dMe stars that show particularly strong Balmer emission, but no flare activity, as have some other authors.

Evolutionary State and Models

Compared with the UV Ceti stars, which are close to the Sun, the number of known flash stars is far greater, amounting to about 1000. A list given by Haro (1968) shows the range in spectra to be K0–M6. The main finding – that in the youngest clusters and T associations, flash stars occur even among K-type stars – has been confirmed by more modern research. Among the flash stars there are particularly frequent indications of superimposed RW Aurigae-type variations. The position of the flash stars on the Hertzsprung-Russell diagram is very similar to that of the very young stars, apart from differences because of the lower mass. On the other hand, UV Ceti stars in the solar neighbourhood have absolute luminosities nearly in accordance with their spectral class, and only slightly above normal main-sequence luminosities. This difference is now seen as being age-dependent, just as the flare stars in individual clusters and T associations (of different ages) have different positions on the Hertzsprung-Russell diagram. Another age effect was found a long time ago in the kinematic properties of *field dMe stars* – potential, if not actual, flare stars – when compared with those of dM stars (without emission features) in the solar neighbourhood: while the space motions of normal red dwarfs do not significantly differ from the average values found in the solar neighbourhood, field dMe stars are observed to have a lower space velocity relative to the Sun, a smaller space-velocity vector at right-angles to the galactic plane, and less scatter in their velocities. It was concluded from this behaviour that the dMe stars in the vicinity of the Sun have lost less of their individuality than the general run of normal M-type dwarfs. The former therefore appear to be rather younger than the latter.

Poveda (1964) made a significant suggestion on the *evolutionary state* of the flare stars. He suggested that they are objects which, during their pre-main-sequence contraction, still have fully-convective interiors. Modelling shows that the earliest spectral class in which this can happen is K1, and that these stars are of luminosity class IV. Very low-mass stars ($< 0.1\mathfrak{M}_\odot$) will still be fully-convective close to the main sequence (spectral class M5 or later). Despite their considerable age (10^9 years), their flare activity shows that they have not advanced very far in their evolutionary development.

The *physical mechanism* that is directly responsible for the flares still remains uncertain. Gershberg's comprehensive monograph (1970) alone listed and described 10 attempted explanations, although some of these are now only of historical interest. In most cases magnetic fields play a part, for example in the ideas put forward by Ambartsumyan, who proposes radiation of relativistic electrons, and in those of Gurzadyan, who suggests that scattering of photons by fast electrons explains the peculiarities found in the radiation from the flares. Both assume that the outbursts are basically caused by a special form of protostellar material, which exists in the interior, and now and then reaches the surface of the star. However, the very existence of such material is purely hypothetical. Gershberg advocates a theory borrowed from those explaining solar activity, where there is no "bright spot" on the surface of the star, but a region of hot, highly-ionized gas that forms above the stellar atmosphere. But this attempted explanation and, above all, the parallels drawn by the author from solar physics are disputed.

Comments on the Statistics

Several flare stars can be observed with small telescopes: UV Cet (13^m-7^m), AD Leo ($9\overset{m}{.}5-9\overset{m}{.}0$), EV Lac($11\overset{m}{.}5-9\overset{m}{.}5$). Certainly monitoring these stars requires great patience, a complete lack of bias, and a lot of experience as notable outbursts are only too infrequent. This fact, and the short duration of the outbursts are, moreover, the reasons for the low discovery probability.

Both the absolute magnitude, and in the majority of cases the apparent magnitude, are very faint in these stars. One of these stars may lie only just above the limiting magnitude reached by a photographic plate in a one-hour exposure. If during the exposure it happens to undergo one of its (all too rare) outbursts lasting a few minutes and amounting to about a magnitude, it is hardly to be expected that it will be discovered from such an event. On the other hand, if a large telescope is used, permitting a much shorter exposure time, then the field is so small that again the chance of finding one of these objects is reduced. Most success has therefore been achieved in clusters and T associations by multiple exposures on one plate. UV Ceti stars in the solar neighbourhood can not be covered by this technique. Here, however, specific photographic or photoelectric monitoring of dMe objects can be employed to discover flare stars. Once again, the statistical picture is, in many respects, both misleading and intractable.

3.4 Hot Variables with Extended Shells

For the sake of simplicity we discuss three types of variables together in this section, even though, as far as their origin is concerned, it is not certain that they have anything to do with one another. However, they cannot be very easily grouped with any other classes. Their characteristic feature is an extended shell around the star.

3.4.1 S Doradus-Type Variable Supergiants

This classification was introduced in the middle seventies (KUKARKIN et al. 1974). They are the objects earlier known as P Cygni stars. It was decided to redefine the classification because the P Cygni phenomenon (emission lines

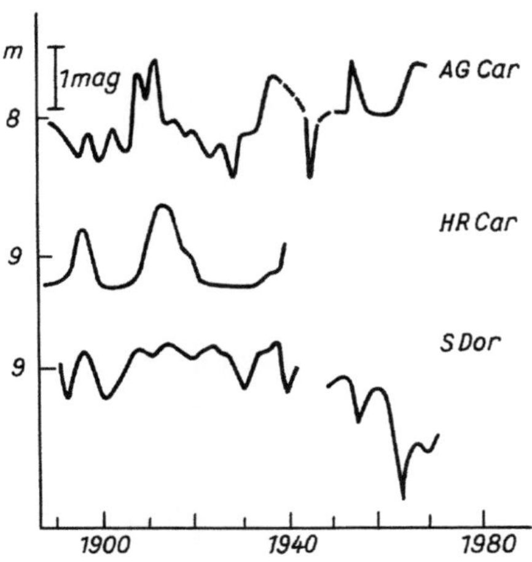

Fig. 108. Light-curve of three S Doradus variables, from various sources and slightly simplified (SHAROV 1975)

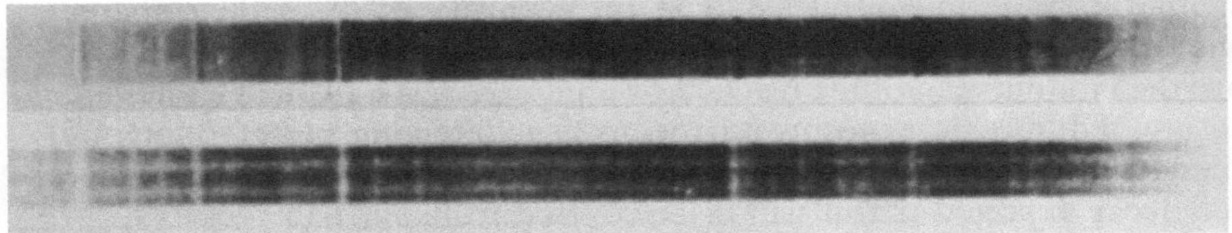

Fig. 109. The spectrogram of P Cyg (*above*), compared with a normal B2 supergiant (χ^2 Ori). From the MORGAN, KEENAN and KELLMAN spectral atlas

with absorption components on the short-wave side) is a purely spectroscopic effect present in many other stars. The characteristic feature of S Doradus variables is *slow, irregular variation in brightness* with typical *time scales* of *years* or even *decades* (Fig. 108). Even in its normal state the luminosity is very high ($M \approx -8 \pm 2$); spectral classes are B or A, with the corresponding colour index. The spectral lines, when observed, show the P Cygni profile. *P Cyg* itself has often been discussed in detail. It must have brightened by about 3 mag between 1597 and 1602. It stayed at maximum, 3^m, for a few years and then declined to about 6^m. A second maximum probably occurred in 1655. In the following centuries the star showed many small-amplitude fluctuations. It is now fairly constant at 5^m. The *spectral class* is cB1peq [(*c*) *high luminosity*, (*p*) *peculiarities*, (*q*) *violet-shifted absorptions*] – Fig. 109. The catalogues by KUKARKIN et al. (1974, 1976) contain the following 8 stars that are regarded as definitely of this class:

AE And	HR Car
AF And	η Car
Z CMa	P Cyg
AG Car	S Dor

The inclusion of Z CMa is doubtful, owing to its much lower luminosity.

It is worth noting that *S Dor* is located in the Large Magellanic Cloud. With its absolute magnitude of $M_v = -9\overset{M}{.}2$ it is one of the brightest stars known (apart from novae and supernovae in outburst). Its *mass-loss* due to the conversion of mass into radiation amounts to around 10^{20} *tonnes* $= 1/60 \, \mathfrak{M}_{\ooalign{$\odot$}} = 5 \times 10^{-8} \mathfrak{M}_{\odot}$ per year. AE And and AF And are members of the Andromeda Galaxy M31. In this galaxy, as well as in M33 and NGC 2403, there are altogether about another dozen more or less definite "Hubble-Sandage variables" of this type, according to the lists given by SHAROV (1975) and HUMPHREYS (1978).

It is known that the *P Cygni profile* arises in a *stationary shell* formed by *gas flowing away from the star*, where the *absorption lines* are produced in the portion of the shell that is moving in the *direction of the observer*. A recent analysis of the line profiles in P Cyg by NUGIS et al. (1978) gave a mass loss of about $10^{-4} \, \mathfrak{M}_{\odot}$ per year. If this extremely high value is associated with the source of the variations, as assumed, for example, by STOTHERS and CHIN (1983) after a critical discussion of six other conceivable causes for the changes in brightness, it is easily understandable why only a few objects of this sort are known, despite their high luminosity. They must pass through this evolutionary state very rapidly indeed, before too much mass is lost. Quite apart from this, they are very massive objects ($50\mathfrak{M}_{\odot}$), which, in any case, are very rare.

The *physical processes* that lead to the formation of such enormous stellar winds are at present as poorly known as the cause of the variability. What can be established is that these objects are in an evolutionary stage which is carrying them away from the main sequence on the H-R diagram.

In an interesting statistical investigation, using 2420 individual Geneva magnitude observations of 327 supergiants of all spectral classes, MAEDER (1980) established that there was a subsidiary maximum of about 0.07 mag in the mean microvariability amplitude among "normal" early BIa stars, and that the mass-loss rate in B and A supergiants increased sharply with this amplitude. Obviously the recently-established group of the weak and non-radially (Sect. 2.3) pulsating B supergiants (cycle-length a few days, amplitude $\leq 0.05\,$mag) makes some contribution to the statistical result, as pointed out by PERCY (1981), for example. The evolutionary connection with the S Doradus type, however, remains uncertain. On a historical note, it is worth recalling that GUTHNICK and PRAGER (1915) were the first to photo-electrically establish a weak, quasi-periodic variability in the A-type supergiant Deneb (α Cyg), which has now given its name to the whole group.

CHENTSOV (1981) looked for objects in the transitional region between the constant, high-luminosity, main-sequence stars and the strongly variable S Doradus objects. A candidate might be V 4029 Sgr for example, spectrum B9eIa, whose luminosity, spectral-line profile and faint variability appear to be appropriate (CHENTSOV 1980).

It must be borne in mind that strong radiation pressure also plays a part in maintaining mass loss, and that very massive stars are essentially unstable (e.g. M. SCHWARZSCHILD and HÄRM 1959).

η *Car* is an extraordinary variable. Its nature and evolutionary state are still very uncertain. The variability of this occasionally very bright object has been known since the 17th century. In 1677 HALLEY observed it at magnitude 4 from St Helena. PAYNE-GAPOSCHKIN (1957) gave the known maxima and minima, reproduced in Table 40. It will be seen that at its extreme maximum the object was almost as bright as Sirius. During this very bright phase it had an *absolute bolometric magnitude of* -13^{M}. At present it is one of the *most intense infrared sources* in the sky.

Table 40. Behaviour of η Carinae

Minima	Magnitude	Maxima	Magnitude
1826:	6^{m}	1827	$1^{\mathrm{m}}_{\cdot}2$
1838	1.5:	1843	$-\,0.8$
1854	1:	1856	0.3
1869	7.0	1871	6.6
1886	7.6	1889	6.7
1901	7.8	1952	6.5

The very complicated spectrum with its numerous emissions has been the subject of several classic studies, an example being that by THACKERAY (1967), who identified a large number of lines of [Fe II] and [Ni II]. The *mass-loss rate* amounts to about $7.5 \times 10^{-2}\,\mathfrak{M}_\odot$ per year at present, so that in a few decades a considerable fraction of the mass of the object has escaped as ejecta. The amount of energy necessary for this comes close to the values that are observed in supernovae. ANDRIESSE and VIOTTI (1979) established that circumstellar dust is being steadily produced in the expanding gas shell. The object consists of a diffuse core, which is surrounded by an irregular ellipse of nebulosity (the so-called Homunculus Nebula), that is obviously involved in the changes in brightness (THACKERAY 1953).

It is not clear whether η Car is a pre- or post-main-sequence object – if these concepts can have any real significance in an object with a mass of $150\,\mathfrak{M}_\odot$, as has been proposed.

3.4.2 γ Cassiopeiae Stars

These objects, also classified as variable *Be and shell stars* (in the narrower sense), lie close to the main sequence in the Hertzsprung-Russell diagram. However, while the rates of mass-loss in the S Doradus objects, discussed in the previous section, are so great that escape velocity is reached (mass is being lost to interstellar space), this is not generally the case in the γ Cassiopeiae stars. Here the material gathers close to the photosphere, forming either a *disk of gas* in the equatorial plane (Be stars) or else a *shell* that completely encloses the star. A star may change from one form to the other.

The *variations in brightness* in these objects are generally small, and the amplitudes can often only be determined photoelectrically to any degree of accuracy. Long, flat, irregular waves with lengths of about fifty to several hundred days and a tendency to form minima, are typical. γ Cas itself seems to be the most active variable of this sort (amplitude 1.4 mag, Fig. 110). Another well-known object is BU Tau (Pleione in the Pleiades); others are X Per, X Oph, φ Per, o And, EW Lac and μ Cen. Spectral class and luminosity are mostly given as BIVpe. The spectral lines (the Balmer series) show a characteristic shape (Fig. 111), which is variable (Fig. 112), and formed by line emission from the disk or shell and the *very wide absorptions* from the rapidly-rotating central star. It is the centrifugal forces, arising from this rapid rotation that are responsible for the formation of the shell. In both γ Cas and BU Tau the formation or thickening of the shell is observed spec-

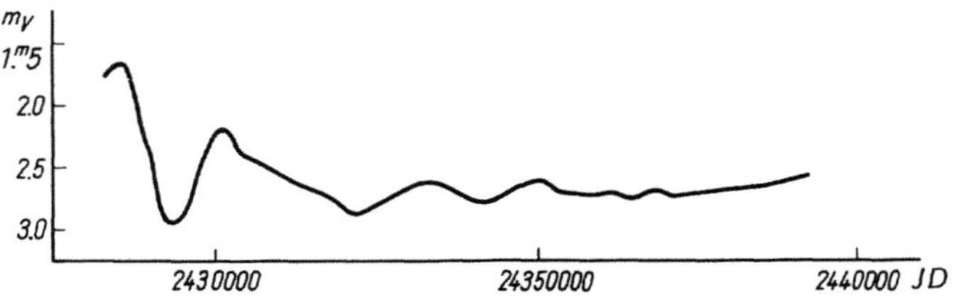

Fig. 110. Light-curve of γ Cas

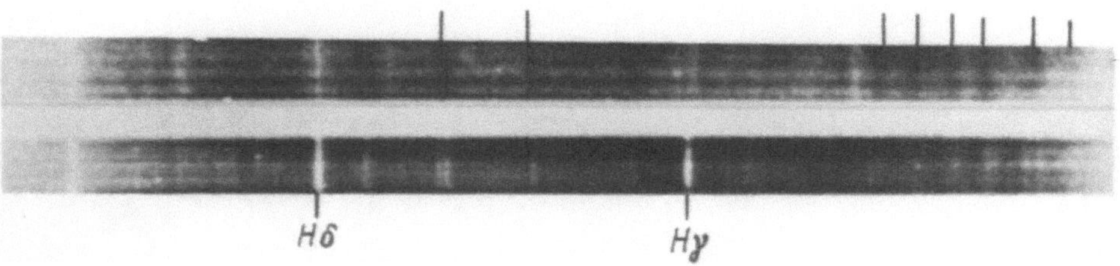

Fig. 111. Spectrogram of the γ Cassiopeiae star φ Per (*above*), compared with an A2 supergiant (α Cyg). Note the complex structure in the hydrogen lines and the presence of emission lines (marked at top). From the MORGAN, KEENAN and KELLMAN spectral atlas

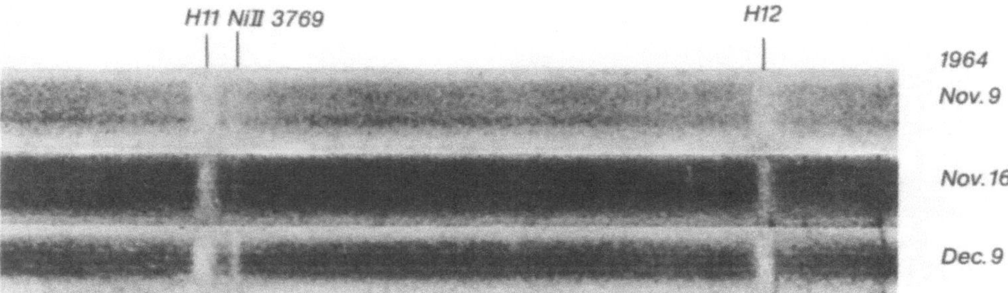

Fig. 112. Abnormal short-term variation of line profiles in the Be star ζ Tau (after UNDERHILL 1966)

troscopically at the time when there is increased photometric activity (GORBATSKIJ 1949, MERRILL 1952). The ejection of a shell by Pleione led to a decrease in the magnitude of the star in years immediately afterwards (SHAROV and LYUTY 1976). STEPINSKI (1980) has theoretically estimated the effect of changes in the dimensions of the gaseous disk in Be stars. In particular the inner ring radius was assumed to vary. As it becomes greater there is first of all an increase in brightness of the system, due to the alteration in the degree of visibility of the disk, and later, with a decrease in the surface area of the ring, a decrease in brightness. STEPINSKI does not discuss the physical processes that are the cause of the changes.

SHAROV and LYUTY (loc. cit.) produced a light-curve of the long-term behaviour of Pleione (Fig. 113), which resembles that of the strange star *XX Oph*, although with a smaller amplitude. The variation found in XX Oph, which has a Beqp spectrum, just cannot be classified anywhere at present. The normal magnitude is bright, and more or less constant, and the variations consist of fades that may be as deep as 1 mag. They occur at irregular intervals, have irregular shapes and may last a year or more. The light-curve in Fig. 114, covering the period from 1890–1940, is taken from a discussion by PRAGER (1940). A little later GAPOSCHKIN (1946) noted that a 15-year standstill followed the 1931 minimum, the next minimum only occurring in 1946. The light-curve for the subsequent years also appears to be of a rather different nature (BEYER 1977). Part of the spectrum's peculiarity (p) is

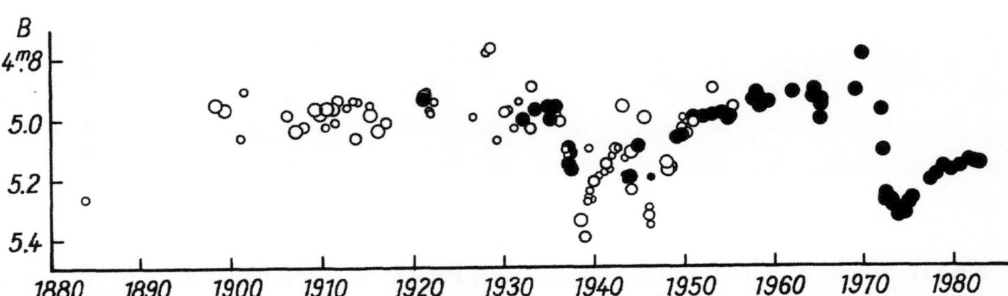

Fig. 113. Light-curve of the γ Cassiopeiae star BU Tau, compiled from various sources. The size of the circles corresponds to their weighting; filled circles represent photoelectric observations (SHAROV and LYUTY 1976, with additions from HOPP et al. 1982)

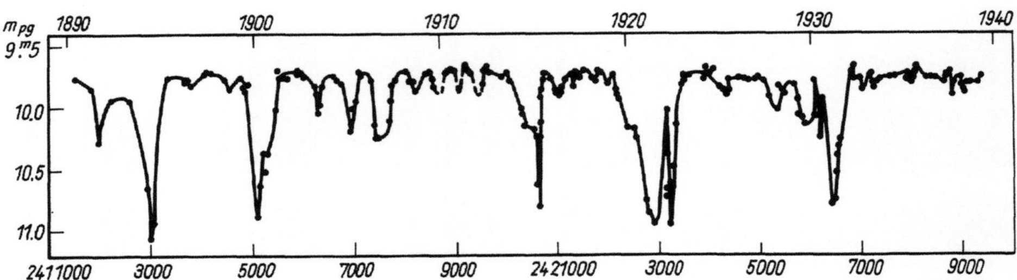

Fig. 114. Photographic light-curve of XX Oph (after PRAGER 1940)

because emission lines (e) of iron occur with a purity that is very difficult to obtain in a laboratory. Because of this, XX Oph has been called "the iron star". The star is sustaining a heavy mass-loss (q) and for this reason it has occasionally been classed with the P Cygni (or S Doradus) stars. However, signs of an extremely high luminosity are unknown. (Could the object be analogous to Z CMa?)

Recently the question of a possible binary nature for the Be stars has been raised, especially by astronomers in Czechoslovakia. They explain the observed properties by evolutionary mass-exchange between the components (e.g. HARMANEC and KŘIŽ 1976; and HARMANEC 1982). In this context we can also point to the observation of hard X-rays from 12 Be stars; this radiation can be explained (RAPPAPORT and VAN DEN HEUVEL 1982) by the presence of a compact companion (a white dwarf or neutron star), like those in other X-ray binaries (Sect. 3.1.7). Direct confirmation for this hypothesis might have been expected from work on the bright variable γ Cas, which has been well-observed. None has yet been forthcoming.

Many shell stars do not vary.

γ Cas illuminates a small reflection nebula. This could well be the case in other objects of this type and increases the possibility of confusion with extremely young Be/Ae variables (Sect. 3.3.2), as HERBIG (1960) has already pointed out.

3.4.3 Variability in Planetary Nebulae or Their Central Stars

Planetary nebulae are luminous gaseous nebulae, generally with a regular shape, frequently spherical, and a diameter of a few tens or hundreds of astronomical units. In a telescope many appear similar to planetary disks.

At present there is hardly anything that can be said about the form of their variability that is generally valid. Very few well-established cases are known, although in examining lists of planetary nebulae that also have variable star designations one gains the opposite impression at first. For example, such a list by ACKER and MARCOUT (1977) contains 26 variables that are given as definite, and 6 possible cases. A closer examination shows, however, that there are a number of false identifications, and that quite a few variables with abnormal spectra (Me-type Mira stars, Z Andromedae stars, etc.) have been incorrectly reported as planetary nebulae. BOND (1976) independently came to the same conclusion. An example of a planetary nebula, incorrectly-classified on the basis of its spectrum, is V 976 Aql, which GESSNER (1982 a) has recently determined to be a normal Mira star with $P \approx 1$ year.

According to the information in the GCVS and its 3 Supplements (KUKARKIN et al. 1969, 1971, 1974, 1976), the following cases appear to be well-established:

AE Ara
V 1016 Cyg
V 1329 Cyg
FG Sge
V 2416 Sgr

HM Sge can now probably be added as CIATTI et al. (1978), KWOK and PURTON (1979) and BALAZS (1980) all see this as a young planetary nebula that is just being formed. The variability was discovered by DOKUCHAEVA (1976) and has been examined in greater detail by WENZEL (1976) and others. In 1975–1976 the object brightened by 6 magnitudes within a few hundred days, and then began a gradual decline. The light-curve around maximum (Fig. 115) chiefly resembled that of an extremely young FU Orionis-type star (Sect. 3.3.2) and it was only the emission spectrum that showed the relationship to the planetary nebulae. KWOK and PURTON seek to explain the sudden brightening by a shock-wave. This occurs if the hot core of the red giant that is believed to be the precursor is exposed by gradual mass-loss. The mass that is lost forms a gas and dust shell, in which the shock-front causes ionization and an increase in brightness. A similar outburst was seen in V 1016 Cyg in 1963–1964.

An unusual case, which resembles many of the extremely young variables, is the central star in the bipolar planetary nebula NGC 2346. Since December 1981 this has shown cyclic variations with a period of about 16 days and an amplitude of more than 2 mag (KOHOUTEK 1982, MARINO and WILLIAMS 1983). Previously it showed no significant variability, as shown by Harvard and Sonneberg Observatory photographs as far back as 1899 (SCHAEFER 1983

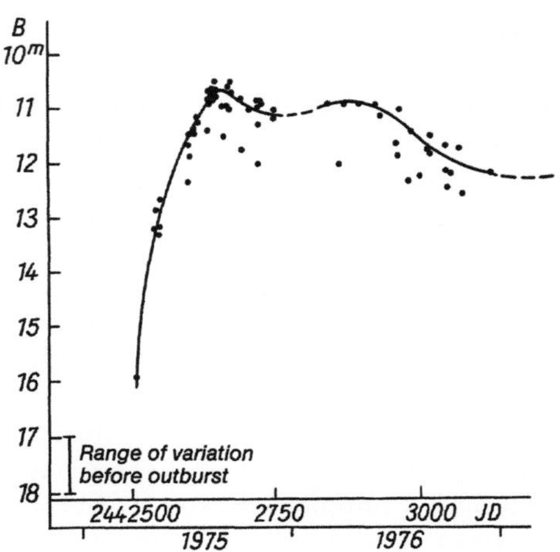

Fig. 115. Outburst of HM Sge in the blue spectral region (after CIATTI et al. 1978). The scatter is predominantly due to instrumental effects

and LUTHARDT 1983). MENDEZ et al. (1982) have suggested, as a preliminary hypothesis, that the star is eclipsed by a dust cloud passing in front of it, so further co-operative observations of this interesting object are urgently required if this is to be confirmed.

FG Sge, already mentioned above, is unique. Its variability was discovered in 1943 by HOFFMEISTER, who considered it to be a star. It is actually the central region of a planetary nebula, which was independently announced by HENIZE (1961). As RICHTER (1960) first recognized, this object's peculiarity lies in the fact that from 1890, when photographic observations began, until about 1967, it grew steadily brighter (Fig. 116). The increase amounted to about 0.5 mag every 10 years, so it rose about 4 mag above its initial brightness of 13^m2. Maximum in the *B* region occurred in 1967; in *U* it had already happened in 1962, but in *V* it only came in about 1970. The cause of this unusual photometric behaviour is to be found in major changes in the spectrum. At first the spectrum consisted of a continuum with absorption lines, the first members of the Balmer series of hydrogen in emission, and the

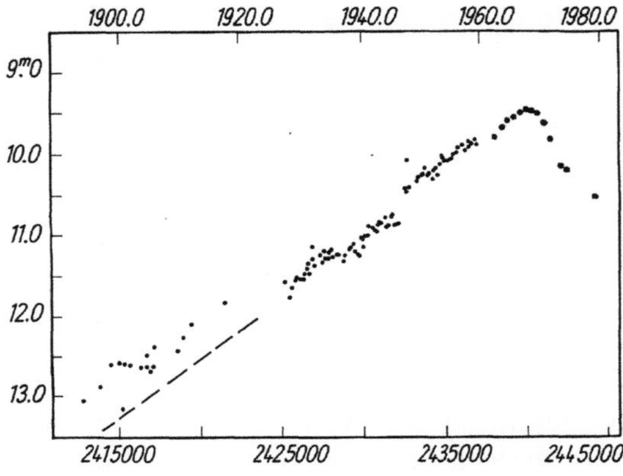

Fig. 116. Secular increase in FG Sge. Individual observations prior to 1920; after 1960 yearly photoelectric (*B*) means. *Broken line* mean slope after reduction to allow for a companion star (after RICHTER 1960 as well as WENZEL and FÜRTIG 1967, with additional Sonneberg measurements)

variable spectrum from a shell that had been ejected by the star. With the continuous change in brightness and in the colour index there was a variation in the spectral type: B4I at 1955.8, A5Ia at 1967.5, F5Ip at 1972.6, G2 at 1975.5 (from a compilation by WHITNEY 1978, where individual references are given). Both the emission features and the indications of expansion disappeared. In 1967 lines of singly-ionized rare earths appeared, which had become abnormally strong by 1972 (showing about 25 times the solar abundance).

WITTMAN (1974) and KRAFT (1974) give clear descriptions of how the "secular" behaviour of FG Sge is a "test case" for stellar evolution. The second author compares the importance of the object to the significance of the Rosetta Stone for the deciphering of hieroglyphics. It is assumed that the object is in a flash phase. Being at an advanced stage of evolution, nuclear energy production occurs no longer in the centre of the star but in a thin spherical shell, where 3 He^4 nuclei are converted into C^{12}, which is fed outwards. At a certain point in its evolution instability arises, leading to a drastic increase in temperature and luminosity, and to the formation of a convective shell. The latter extends out as far as the still hydrogen-rich outer layers. In this way C^{12} is mixed with protons, C^{13} being formed. Interaction with α particles produces O^{16} nuclei, neutrons being released. By means of the s-process ("s" for "slow neutron capture") the neutrons take part in the formation of heavy elements from those of the iron group (Fe, Ni, etc.). For example, rare earths such as Ba, etc. are produced. The theory outlined can account very well for the formation of the planetary nebula in several stages, as well as for the abnormal abundance of elements and the movement on the Hertzsprung-Russell diagram that is linked with the change in luminosity. It is assumed that the object was initially a star of a few solar masses. Other groups of theoreticians, however, describe the object as being completely atypical, e.g. by SCALO (1981) in his review of mixing processes in red giants.

Apart from the strong increase in brightness, FG Sge shows, among other things, secondary waves that amount to a few tenths of a magnitude. This was first noted by WENZEL and FÜRTIG (1967) in extensive photoelectric measurements. Later, more observations by these and other authors showed that the cycle-length of these waves increased considerably from 15^d in 1962 to 108^d in 1979 (JURCSIK and SZABADOS 1979). This can be explained by the expansion of a pulsating atmosphere. More detailed analysis of the light- and velocity-curves shows that the physical behaviour may more closely resemble that found in the Mira stars than that in the classical δ Cephei stars, and that the radius of the pulsating "surface" amounted to about 200 solar radii in 1978 (MAYOR and ACKER 1980; see also WHITNEY 1978).

In these series of photoelectric measurements of FG Sge, additional rapid, low-amplitude changes are found with a time scale of hours and amplitude < 0.1 mag. Similar variations in a number of other central stars in planetary nebulae have been covered by various authors. STOTHERS (1977) gives a summary of results, according to which fluctuations of the order of 0.01 mag and with cycles as low as seconds are supposed to have been detected (ALEKSEEV 1973). STOTHERS (loc. cit.) tries to explain pulsations of that order of magnitude by calculating modes for a Kappa Mechanism (Sect. 2.1.2) that operates in the regions where the CNO-elements are ionized. He thus takes as a basis a star in an advanced evolutionary phase where the outer layers are thrown off (as just described) to form a planetary nebula, leaving behind an object with a greater luminosity and a relatively low mass ($1\mathfrak{M}_\odot$). However, both the observational and theoretical bases are too weak for this phenomenon – which may be regarded as of minor importance – to be discussed any further.

The formation of a planetary nebula does not solely occur in individual stars, but can also take place in binaries. From the work by BOND (1978, 1980) we know two close *eclipsing binaries* that are the central stars in the planetary nebulae Abell 46 and Abell 63. The physical properties of the two systems are strikingly similar. The first object (*V 477 Lyr*) has an orbital period of $0\overset{d}{.}472$; the second (*UU Sge*) has $P = 0\overset{d}{.}465$ and consists of an O-type subdwarf and a K-type dwarf. The systematic investigation of central stars for signs of a binary nature or variability is a very topical field of research, and further interesting results should be obtained in the near future. Many authors have suggested a close relationship with the symbiotic stars (Sect. 3.1.6). However, the fact that in the latter the cool component is a giant, and not a dwarf as found in both the cases just mentioned, complicates the situation still further.

3.5 R Coronae Borealis Stars

Light-Curves and Position on the Hertzsprung-Russell Diagram

This group of stars is very small. The variations in the prototype R CrB are very characteristic of them all, and its luminosity is very high. The star is normally at *maximum light*, but this is interrupted by *deep minima* with very irregular behaviour. There are short fades lasting a few weeks, but also minima that continue for several years. The latter are generally interrupted by secondary brightenings which do not completely reach normal light. A very instructive, slightly schematic, uniform light-curve for R CrB since 1844 (Fig. 117) is given by ZHILYAEV et al. (1978) in a comprehensive discussion of this star. A similar light-curve had already been published by MAYALL (1960) from the intensive visual observations made over several decades by members of the American Association of Variable Star Observers (AAVSO). The *amplitude* of R CrB can be as much as 9 mag, $5\overset{m}{.}8 - 14\overset{m}{.}8$ visual. A decline of 6–7 mag is often completed in 30–35 days, but the rise, particularly in the upper portion, generally takes longer. All well-investigated stars of this type exhibit the spectra of *hydrogen-poor, carbon-rich objects* (Fig. 118). We should therefore accept this property as a characteristic feature. This makes it easier to exclude falsely-identified variables, especially those that with further, more detailed study would have proved to belong to other well-known classes. FEAST (1975) gives 17 as being the number of objects definitely

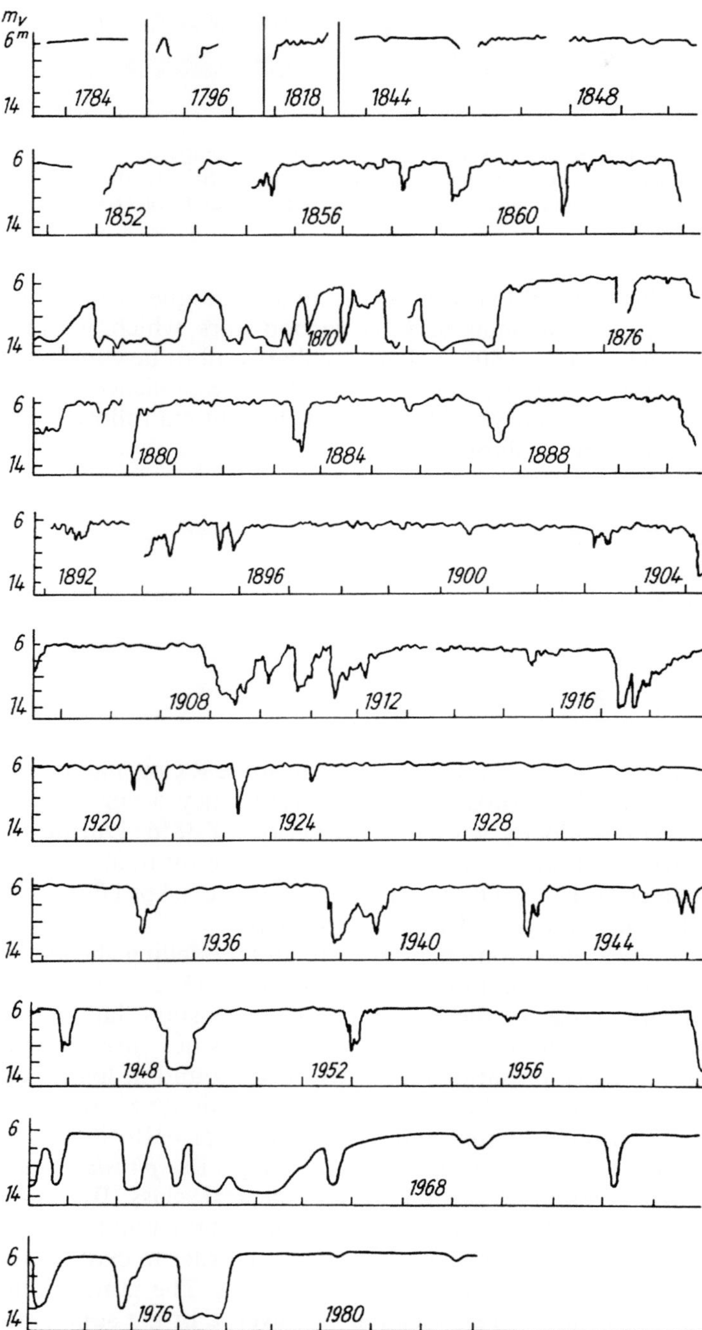

Fig. 117. Visual light-curve of R CrB from 1784 to 1982. The curve was compiled, from the observations given by many authors, by MAYALL (1960) and ZHILYAEV et al. (1978) up to and including 1956, and from 1957 by the present authors

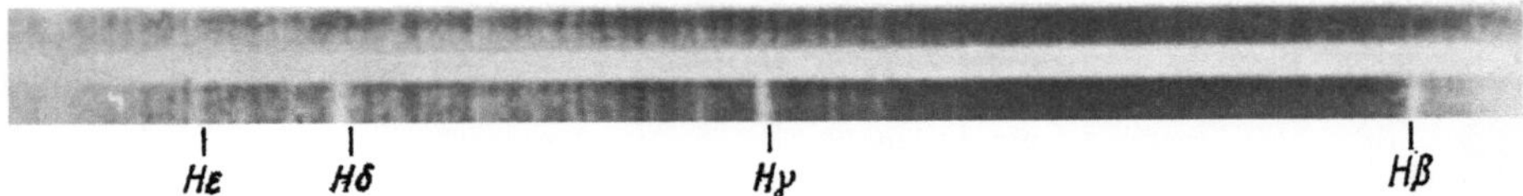

Fig. 118. Spectrogram of the R Coronae Borealis star SU Tau (*above*), compared with a normal F5 supergiant (α Per). Note the almost complete lack of hydrogen lines in SU Tau; the metal lines are of similar strength in both stars. From the MORGAN, KEENAN and KELLMAN spectral atlas

known to belong to this class. BIDELMAN (1979), in a catalogue of luminous hydrogen-poor stars, lists 21 strongly variable carbon stars, which, with the exception of the peculiar object V 605 Aql, are probably all to be considered as R CrB stars (Table 41). Much earlier STERNE (1934) established from a statistical analysis of the light-curve of R CrB that the minima followed one another in a "truly random" fashion.

Table 41. R Coronae Borealis stars

S Aps	WX CrA	SV Sge
U Aql	R CrB	RY Sgr
(V 605 Aql)	V 482 Cyg	VZ Sgr
XX Cam	W Men	GU Sgr
UV Cas	Y Mus	LR Sco
UW Cen	RT Nor	SU Tau
V CrA	RZ Nor	RS Tel

Secondary quasi-periodic variations are found in some R Coronae Borealis stars. The best-observed are those in RY Sgr where they occur in luminosity, colour and radial-velocity with a mean period of $38^{d}6$. During the principal minima these fluctuations in brightness may amount to as much as 1.5 mag (ALEXANDER et al. 1972). Other similar objects are S Aps ($\overline{P} \approx 113^{d}$) and UW Cen ($43^{d}4$) – KILKENNY and FLANAGAN 1983.

This is thought to be a pulsation phenomenon, and modelling shows this to be similar to those in Population II pulsating stars (Sect. 2.1.2). The spectral and luminosity classes of the R Coronae Borealis stars (late F-type supergiants) match this; all well-established luminosities are around $M_v = -4 \pm 1$. Recently weak pulsations have also been found in hydrogen-poor helium stars, which do not show R Coronae Borealis-type variations. V 652 Her has been well-investigated and has a period of $0^{d}1079950$, and an amplitude in the V region of about 0.07 mag. Its *physical data* are: $M_v = -0.3$, $R = 1.6 R_{\odot}$, $T_{\mathrm{eff}} = 25\,500$ K, spectral class B1, mass $\mathfrak{M} = 0.9\,\mathfrak{M}_{\odot}$. HILL et al. (1981) who published these data point out the necessity for further research on other similar objects in order to clarify their evolutionary relationship to the R Coronae Borealis stars. The same applies to objects like the helium star PV Tel, which is, however, a bright supergiant ($M_v = -4.5$). It is evidently weakly variable (≈ 0.05 mag) on a time scale of days or years (WALKER and SCHÖNBERNER 1981, WALKER and KILKENNY 1980).

We also encounter the sort of variation discussed in this section, together with a lack of hydrogen or an over-abundance of carbon (or both), in certain hot stars. Both MV Sgr (a helium star) and V 348 Sgr (possibly related to planetary nebulae) have been well-investigated.

Model and Evolutionary State

LORETA (1934) and O'KEEFE (1939) first proposed the theory that the minima of the R Coronae Borealis variables were caused by a *cloud of* (carbon) *particles darkening the star*. This is still the most plausible explanation, especially since it has been realized that the abnormal spectra are due to a true anomaly in the abundances. The *infrared excesses* found in these variables – observed by STEIN et al., LEE, FEAST, GLASS and others – are an important characteristic. They are the result of thermal radiation from a (structured) *circumstellar dust shell*, which also exists outside the minima.

The spectroscopic changes during the minima are complex and difficult to understand. The alteration of the absorption-line spectrum during the decline into a chromospheric-like emission spectrum with the same lines (HERBIG 1958a, PAYNE-GAPOSCHKIN 1963, ALEXANDER et al. 1972), is very characteristic. This has been explained by FEAST (1975), one of the specialists on these stars, by the ejection of a cloud of particles in the direction of the observer, in such a manner that a significant portion of the photosphere is hidden, just as in a solar eclipse. KRELOWSKI (1975), on the other hand, describes a model with a spherically-symmetrical shell of gas ejected by the star, in which dust condenses.

Despite there being many starting points for theoretical models, the evolutionary stage of the R Coronae Borealis variables is still not definitely established. Modelling by TRIMBLE (1972), who considered high-luminosity, low-mass ($1-2\,\mathfrak{M}_\odot$) pulsating helium stars, indeed led to the idea that RY Sgr (for example) might be a member of the old disk population in the Galaxy, related to the W Virginis stars, and at an advanced stage of evolution. Instabilities were also found, which could be the cause of the mass ejection. Nevertheless, some doubts remain, especially as BIERMANN and KIPPENHAHN (1971) found in their models that the pulsations more closely resembled those of the Mira stars. The observed systematic period changes, first noted by PUGACH (1977), lie between $-0\overset{d}{.}19$ (S Aps) and $+0\overset{d}{.}003$ per cycle (UW Cen) – KILKENNY and FLANAGAN (1982). The latter (see also KILKENNY 1982) tried to explain these by considering SCHÖNBERNER's evolutionary models (e.g. 1977), that is by highly evolved stars with compact C/O cores and He exteriors. However, the situation is unsatisfactory in just the same way as with the progressive period-changes in the RR Lyrae variables (Sect. 2.1.3). Moreover, SCHÖNBERNER's models are also disputed. A brief list of work carried out on "extreme helium stars" in the last two decades is given by LYNAS-GRAY (1981).

Many authors also refer to the connections between these stars and the planetary nebulae, the Wolf-Rayet stars and the novae.

3.6 γ-Ray Bursters

The search for optical counterparts of the γ-ray bursters has been the source
of so much discussion in astronomical literature in recent years that the
matter must be briefly mentioned here. Cosmic γ-ray bursts at wavelengths
between 10^{-3} and 10^{-2} nm (corresponding to a photon energy of about
$0.1-1$ MeV) were first discovered in 1967 by the Vela satellite detectors, and
have since been observed by an increasing number of other spacecraft, the
Prognoz and Venera series, for example. As the detectors are not directional,
the position of a source has to be determined from the time differences
between observations made by various satellites. If possible these should be
widely separated and in suitable orbits. If only two spacecraft show a coincidence, then the time difference (i.e. the difference between the signal's travel
times to the two detectors) defines a circle on the celestial sphere, on which
the source must lie. If more than two satellites are involved the required
position is further refined by intersections of further circles. "Error boxes"
are thus obtained that may be examined for *optical counterparts*. This procedure only works because the bursts – or at least certain characteristic
features – are *exceptionally short*, lasting between 0.1 s and a few seconds.

KLEBESADEL et al. (1982) have published a catalogue of 111 gamma-ray
bursts recorded up to 1979 June 13. The work involved 14 scientists from four
important space-research institutes: Los Alamos National Laboratories
(U.S.A.), Goddard Space Flight Center (U.S.A.), Centre d'Etude Spatiale
des Rayonnements (France) and Institut Kosmicheskikh Issledovanij
(U.S.S.R.). The number of newly announced bursts continues to increase.

As yet the search for optical (i.e. photographically recorded) counterparts
has had little success, despite being carried out systematically in the plate
archives at Harvard, Ondřejov, and Sonneberg Observatories and in Institutes in the Soviet Union, among others (albeit with varying persistence). An
identification of the 1979 March 5 burst with the *N 49 supernova remnant* in
the Large Magellanic Cloud (e.g CLINE et al. 1982) has not gone unchallenged
(see PEDERSEN et al. 1983). There has been a lot of discussion about the
star-like image discovered on a Harvard plate by SCHAEFER (1981), which lies
in the error box of the 1978 November 19 burst. The plate is the fourth of
six identical 45-minute exposures made on 1928 November 17. If the "optical" outburst had a duration of only 1 second, like the γ-ray bursts – which
is purely hypothetical – then the object that SCHAEFER discusses must have
had a momentary brightness of about $m_{pg} \approx 3^m$. On high-resolution CCD
images (Sect. 8.1.2) there are at least three stars of magnitude $\approx 24^m$ at this
position, all of which are suspected of being possible sources of the variations
– for example, by PEDERSEN et al. (1983) and SCHAEFER and BRADT (1982).
Workers experienced in routinely interpreting photographic plates warn that
star-like plate faults can occur, so the reality of SCHAEFER's outburst and thus
the amplitude of ≈ 21 mag are not free from suspicion. It should also be
noted that there are four quasi-stellar radio sources and one X-ray source
either inside or very close to the error box.

We shall not discuss the theoretical – and, in part, highly speculative – attempts that have been made to explain the gamma-ray bursts.

3.7 Other Types of Object

We include under this heading types of variation that can be said to have some relationship to one another, but which have very different astrophysical and evolutionary characteristics. They are stars whose variability – at least according to reasonably well-established hypotheses – are probably caused by restricted centres of activity ("spots"), although these are almost certainly of very different origins.

3.7.1 BY Draconis Stars and Similar Objects

BY Draconis Stars

Stars of this class were first officially differentiated from other groups of variables (e.g. the flare stars) at the beginning of the seventies when KUKARKIN et al. (1971) defined them as follows: "Emission stars of late spectral classes showing periodic light variations with variable amplitudes (from 0.3–0.5 to 0.0 mag). The form of the light-curve varies. Periods are usually from a fraction of a day to a few days." Typical light-curves are given in Fig. 119.

For a time BY Draconis itself, a K4eV star (magnitude $8^{m}.3$ visual), was thought to be a flare star until the intensive photoelectric observations by CHUGAINOV (1966 etc.) demonstrated the quasi-periodic and continuous nature of the changes in magnitude. KRON (1952) had already found that in the eclipsing system YY Gem (which consists of two M1eV stars) a sine-wave variation was superimposed on the geometrical changes. He ascribed this to an uneven brightness distribution on the surface of the rotating star. *Spot models* of this sort still offer the most plausible explanation of these stars' photometric behaviour.

A whole series of models with a wide range of spot temperatures and areas have been calculated by TORRES and FERRAZ-MELLO (1973) for example, as well as by FRIEDEMANN and GÜRTLER (1975). The position of the spot on the surface of the star and the orientation of the rotational axis relative to the observer have to be taken into account. The first pair of authors found that a typical light-curve was satisfactorily explained by a spot covering 5–20% of the visible hemisphere and 500–1500 K cooler than the photosphere. OSKANYAN et al. (1977) have published a detailed analysis of the variability of BY Dra itself. They also consider bright spots and come to the conclusion that the active regions primarily exist on the star's rigidly-rotating polar cap. HARTMANN and ROSNER (1979) discuss in detail the difference between the amount of energy radiated from cool spots and from an undisturbed portion of the photosphere that is of equal size and has an even surface brightness.

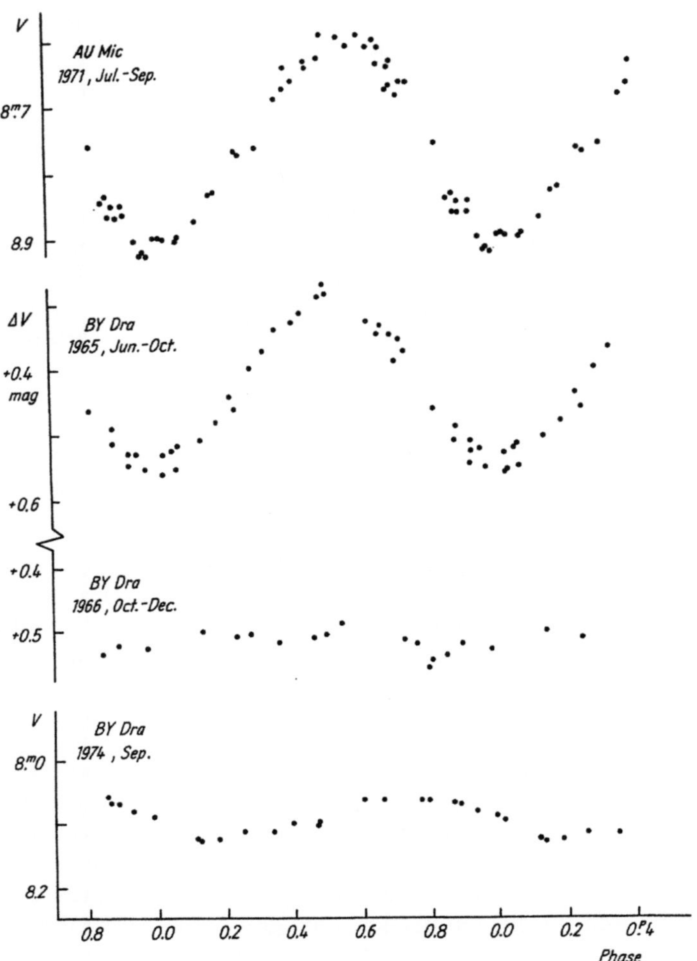

Fig. 119. Light-curves of AU Mic (Torres and Ferraz-Mello 1973) and BY Dra (Chugainov 1973 and 1976); the periods taken as a basis were $4^\mathrm{d}\!8657$ for AU Mic and $3^\mathrm{d}\!836$ for BY Dra (after Rodono 1980)

However, as indicated in the definition given above, the photometric behaviour of these stars is very complex. Chugainov (1973) found that the amplitude of BY Dra in *B* decreased from 0.45 to 0.05 mag between 1965 and 1972, while the period altered from $3^\mathrm{d}\!84$ to $3^\mathrm{d}\!79$. A similar long-term variation occurred in the emission lines in YY Gem, which were strong in 1920–1926, but not conspicuous in 1950 (Struve et al. 1950). Analysis indicates that the star's tendency to produce spots varies on a time-scale of years or decades, whereas the lifetime of the active regions is only of the order of months. The *analogy with solar activity* is very obvious, especially if differential rotation like that of the Sun is present which, through a shift in the position of the spot towards the pole or the equator, might explain the change in primary periods observed in BY Dra and other stars. Astrophysicists who see similarities between the phenomena shown by flare stars and solar activity find no basis for differentiating between BY Draconis stars and UV Ceti stars, especially as occasional flares can occur in the former type (GERSHBERG and SHAKHOVSKAYA 1974).

RODONO (1980) has published a review on the subject of solar-type "stellar activity". He attempts to bring together the phenomena found in BY Draconis stars, flare stars, and binary stars of the RS Canum Venaticorum type, as well as the chromospheric and coronal activity that certain stars exhibit in the far ultraviolet and in X-ray emission. There is an apparent similarity between BY Draconis stars and the post-T-Tauri stars (end of Sect. 3.3.2), and future observations should show whether this is real.

RS Canum Venaticorum Stars

The RS Canum Venaticorum class (RSC stars) was defined in 1975 by HALL, who has published a number of papers on the subject (for example HALL 1972, 1976; HALL et al. 1979; EATON and HALL 1979). There is no doubt that these are *binary systems*, where the cool component is a late G or early K subgiant, and the hotter component an F or G star of luminosity class IV or V. These characteristics differentiate the objects from the BY Draconis variables. As the generally accepted model for the RSC stars is also based upon the presence of spots on the cool component, the term "BY Draconis syndrome" is also occasionally applied to these stars. *Superimposed* on the eclipsing light-curve there is a *wave* that has an amplitude of up to 0.2 mag. The characteristic feature is that this wave exhibits a shift, normally backwards with respect to the eclipse curve, i.e. towards smaller phases (Fig. 120). If the overall rotation is assumed to be synchronous (orbital period = mean rotational period, as with the Moon in the Earth-Moon system), presumably the spots predominate in a surface zone that is rotating faster than the average. Such differential rotation occurs in the Sun's equatorial region. As the effect has only been known for a short time, the cycle-length of the shift – between the times when the wave's maximum occurs at the same phase of the eclipse curve – has only been approximately determined. In RSCVn it amounts to about 10 years. As a *beat period* P_b is produced between the *orbital period* P_0 and the *actual period* of the wave P_1, the equation derived for the double-period δ Cephei stars applies:

$$\frac{1}{P_1} = \frac{1}{P_b} + \frac{1}{P_0}.$$

For RSCVn ($P_0 = 4\overset{d}{.}8$) one obtains $P_1/P_0 = 99.87\%$, which is a much smaller deviation of the equatorial rotation rate from the mean than that found in the Sun. The effect of the travelling-waves makes itself apparent as increased scatter in mean light-curves compiled from observations made at various times.

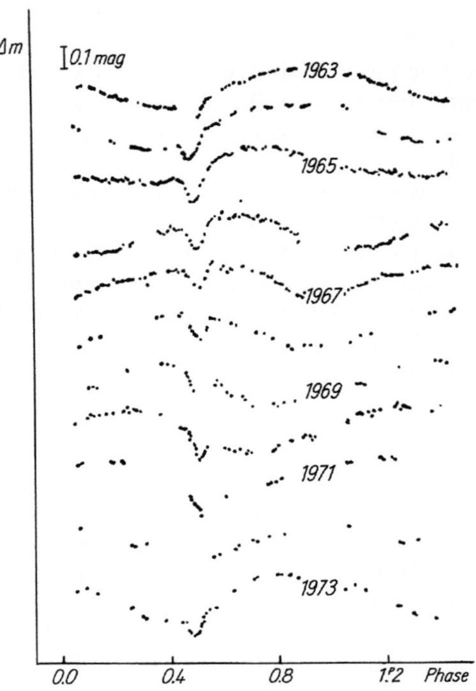

Fig. 120. Light-curves of RSCVn for the years 1963 to 1973, showing the subsidiary wave, which shifts to smaller phases; points at primary minimum (phase 1.0) have been omitted (after CATALANO and RODONO 1967)

Apart from the prototype, well-known objects in this class are AR Lac (whose eclipsing variability was first discovered by LEAVITT in 1907 on Harvard plates), RT Lac, SS Boo and RW UMa. HALL (1976) included 24 variables as members, but the total seems to be rising rapidly.

There are a number of further indications of activity in the RSC stars: *flare activity* similar to that of the flare stars (Sect. 3.3.3) (see PATKÓS 1981 for example), strong Ca II *emission lines*, high *UV-excess*, non-thermal *radio frequency outbursts*, and variable *X-ray radiation*. The latter can even be used to detect the numerous RSC systems that do not show eclipsing light-curves, owing to the high inclination of their orbital planes, and which are difficult to discover in the optical region because of their low amplitude (which is now identical with the amplitude of the wave). Examples are UX Ari, V 711 Tau and σ CrB. (See also the penultimate paragraph of Sect. 3.1.7.)

Magnetic field effects play a part in the models, together with active regions, *chromosphere* and *corona* – just as in the Sun – (e.g. ROSNER et al. 1978). The presence of rapid rotation and significant outer convection zones supports this view.

It is not yet possible to explain unambiguously either the origin and the evolutionary state of the RSC stars, or the physical processes that cause mass motions and outbursts of radiation that are enormous when compared with those on the Sun. This is because it has not yet been possible to obtain photoelectric and spectroscopic observations (as well as those from space) to a sufficient extent in the few years since the phenomenon was discovered.

Even the important conclusion that a particular range of longitudes in the star's equatorial zone is subject to spots for years or decades, while the other hemisphere is essentially free of them, appears to be insufficiently explained. Comprehensive references can be found in the detailed papers by RODONO (1981) and RÖSSIGER (1982).

3.7.2 Pulsars

The name "pulsar" derives from the term "pulsing (radio) star", originally used for these objects, and is retained for convenience. We now know that the observed cyclic variability in the radio emission from these objects is not caused by pulsation in the stars, but rather by their *rapid rotation*. Beginners should not confuse pulsars with the pulsating variables described in Chap. 2.

When MANCHESTER and TAYLOR (1981) drew up their list, containing observed and derived properties of all known pulsars, 330 objects were included. This would not have been sufficient to justify their inclusion in a book on variable stars, if our earlier definition of "variability" is observed. However, at present at least two radio pulsars are also active in the photographic and visual regions, and can thus be observed "optically" as pulsars. These are *CM Tau* and *HU Vel*. *Pulsar periods* are exceptionally short and fall, so far as we yet know, between 0.00156 s (4C 21.53) and a few seconds. Very short periods are found in CM Tau (0.033 s) and HU Vel (0.089 s). Between these two lies the binary radio-pulsar PSR 1913 + 16 (0.059 s) but, as with several other objects, all attempts to observe it at optical wavelengths have given negative results (e.g. NATHER et al. 1977). There are theoretical models that predict that the optical luminosity is proportional to P^{-10} (e.g. GINZBURG and ZHELEZNYAKOV 1975). This would explain why only the shortest-period pulsars are visible in the optical region; the time-averaged apparent magnitude of PSR 1913 + 16 would be fainter than 26^m, below the limit of detection. But 4C 21.53 may be of a somewhat different nature.

Close binary stars are known where one component is a radio or X-ray pulsar (Sect. 3.1.7). In the optical region, the presence of a pulsar in such systems can generally only be determined indirectly, by the effect of its X-radiation on the system's second component, for example.

Because of the extremely short periods, discovery and measurement of pulsar activity in the optical region is naturally only possible with special equipment. The photoelectric integration must be synchronized with the periods known from radio-frequency observations. This was the method used to discover the optical variability of CM Tau early in 1969 (COCKE et al. 1969; NATHER et al. 1969; LYNDS et al. 1969 – among others) and of HU Vel early in 1977 (WALLACE et al. 1977).

The Taurus pulsar is the south-preceding (= southwest) component of Baade's Star, the well-known binary object in the centre of the Crab Nebula. The maximum visual magnitude of the pulses, estimated by the discoverers, is 15^m and the time-averaged magnitude is 18^m. There is also a subpulse with a total energy 55% of that of the main pulse. The half-intensity width of the

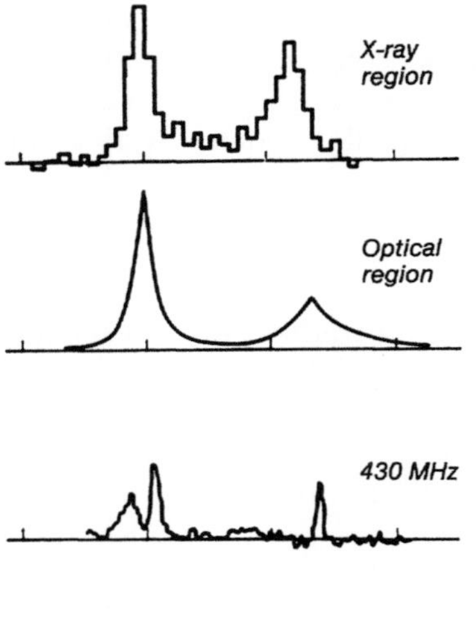

Fig. 121. The shape of pulses in the Crab pulsar CM Tau at optical, X-ray, and various radio wavelengths (after DAUTCOURT 1976)

main pulse is 1.8 ms and that of the subpulse is 3.1 ms (Fig. 121). Baade's object and the Crab Nebula are known to be the remnants of the 1054 supernova, and the former has had the variable star designation CM Tau for a long time. The fact that it is a pulsar was discovered at the end of 1968 as the result of a specific search – the first radio pulsar, PSR 1919 + 21, having been found accidentally early in 1968. The Crab pulsar also radiates pulsed soft and hard X-rays, as well as γ-rays – in other words it emits photons over practically the whole electromagnetic spectrum.

HU Vel and the radio and X-ray pulsar PSR 1509 − 58 are also in the centres of nebulae that are probably supernova remnants.

It is now thought that, at least in certain cases, supernova explosions may leave behind rapidly rotating neutron stars (Sect. 3.2). The reduction of the stellar radius to about a millionth of its previous size raises the magnetic field strength from the previous weak value of ≈ 1 Oersted to $\approx 10^{12}$, while conservation of angular momentum increases the rotational rate, so that the period becomes a fraction of a second. For physical reasons – the discussion of which would lead us too far – a stream of charged particles is ejected at almost the speed of light along the magnetic axis, which does not coincide with the rotational axis. (The star is an oblique rotator.) By means of the synchrotron effect this stream of particles produces electromagnetic radiation primarily in the forward direction. It is essentially this "searchlight beam" that gives rise to the pulsar phenomenon, as it rotates with the star and intermittently encounters the Earth, giving a lighthouse effect. The effect is not visible on the Earth if the line of sight does not periodically coincide with the axis of the beam – which itself sweeps out a cone in space. This probably explains why no pulsars are observed in many gaseous supernova remnants.

The phenomenon is physically much more complicated than outlined here. There are also still a number of unanswered questions. A comprehensive description, generally understandable by anyone, has been given by MANCHESTER and TAYLOR (1977) and by F. G. SMITH (1977).

3.7.3 α^2 Canum Venaticorum (Magnetic) Stars

Because of its length, the name of the prototype, α^2 CVn is only rarely used to describe these "magnetic stars in a restricted sense". Frequently one simply speaks of "magnetic variables". This term has recently been carelessly used for the AM-Herculis objects (polars), which also have strong magnetic fields (Sect. 3.1.4), despite it having been already applied to the α^2 CVn stars. The reader is expressly warned about this source of confusion.

The variability of the α^2 Canum Venaticorum stars is very small. It is generally less than 0.1 mag and can thus only be established by photoelectric measurements. The members of this class may be recognized by the following characteristics, among others:

1) Very strong, extensive magnetic fields with typical field strengths of $10^3 - 10^4$ Oersted (from the Zeeman splitting of the lines). In the literature the incorrect unit "Gauss" (magnetic flux density – cgs) is frequently encountered. In order to ensure that the numerical values are those given elsewhere in astronomical literature, we use the term "Oersted"(magnetic field strength – cgs), even though it is no more legitimate. The Earth's magnetic field ≈ 0.5 Oe; sunspots $\approx 10^3$ Oe.
2) Exceptionally strong spectral lines of certain elements, with excess abundances for iron-group elements (by a factor of $10-100$), for Sr, Y and Zr (by a factor of 1000) and for the rare earths (by a factor of $300 - > 1000$). This has led to the use of the term "peculiar A star" (Ap stars).
3) Variability of the magnetic fields, spectra and luminosity. Typical periods are $5-9$ days, but shorter an much longer periods are also found; a small group has periods of a few years. This has given rise to the term "spectrum variable", which is not recommended.

Recently, a subgroup has been distinguished, the "SX Arietis stars", where the characteristic feature is a higher temperature (spectra B0p–B7p), and for which the rather inappropriate name "helium variables" is sometimes used.

The Am stars ("metal line stars") are occasionally grouped with the Ap stars. The two groups lie close together on the H-R diagram, slightly above the main sequence. According to recent research, the Am stars merely appear to be a less extreme form of the Ap stars, as regards element abundances. In

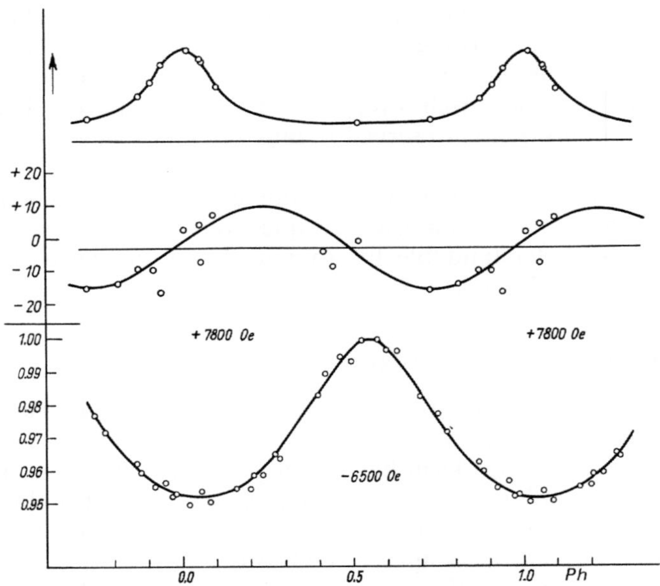

Fig. 122. Periodic features in the magnetic star CS Vir (after STIBBS). *Top*: intensity of the Eu II line; *centre*: radial velocity (km/s); *below*: light intensity. Three magnetic field strengths in Oersted (Oe) are also included

addition, they generally show no detectable magnetic field, and no luminosity or spectral variations.

The variations in the observed properties, mentioned under 3, mostly take place in synchronism (Fig. 122), which immediately shows that all these anomalies have some physical connection. The existence of subsidiary, very short-period and very low-amplitude irregularities (of the order of 0.001 mag) cannot be ruled out, particularly as photometric and spectroscopic data are so difficult to obtain, except for the brighter objects. WEISS et al. (1976) give a detailed description of many theoretical and practical aspects. In the same publication SCHÖNEICH and STAUDE also interpret their 10-colour photometry in terms of a spot model, where the relationship between temperature and depth differs from that in an undisturbed atmosphere, owing to the effects of the magnetic field. The highest layers of the spot are more than 2000 K cooler than the surroundings. The variation in light and in the measured magnetic field strengths are produced by the rotation of the star.

We will not discuss the controversial question of the origin of the magnetic field, nor its structure and location relative to the rotational axis. Reference should be made to the specialist literature on these points.

The Ap star V 816 Cen (PRZYBILSKI's star) appears to be a special case, where a field of −2200 Oersted and an F0 spectrum were found. The latter alone would suggest that it should be considered a δ Scuti star (Sect. 2.1.4), were it not for the fact that the rare earth abundances are several orders of magnitude greater than normal. The star shows luminosity changes with a period of 12.141 minutes and the particularly low amplitude of 0.006 mag in B. There do not appear to be long-term luminosity, spectral or radial velocity variations. At present this variable cannot be classified anywhere, but this may well be only because similar cases are exceptionally difficult to discover.

4. Eclipsing Stars

4.1 General

Estimates show that about a quarter or even as many as half of all stars in the Galaxy are binaries. When these have an appropriate orbital orientation eclipsing light-curves may be visible. In known eclipsing stars the two components are generally fairly close as in wider pairs the probability that eclipses will be seen from Earth is very low (Fig. 123).

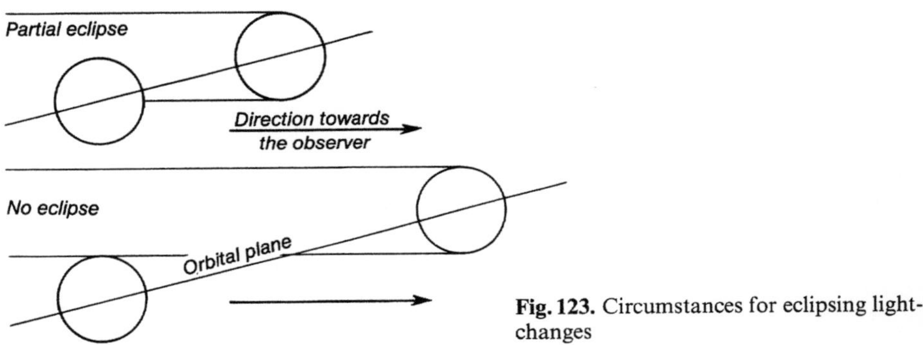

Fig. 123. Circumstances for eclipsing light-changes

The eclipsing stars have a somewhat unusual situation among variable stars in that many authors do not regard them as being true variable stars and class them as "optical variables" as distinct from stars that vary physically. They will be given full consideration here, however, as current knowledge shows that an exclusively eclipsing variation is very rare.

In most close binaries physical variability is induced by mutual interaction between the components. This is particularly marked in the eruptive binaries, where the eclipsing variation (when it is present) is often hardly detectable beneath the strong changes in luminosity caused by physical processes. Moreover, in many eruptive binaries it is the eclipse of a disk of gas or dust that is observed, rather than an eclipse of one of the stellar components.

Neither do most of the "classical" eclipsing stars show purely geometrical luminosity variations. Even in the "detached" systems the gravitational and magnetic fields and the electromagnetic radiation from the two components affect one another to a great extent. Tidal deformation, reflection effects, the formation of a general dust and gas shell, exceptionally strong photospheric and chromospheric activity, as well as ejection of mass may all take place. These processes, which can be partly followed spectroscopically, produce periodic or non-periodic variations in the eclipsing light-curve, luminosity outbursts, and period changes.

As a result it is therefore not possible to draw a clear dividing line between the non-eruptive doubles and the eruptive binaries that have been discussed elsewhere.

We shall define "eclipsing stars" as being all objects where the *geometrically produced variations* dominate those with *physical origins*.

Finally, it should be mentioned that the observational methods used for the eclipsing stars are identical to those employed for physical variables.

4.2 Geometrical Properties

Eclipsing variables are, as a rule, *spectroscopic binaries*, where both components are visible in the spectrum, and the orbital motion can be recognized by the Doppler effect on the spectral lines. Naturally eclipses will not be observed in all such systems, but only when the observer's line of sight is not too strongly inclined to the orbital plane. Among systems with the same inclination and otherwise similar dimensions, those with the smaller separation will be more likely to show eclipses, as Fig. 123 will show. Here, the inclination of the line of sight to the orbital plane is assumed to be 15° and for simplicity the components are taken as being the same size. In the upper example a partial eclipse takes place, but in the lower there is only a graze. If an eclipse is to be observed, then the angle between the line of sight and the orbital plane must decrease as the separation increases, other things being equal. This fact has a great effect upon statistics. Systems with long periods are only rarely recognized, whereas objects with rapid variation, in particular the type known as contact systems, are especially favoured.

The relative size of the components has as great an effect on the observed phenomena. Fig. 124 illustrates the various limiting cases. Here the separation and mass are taken as being constant. The shape (known technically as a lemniscate) represents a cross-section of the equipotential surface through

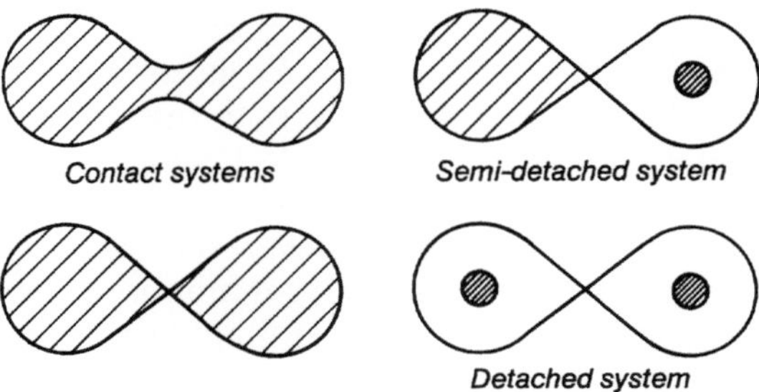

Fig. 124. Principal types of spectroscopic binaries

the centres of both stars and at right-angles to the orbital plane. The equipotential surface should be regarded as a stability boundary (= Roche boundary). There are stars that completely fill the volumes within these boundaries. The figure illustrates a *contact system* with two such stars, which often have a common envelope and exchange mass through gas streaming. They are in contact and are tidally deformed. In the second type (*semi-detached system*), one of the stars is smaller, and its surface lies well inside the boundary surface. In the third type, both members are *detached*. These three somewhat schematically described cases are all actually found to exist.

The geometrical properties have been extensively investigated by KOPAL (1978, 1979).

4.3 Classification

The way in which the light changes occur under these different circumstances may be easily described. If we consider a contact system with components of approximately equal size and brightness, then in a single 360° rotation two maxima and two minima occur. As stars A and B are approximately equal, it makes little difference whether A is in front of, and eclipses, B, or vice versa. The two minima m_1 and m_2 will have approximately the same depth, and the curves will be very similar. In addition the light-curve will not show any constant light. This property is typical of contact systems. Variables of this type are known as *W Ursae Majoris* stars (Fig. 125), *EW* in the GCVS. Detailed investigation shows that both components lie close to the main sequence, are of almost the same brightness, but have somewhat different masses (the mass ratio is about 2 : 1 on average). The orbital period is < 1 day. Most W Ursae Majoris stars are only approximately contact systems. The physical and evolutionary properties of the W Ursae Majoris stars have not yet been the subject of sufficient research. Models for these systems have been discussed in a review by SAHADE and WOOD (1978, p. 34).

A contact system with members having unequal surface brightness also has a light-curve without periods of constant light; however, the minima will be of different depth. The deeper minimum corresponds to the eclipse of the brighter star by the fainter. Such pairs are the *β Lyrae stars* (Fig. 125), *EB* in the GCVS.

Detached and semi-detached pairs, whatever the type of components, can be recognized by the light-curve having a horizontal portion – a bright "normal light". These are the very numerous *Algol* stars, *EA* in the GCVS.

If the components have identical brightness and size, then the primary and secondary minima will have the same depth, as in a W Ursae Majoris star. When classifying a newly-discovered star of this type one runs the risk of taking the half-period for the true one. As a rule, however, one star is of lower surface brightness than the other, and the secondary minimum, when the fainter star is eclipsed by the brighter, is much shallower than the primary minimum. As an example, Algol itself may be taken, where the secondary

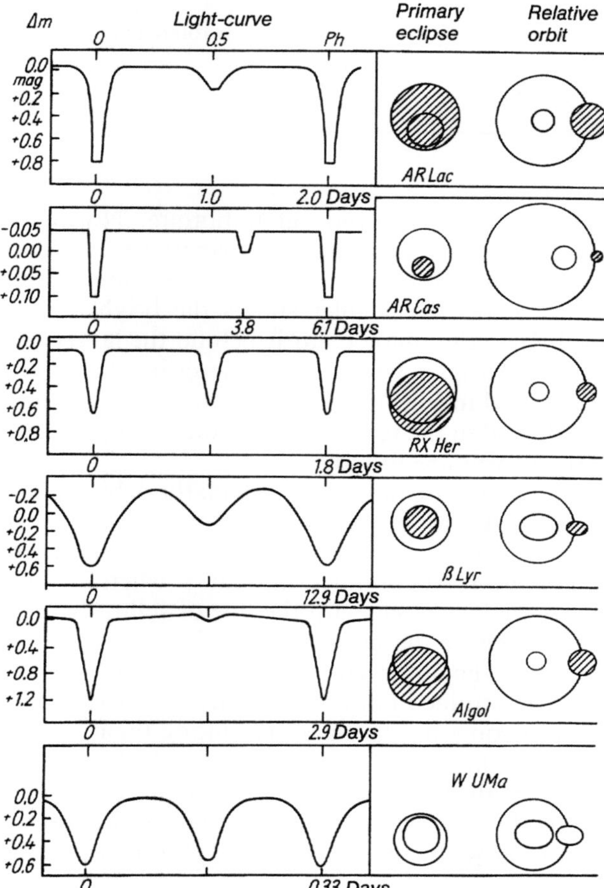

Fig. 125. Various types of eclipse light-curves. The right-hand side shows the configuration at the time of the eclipse of the brighter star by the fainter, and the corresponding orbit at a reduced scale. After STRUVE (1962), with additions

minimum only has a depth of about 0.1 mag. In many systems it is essentially missing, as if the companion did not contribute to the system's overall magnitude.

The shape of the light-curve in Algol stars differs considerably from one object to another. It is determined by the relative size and surface brightness of the stellar disks, the ratios between the orbital radius and the stellar radii, and the position of the orbital plane relative to the line of sight. Moreover, it is affected by whether or not the eclipse is central. With a central eclipse of two similar stars the minimum is sharp, and the intensity is reduced to a half of normal light, which corresponds to a magnitude difference of 0.75 mag. If the eclipse is not central then the amplitude is reduced and minimum light is no longer sharp. In both cases two similar minima m_1 and m_2 occur, generally separated by half the orbital period. When one component is considerably smaller and fainter than the other, but has the same brightness per unit surface area, we have two similar minima, both of correspondingly

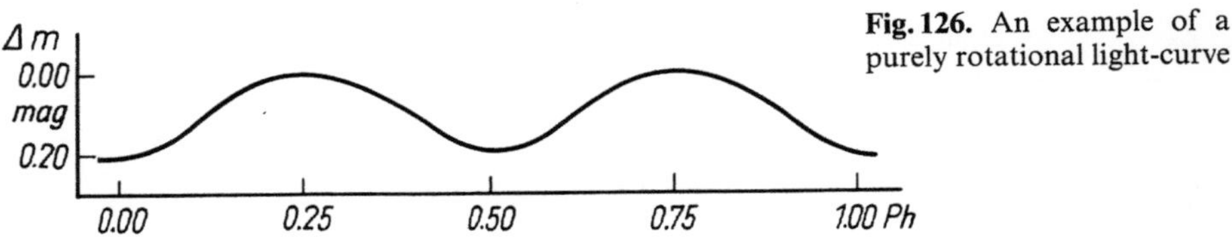

Fig. 126. An example of a purely rotational light-curve

smaller amplitude and with constant magnitude at minimum. Normally, however, one of the two stars has a much lower surface brightness. It is only in this manner that we can explain the observed large amplitudes, amounting to several magnitudes. In most cases the smaller star is the brighter, and primary minimum is observed when the smaller star is behind the larger. An extreme example of this type is ζ Aurigae, which will be described later. The shapes of the light-curves illustrated in Fig. 125 show a great range of variation, both in the total amplitude and in the relative amplitudes of m_1 and m_2. Stars having every intermediate form are found.

A small group of stars remains to be mentioned, in which the orbital plane is so aligned that the components do not eclipse one another. Nevertheless, a very small variation in magnitude may occur, if both stars are distorted into ellipsoids by their mutual gravitation. At conjunction, that is, when the long axis of the ellipsoids forms the smallest angle with the line of sight, then the apparent area of the system, and hence the brightness as observed from the Earth, is at a minimum (Fig. 126). The observed amplitudes mostly lie below 0.1 mag, and can thus only be determined photoelectrically. The total number of cases known is small, but some bright stars are among them: ζ And, b Per, o Per, π^5 Ori, α Vir. In ζ And (amplitude 0.16 mag) and α Vir (amplitude 0.07 mag) there may also be a small partial eclipse. The true total of these *ellipsoidal variables* is undoubtedly much larger than shown by the discoveries that have been made up to now. In the GCVS these objects are designated *Ell*.

Table 42. General properties of the three photometric classes of eclipsing variables

	EA	EB	EW
Prototype	Algol	β Lyrae	W Ursae Majoris
Period (days)	> 0.4	> 0.4	$0.2 - 1.0$
Spectrum	O6 – M1	B8 – G3	F0 – K4
Total ($< 12^m$)	≈ 1000	≈ 200	≈ 100

Table 42 shows the distribution of eclipsing variables in the three photometric classes that have been described (see also HAZLEHURST 1976).

Known *spectra* of Algol stars cover the range from O6 (V 444 Cyg) to M1 (YY Gem). They show a narrow maximum between A1 and A5, whereas the β Lyrae stars and the W Ursae Majoris stars have a wide maximum. As might be expected, the W Ursae Majoris stars form a relatively uniform group (although some authors suggest that there may be two sub-types). The Algol stars, on the other hand, show a much greater range of variation in their properties. The totals change rapidly as new objects are always being found. In fact, systematic searches of star fields always reveal new eclipsing stars.

Even among the brighter eclipsing variables there are quite a number where little is known apart from the fact that they vary, and their class. Even amateurs only equipped with small telescopes ought to be able to tackle these.

KREINER and TREMKO (1978) published a list of 28 β Lyrae stars in the northern hemisphere (of which 27 are brighter than magnitude 11), for which future observations seems desirable for various reasons (earlier neglect, variable periods or light-curves, physically interesting stellar components).

4.4 Analysis of the Light-Curve

The light-curve of β Per (Algol) shows primary and secondary minima, and between the two minima a rise and decline of about 0.1 mag (Fig. 125). This is a *reflection effect:* the faintly luminous companion is illuminated by the primary star and disappears behind the primary at secondary minimum, shortly before it reaches its full phase as seen from the Earth. Figure 127 shows the relative dimensions found in the Algol system. It is inferred that the side of the faint companion that is turned towards the bright primary is brighter than the averted side. The dashed circle shows the position of the companion during the partial eclipse at primary minimum.

Fig. 127. Relative sizes in the Algol system (after STEBBINS)

An additional increase in brightness is observed in systems with very eccentric orbits at the times of minimum separation between the components. In such cases additional excitation by UV radiation probably occurs as well as the reflection effect. This phenomenon is called the *periastron effect.* Objects are conceivable in which the periastron effect occurs, but which have no eclipsing variation. $KQ\,Pup(4\overset{m}{.}9 - 5\overset{m}{.}2)$, otherwise regarded as being related to VV Cep (see below), seems to be just such an object.

A further perturbation of the light-curve can be ascribed to the fact that the stellar disks are not of even brightness. This is the *limb-darkening effect,* well-known on the Sun. In most cases its amount can only be estimated, but luckily the effect upon the final result is not very great.

If the light-curve is corrected in the manner briefly indicated here, then we obtain a "rectified light-curve", which can now be used for determining information about the size of the components and their motion in the system (the system constants).

The foregoing shows that a good light-curve is capable of giving a considerable amount of information. We have the period P, the duration D of the primary minimum, the duration d of any constant brightness at minimum,

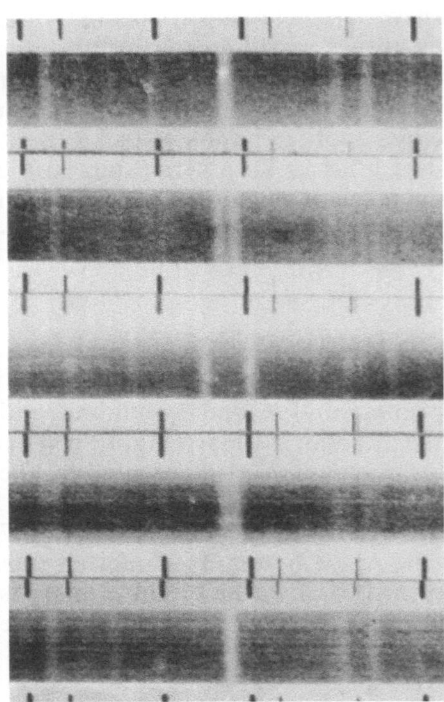

Fig. 128. Variable line-splitting in the β Aur eclipsing system, a Doppler effect caused by the orbital motion of the components (after O. STRUVE)

the corresponding values for the secondary minimum, the amplitude of both minima and the course of the rise and decline. These values follow from the photometric observations alone, and photoelectric photometry enables them to be determined to a precision of a hundredth of a magnitude. In addition there are the observations in various colours, and spectroscopic observations including the Doppler effects arising from orbital motion (Fig. 128). Moreover, as many bright objects are available, the crucial factors may be determined with a high degree of accuracy. From these, methods have been established for orbital calculation in spectroscopic binary stars in general, and in eclipsing systems in particular. As a result these objects are of great significance for astronomical research; a large part of our information about stellar properties has been obtained from research into eclipsing systems. A detailed description of the theory of eclipsing stars and methods of determining the orbits can be found in the work by SCHILLER (1923, p. 288). RUSSELL (1912) was the first to solve the problem. SHAPLEY (1915) and FETLAAR (1923) should also be mentioned. Later SCHNELLER (1949) developed RUSSELL and FETLAAR's work further, and simplified the calculations. In addition he made tables available, as HETZER had earlier. A later work is that by BINNENDIJK (1960). Further information about the method of determining orbits can be found in the book by TSESEVICH (1971). Comprehensive works on the analysis of light-curves have been published by KOPAL (1978, 1979).

4.5 Period Changes

It might be assumed that, as the variations in eclipsing stars are caused by a mechanical process, the details relating to the stability of the periods would be quite obvious. Strangely this is not the case. However, before we enter into a description of individual cases we must deal with period-changes that have a mechanical explanation.

4.5.1 Periodic Changes

If the orbit of an eclipsing system is strongly eccentric, the secondary minimum does not, in general, fall in the centre of the interval between the primary minima (Fig. 129), unless the line of apsides points towards the observer. The asymmetric position of the secondary minimum is therefore a sign of considerable eccentricity in the orbit. An extreme case of this is DI Her, where $P = 10^{\text{d}}550$ and secondary minimum is at phase 0.768 (Fig. 130).

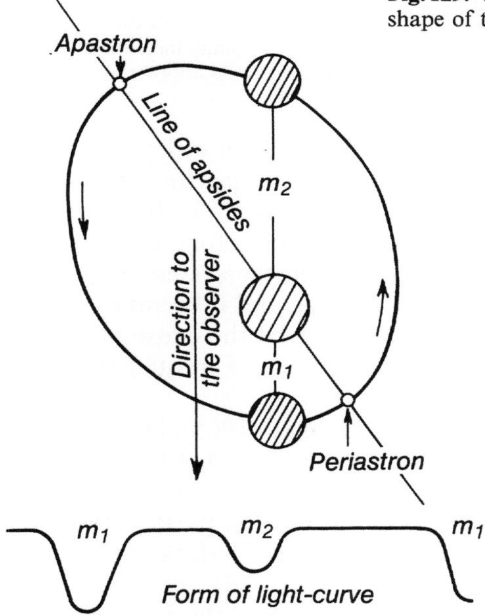

Fig. 129. The effect of great orbital eccentricity on the shape of the eclipsing light-curve

However, there are cases in which the line of apsides is rotating. The secondary minimum then oscillates about its average position with the *apsidal rotation* period (Fig. 131), and its period undergoes cyclic changes. The same also applies to the primary minimum (Fig. 132). The formula for representing the times of minima then contains a sine term, whose coefficient has opposite signs for the two types of minima.

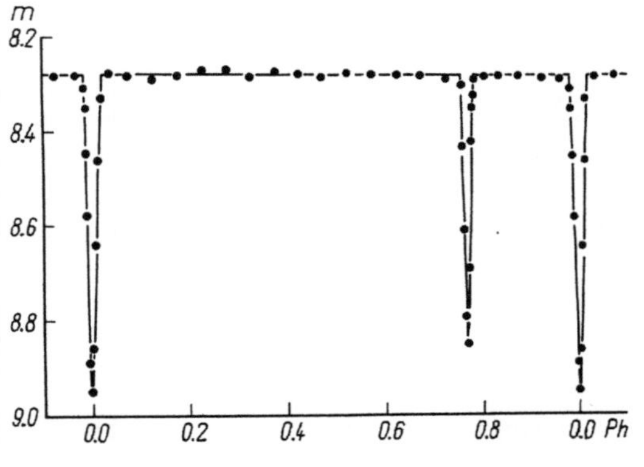

Fig. 130. Light-curve of DI Her (an example of great orbital eccentricity; after JACCHIA)

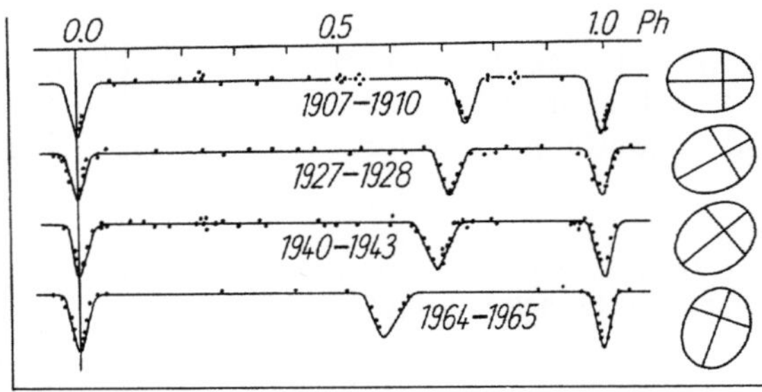

Fig. 131. RU Mon. Progressive change in the light-curve due to rotation of the line of apsides (after MARTYNOV 1971). For the purposes of illustration, the primary minima have been drawn beneath one another

A further periodic term can result from the finite speed of light, that is if the eclipsing system as a whole describes an orbit in the *gravitational field of a third body*. The effect resembles the light-time effect produced by the Earth's orbit around the Sun, described in the introductory chapter. The observed period is longer than the true period when the system is receding from us (as a result of motion about the third body), and it becomes shorter as the system approaches us. A third body can also be the cause of apsidal rotation, similarly to the way in which the planets cause mutual perturbations.

A special study of apsidal motion is given in KOPAL (1965). He deals with the theory and the observed effects, and stresses the great significance which these close binaries have for research on stellar interiors, since the gravitational effects, which are primarily produced by the densest parts of the stars, do not undergo any form of absorption. In his paper 21 stars are given as examples.

Recent work on the problem of apsidal motion and the influence of a third body on the epochs of minima has been given by MARTYNOV (TSESEVICH 1971, Chap. 9), BATTEN (1973, Chap. 6), SAHADE and WOOD (1978, Chap. 6) and KOPAL (1978). The list of eclipsing systems with fully-investigated apsidal motion, which is given in Table 43, has been compiled from the last source.

A particular form of periodic, or rather nearly periodic, changes, which certainly cannot be explained mechanically, occur in a special class of de-

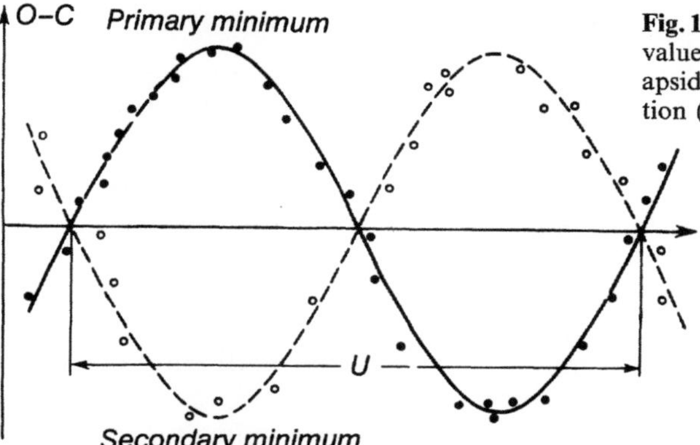

Fig. 132. Periodic changes in the $O-C$ values caused by rotation of the line of apsides. $U=$ period of the apsidal rotation (after MARTYNOV 1971)

Table 43. Eclipsing systems with apsidal motion

Star	Spectra	P	U/P	e	$\mathfrak{M}_2/\mathfrak{M}_1$
GL Car	B3 + B4	$2^\mathrm{d}422$	3800	0.16	1.0
HH Car	B5 + B8	3.2315	75000	0.16	0.9
AR Cas	B3 + A0	6.0665	25000	0.25	0.25
PV Cas	B6 + B6	1.7505	18988		
V 346 Cen	B4 + B6	6.3227	11000	0.20	1.0
XX Cep	A8 + G6	2.3373	10000	0.14	0.22
Y Cyg	O9.5 + O9.5	2.9963	5900	0.14	0.99
MR Cyg	A0 + F7	1.6770	12000	0.05	0.85
V 380 Cyg	B1.5 + B3	12.4256	59000	0.22	0.57
V 477 Cyg	A3 + F5	2.3470	54300	0.30	0.68
HS Her	B5 + A4	1.6374	3450	0.05	0.34
CO Lac	B8.5 + A0	1.5422	10010	0.03	0.82
RU Mon	B9 + A0	3.5847	28900	0.38	0.96
GN Nor	?	5.7034	31000	0.21	1
FT Ori	A0 + A3	3.1504	60000	0.40	0.9
δ Ori	B1 + B2	5.7325	9900	0.08	0.38
AG Per	B5 + B7	2.0287	12900	0.07	0.88
YY Sgr	A0 + A0	2.6285	46000	0.16	0.9
V 523 Sgr	A5 + A5	2.3238	33000	0.18	1.0
V 526 Sgr	A0 + A3	1.9195	27800	0.22	0.8
V 2283 Sgr	A0 + A2	3.4714	59000	0.49	0.7
α Vir	B2 + B3	4.0142	11200	0.15	0.62
DR Vul	B7 + B8	2.2512	6140	0.09	0.95

Here: $P=$ Orbital period of the binary; $U=$ Apsidal rotation period; $e=$ Orbital eccentricity; $\mathfrak{M}_2/\mathfrak{M}_1=$ Mass ratio

tached systems, the *RS Canum Venaticorum stars*, discussed in Sect. 3.7.1. Apart from the orbital period, a second period is active in these objects, which is less than the orbital period by a small amount (in rare cases it is slightly longer), so that a form of "beat" occurs with a cycle-length of several years (see Fig. 120). Physical causes are probably responsible for this. An explanation of the phenomenon (the star spot theory) has been given in Sect. 3.7.1.

4.5.2 Non-Periodic Changes

A new aspect arose when it was realized that contact and semi-detached systems have a common atmospheric envelope and that the atmosphere around at least one component has expanded to the stability boundary, so that *mass loss* and even *mass-exchange* between the components take place. The possibility of sudden period changes by mass-exchange was recognized by WOOD (1950). Even detached systems, in which both components still lie inside the stability boundary, may show considerable mass ejection, as is apparently the case in the RS Canum Venaticorum stars just mentioned. Obviously the circumstances are similar to those we have already described in the eruptive binaries, although here the effects are much smaller. As the component that is accreting the material is not a compact star (a white dwarf or similar object), the amount of gravitational energy released by the deposition of material is not very great. Figure 133 shows as an example the $O - C$ curve of W UMa from 1912–1982.

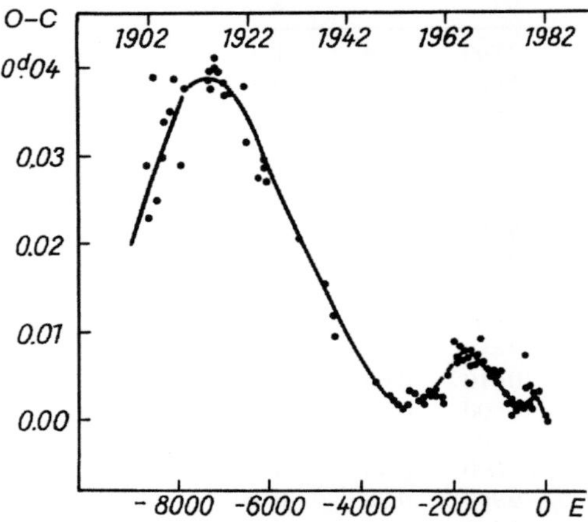

Fig. 133. $O - C$ curve of W UMa (after HAMZAOGLU et al. 1982)

There are numerous publications about mass-exchange and mass-loss in close binary systems. Mention should be made of the work by BATTEN (1973), SAHADE and WOOD (1978, Chap. V) and some contributions in the book by GYLDENKERNE and WEST (1970), where these problems are discussed from a mainly observational point of view. Additionally, KOPAL (1978, Chap. V) mainly deals with the theoretical aspects.

SCHNELLER (1960) had previously tried to find a complete solution to the problem of period changes. He investigated how far changes could be explained mechanically, that is by apsidal rotation, light-time effects, or the existence of further bodies. Table 44 gives the results obtained from investigation of 68 systems. This shows that, on average, the detached systems are the most stable. The author says that only in the rarest cases can period changes be explained by "simple models". "Rather does it appear that short-term, catastrophic processes cause the period changes." SCHNELLER also found that a better representation of the $O - C$ diagrams can, in most cases, be obtained

Table 44. Period variation in eclipsing stars

	Detached [%]	Semi-detached [%]	Contact systems [%]
Constant period	73	47	33
Variable period	19	53	50
Uncertain	8	0	17
Total number of objects	26	30	12

by polygonal functions than by curves. The following extract is also of interest: "At the same time this research shows the importance of continuing observation of as many eclipsing variables as possible. Without the use of the numerous minima determined by amateurs, where the timing has quite sufficient accuracy for investigation of major features, many of the diagrams for the stars discussed here could not have been brought up to date."

According to very recent work (MATESE and WHITMIRE 1983) period changes may be caused not only by mass-streaming, but also by small changes in the structure of the secondary (radius changes of up to 3%, with or without changes in the mass concentration inside the star).

4.6 Statistics

The *amplitudes* of close contact systems of the W Ursae Majoris type, defined above, are mostly about 0.7 mag, if the eclipse is central. With partial eclipses every lesser value is possible down to the limit of detection. The latter point also applies to semi-detached and detached systems, but these may, in theory, have any arbitrarily large amplitude, as it is conceivable that a dark companion could totally eclipse the bright star. No such case has yet been observed, but some very large amplitudes are known. Some ranges that are listed are: RW Tau 8^m0-12^m3 pg., U Cep 6^m6-9^m8 pg., SS Cet 9^m4-13^m0 vis., and WW Cyg 9^m9-13^m7 vis. Amplitudes of more than 3 mag are rare. The extreme case known is probably V 442 Cas with an amplitude of about 5 magnitudes.

For statistical purposes the value D/P, the ratio of the *duration of minimum* to the period, is more important. This value is closely related to the discovery probability, if the generally insignificant secondary minima in Algol stars are excluded. The older average value of 1/7 has been considerably reduced by the discovery of more difficult objects as a result of systematic searches. RICHTER has obtained a value of 0.123 from the Sonneberg Field material, which becomes 0.112 when the discovery probability is taken into consideration. But a number of cases are known with values near 0.02, so that one must examine 50 plates, on average, to discover one minimum. Extreme values must be those of SW Nor with 0.014 and V 1108 Sgr and EE Cep with 0.015.

Extreme periods known in eclipsing stars are as follows:

Shortest periods		*Longest periods*	
AM CVn	18 min	ε Aur	9883 days
GP Com	46 min	VV Cep	7430 days
		V 381 Sco	6475 days
		V 383 Sco	4900 days

The components in AM CVn and GP Com are probably white dwarfs. Here the physical changes in luminosity caused by mass-exchange are so strong that it is very difficult to observe the eclipsing changes. The objects have been discussed elsewhere (Sects. 3.1.3 and 3.1.5) with the eruptive binaries. The 14 eclipsing binaries known to date with periods below 0.17 days are all eruptive binaries. It should be said that periods below 0^d3 are not present in the "true" Algol stars, although W Ursae Majoris stars have a few representatives below 0^d3. The W Ursae Majoris stars with the shortest known periods are given in Table 45. According to recent research by WILLSON et al. (1981) – if these authors have correctly interpreted the luminosity changes – the eclipsing star with the longest known period is R Aqr, the symbiotic Mira star discussed earlier. Here the disk of material surrounding the compact companion eclipses the Mira star every 44 years (≈ 16000 days) for some 6–7 years, during which the Mira-type variations are invisible. KQ Pup may also have one of the longest periods ($P = 9752$ days) but it is not certain whether the changes in luminosity caused by the periastron effect (see above) are accompanied by any eclipse. Among the stars with very long periods there are some very interesting objects; some of these will be discussed later.

Table 45. The W Ursae Majoris stars with the shortest known periods

Star	m_{Max}	m_{Min}	P
AB Tel	13^m4	13^m9 pg	0^d17
BF Pav	12.8	13.4 pg	0.17:
CC Com	11	11.9 v	0.22068
V 523 Cas	11.2	12.1 B	0.2337

4.7 Some Interesting Eclipsing Systems

β Lyrae

The photoelectric measurements of β Lyrae by GUTHNICK and PRAGER (1917) – see Fig. 134 – gave the first indication of the existence of non-periodic changes in eclipsing stars. Since then the β Lyr system has continued to yield interesting facts.

At primary minimum emission is seen from the part of the common atmosphere that surrounds the brighter star, while the atmosphere of the

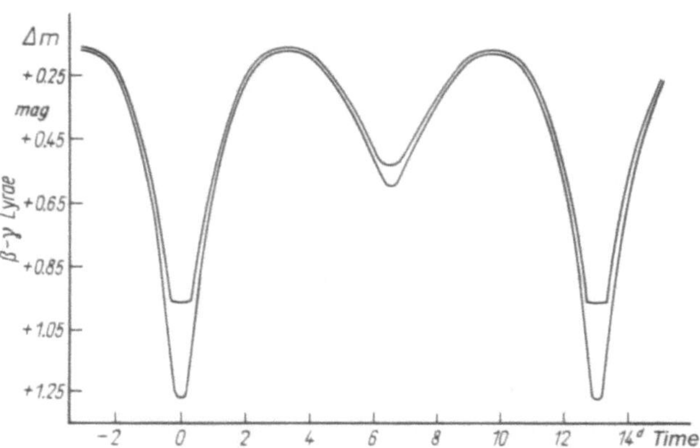

Fig. 134. Photoelectric light-curve of β Lyr after GUTHNICK. *Upper curve*: 1925/26 and 1943/44, *lower curve*: 1915/16

Fig. 135. 32 spectrograms of the eclipsing system β Lyr, arranged according to phase. The spectrograms have been aligned so that absorption lines belonging to the pair's brighter component fall beneath one another (for example the first 3 lines at the left edge). Doppler shifts in other lines then show the relative motion of the shell, which principally surrounds the faint (but massive) star (for example, the bright lines of 388.9 He I, about 3 cm from the left edge, 397.0 H_ε, right of centre, and 402.6 He I at the right-hand edge). The position of the "static" interstellar Ca II line also alters (sharp line alongside the strong absorption line just to left of centre), and thus reflects the orbital motion of the bright component about the system's centre of mass. Lines from the faint component are not visible (after O. STRUVE 1957)

Fig. 136. Model of β Lyr (after SAHADE and WOOD)

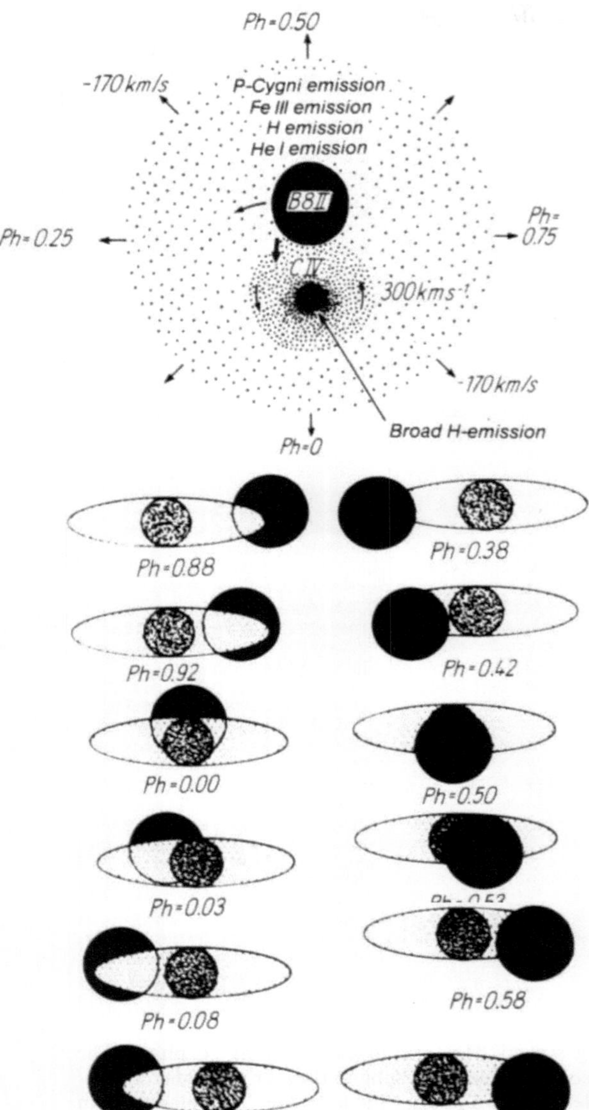

Fig. 137. Schematic picture of β Lyr as seen from Earth, at various phases (*Ph*) in the light-curve (after BROWN and HUANG 1977)

fainter star, which is then nearer to the observer, contributes absorption lines. In the β Lyrae system the fainter component is a massive, low-luminosity star, whose spectrum is difficult to determine (the features range between A7 and B5). The second, brighter, but lower-mass component has a B8 spectrum (Fig. 135).

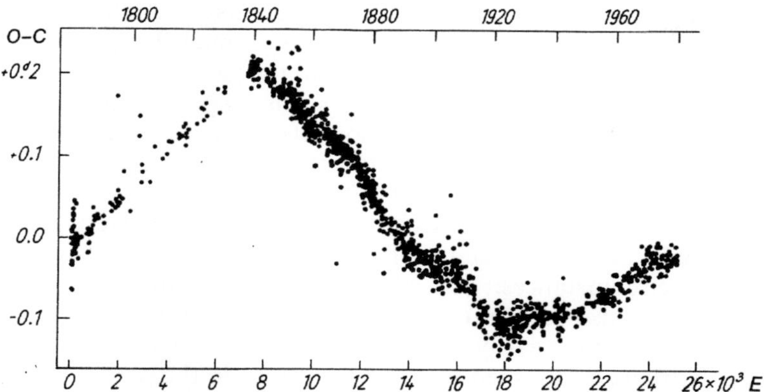

Fig. 138. $O - C$ curve of Algol (after SCHNELLER, with additions by FUHRMANN, Sonneberg)

The masses are high, but uncertain: according to one source they are 2 and 11 solar masses, while another gives them as 13 and 23 solar masses. The current opinion is that the massive component is surrounded by a disk of material, which is the source of the hydrogen and C IV emission lines, and rotates at a velocity of 300 km/s. Material streams from the B8 component into the disk. Both objects are surrounded by a common gaseous shell, expanding at velocities up to 170 km/s, which is the source of further emission lines (Fig. 136). Figure 137 schematically illustrates the course of the eclipses (left at primary minimum, right at secondary minimum). We are still far from a complete understanding of the β Lyrae system.

Algol

Algol may be taken as an example of the various types of period changes (Fig. 138). CHANDLER's highly complicated formula, by which the minima around the turn of the century were predicted, gave an error of about 2 hours even as early as 1915. The situation was similar to that found with the Mira stars, and there was no other solution than to employ "instantaneous elements" for Algol as well. It was not possible, however, to give an explanation for the phenomena. The literature in the last few decades, for Algol in particular, is so extensive that only a short summary can be given here.

As explained previously in Sect. 4.5.1 there are three possible causes for cyclic period changes that can be considered: rotation of the line of apsides, light-time effect, and RS Canum Venaticorum-type behaviour (not found in Algol). Let us first establish what has to be explained. Algol's primary period is $2^{d}\!86731$. EGGEN in 1948 found another three periods: the first of 1.873 years, which had already been deduced from radial velocity measurements by McLAUGHLIN; the second of 188.4 years, which needed to be confirmed; and finally a period of 32 years, which could be ascribed to rotation of the line of apsides. EGGEN's other periods apparently indicated orbital motion, so that the Algol system could consist of 4 stars, whose masses have been given as: 5.0 and 1.0 (the eclipsing pair), 1.2, and 3.8 solar masses. After LUYTEN voiced doubts about the apsidal motion theory, as a much longer period would be required, PAVEL (1949) tried to explain all the variations by orbital motions and light-time differences, for which he needed 4 perturbing bodies. The Algol system would thus have 6 components. FERRARI had already settled on a quintuple system, and supported this theory, about the same time as PAVEL, with further research (FERRARI 1950).

In view of this state of affairs, one could hardly avoid subjectively-based objections. The doubt about the reality of the many-body hypothesis was underlined by two circumstances: first, that phenomena similar to those observed in Algol were also seen in other eclipsing stars, above all in some bright ones that could be most easily studied (examples are β Lyr and λ Tau); and second, that workers established that sudden jumps in the period length of Algol seemed to take place. These could not be explained by any processes based on celestial mechanics.

As a result, we have here in Algol, a semi-detached system, another example of non-periodic changes (see Sect. 4.5.2), produced by the exchange of mass.

Nowadays it is believed that Algol actually consists of just three objects: the eclipsing pair Algol A (B8V spectrum) and Algol B (probably a G to K subgiant) with an orbital period of 2.87 days, and Algol C, which, like the close AB pair, orbits the centre of mass in 1.86 years. In fact, for a short time during primary minimum, numerous faint absorption lines are observed, originating in Algol C. The spectroscopic evidence for Algol B was also obtained recently, with the successful discovery, using the most modern techniques, of the well-known sodium doublet, which can be ascribed to it (TOMKIN and LAMBERT 1978).

All other periodicities are spurious; the changes are non-periodic.

SAHADE and WOOD (1978) liken the introduction of a fourth, fifth or sixth component to the somewhat similar situation that arose long ago, when attempts were made to rescue the Ptolemaic theory by introducing more and more epicycles.

For some years Algol has also been known as a radio and X-ray source. Future investigations will be needed to determine whether these observations indicate some form of "stellar activity" that is similar to solar activity, but much stronger.

A detailed description of the Algol system can be found in SAHADE and WOOD (1978, Chap. 10).

V 444 Cygni and CV Serpentis

The Algol star *V 444* Cyg, discovered in 1940 by GAPOSCHKIN, with a period of 4.2 days and a low amplitude ($8^{m}3 - 8^{m}6$), is a peculiar object. One component is a *Wolf-Rayet star* (*Sp* WN5) and the other is an O6 star. Wolf-Rayet stars (= WR stars) have highly complicated spectra with strong emission lines of He, C, N and O, that indicate an extended shell. These objects include numerous spectroscopic binaries, and the question of whether any single WR stars occur at all remains open. Statistics might provide the answer, if we knew the total number of WR stars that are eclipsing stars. In addition, thorough analysis of the light-curves of WR eclipsing stars in conjunction with spectroscopic observations enables some evidence to be obtained about the physical details of these peculiar objects, which are very difficult to understand. Unfortunately, as yet only 7 WR eclipsing stars are known, and none of these show central eclipses, so that the results are still very uncertain. Nevertheless, it is obvious that the masses are very high. As yet, the best-observed object is V 444 Cyg. According to KRON and GORDON, the O6 star has a diameter 4.5 times as great as that of the WN5 star. The (uncertain) masses amount to 10 solar masses for the WR star, and 26 solar masses for the O star. In addition, there is an inner luminous shell around the WR core, and an outer electron shell (Fig. 139).

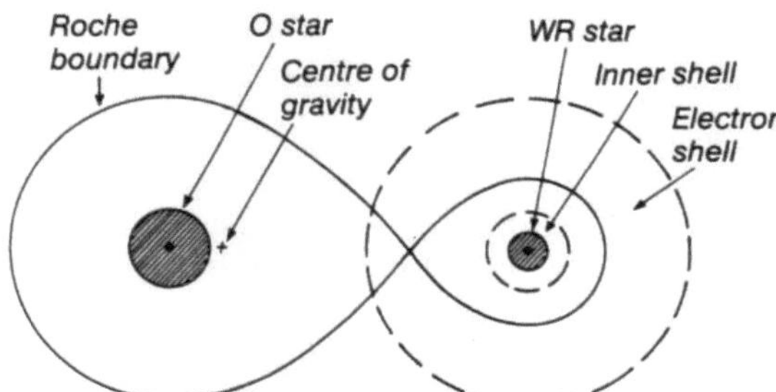

Fig. 139. Model of V 444 Cyg (after SAHADE 1980)

The other six objects of this sort are CV Ser, V 1676 Cyg, V 1696 Cyg, GP Cep, CX Cep, CQ Cep.

CV Ser is an eclipsing star with a period of 29.7 days. Until 1963 the magnitude oscillated (with variable amplitude) between the extremes of 9^{m}7 to 10^{m}4. In 1970, however, no variation at all was detectable! COWLEY et al. (1977) give the possible explanation that in the earlier variations it was not one of the stars that was being eclipsed – it cannot have simply vanished – but some form of bright material between the two stars.

A detailed description of these interesting objects can be found in SAHADE and WOOD (1978, p. 93) and in SAHADE (1980, p. 46); see also TSESEVICH (1971, p. 256) and BATTEN (1973, p. 51).

WR binaries have been known as *X-ray sources* for some time (see e.g. SANDERS et al. 1981).

U Cephei

A summary of the numerous works on this interesting semi-detached system is to be found in SAHADE and WOOD (1978, p. 142). It is assumed that the object, discovered in 1880, consists of B7V and G8III–IV components. There is a total eclipse of the hot star by the cool giant every 2.5 days. The visual amplitude amounts to about 2.2 mag. The light-curve somewhat resembles that of an Algol star, but there are distinct irregularities in the shape of the

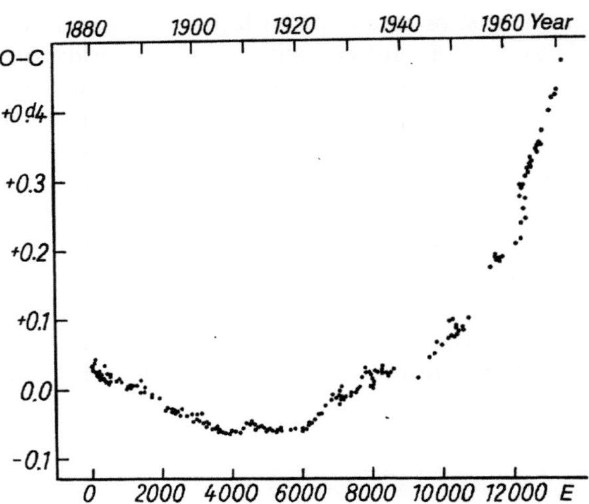

Fig. 140. $O - C$ curve of U Cep (after BATTEN 1973)

curve. The primary minima show irregular changes in both depth and duration. There have been regular observations for more than 100 years (!), and the very interesting $O - C$ curve (Fig. 140) shows that on average the period-length has been increasing steadily. This is a sign of extremely active mass-exchange. The spectra occasionally show emission lines of hydrogen and other elements.

The following model of the system has been established from thorough analysis of the observations. After very strong mass ejection from the giant star, as happens from time to time, a temporary equatorial disk of material of considerable density forms around the B star, and hot spots also occur in the B star's photosphere.

This is probably a system that is going through a very rapid evolutionary phase.

V 471 Tauri (= BD + 16°516)

This eclipsing star, discovered by NELSON and YOUNG (1970, 1976), has a period of 0.52118340 days and is a member of the Hyades cluster. It has a light-curve which cannot be classified in one of the three major groups, the Algol, β Lyr or W UMa stars (Fig. 141). At first sight, it apparently resembles an RR Lyrae light-curve, with a hump on the descending branch. In the visual, blue and ultraviolet spectral regions the amplitudes amount to 0.3, 0.4 and 0.65 magnitudes respectively.

Based on spectroscopic analysis, the authors mentioned suggest that this is an eclipsing system, with a K0 star (0.7 solar masses and 0.8 solar radii) and a hot white dwarf (0.7 solar masses and 1.3 Earth radii). Practically all of the variability in the visual and blue regions arises from the tidally deformed K0 star, one side of which is strongly heated by the white dwarf. A remarkable eclipsing variation is seen only in the UV (Fig. 141). There is no secondary minimum. The decline to primary minimum lasts 55 seconds (!) and its

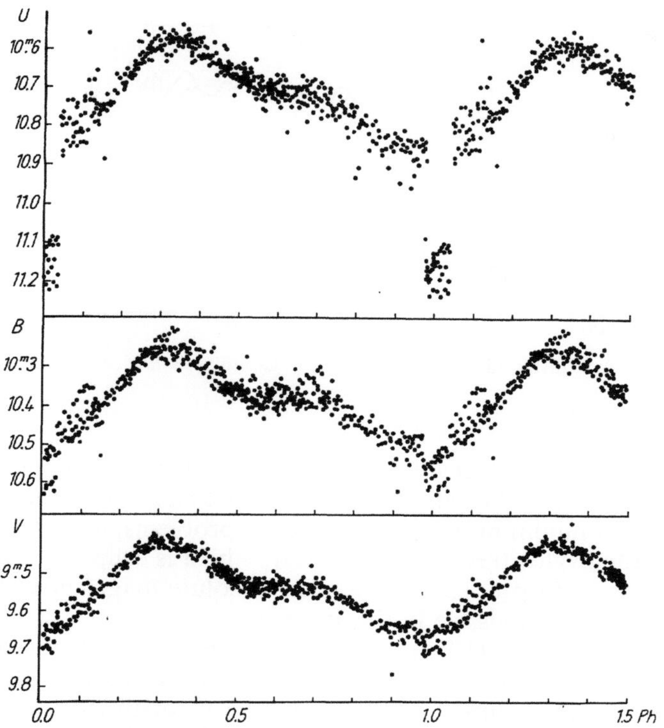

Fig. 141. Light-curve of V 471 Tau in *U* (*above*), *B* (*centre*) and *V* (*below*); after NELSON and YOUNG 1970

duration is 47 minutes. There are irregularities in both the light-curve and the period. In many respects the system resembles that of a U Geminorum star, with the difference that no mass-exchange could be established, and no eruptions. Some authors see the object as a U Geminorum precursor. For further information see HAMZAOĞLU (1981) and RUCINSKI (1981). UU Sge (Sect. 3.4.3), AA Dor and GK Vir (= PG 1413 + 01) are related (PACZYNSKI 1980).

ε Aurigae

The variability of this object was discovered in 1821 by the Quedlinburg clergyman FRITSCH, but it was LUDENDORFF who first established, in 1903, that eclipsing variation was present with the exceptionally long period of 9883 days $\approx$ 27 years. Minima so far observed were in 1821, 1847–1848, 1874–1875, 1901–1902, 1929, 1956 and 1983–1984. No secondary minimum is visible, even photoelectrically. The visual amplitude amounts to 0.63 mag (3^m23 to 3^m86). The shape and width of the mimima vary (Fig. 142). On average, the decline and rise each last 197^d, and the constant light at minimum 360^d, so the overall duration of the minimum $D = 754^d$. Only a few

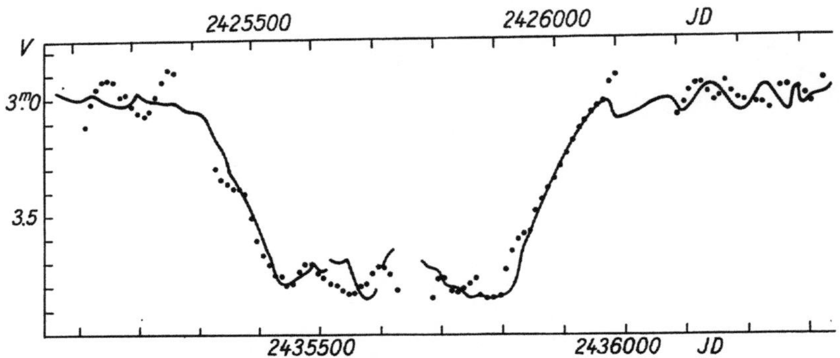

Fig. 142. Eclipse minima of ε Aur (after GYLDENKERNE 1970). *Continuous line*: 1955–1957; *points*: 1928–1930

details can be given from the very extensive literature. Apart from the eclipse, the light from the star shows small oscillations of up to 0.2 mag. The behaviour of the spectral lines at minimum poses a number of problems, including some concerning the radial velocities. The bright star, which is eclipsed, is a supergiant of spectral class F0epIa, said to have an absolute magnitude of $-2^{\text{M}}8$. No second component is visible in the spectrum.

The model described by STRUVE (1953) is very surprising. All the observed light comes from the smaller component. Indeed, according to KUIPER the smaller, brighter star is totally eclipsed, but can be seen through the outer layers of the larger star, being only weakened by about 50%. The smaller component has about 300 times the diameter of the Sun, and the larger at least 3000 times! Even so, this value is probably too small as there are signs that the atmosphere of the large star extends as far as the periastron of the orbit of the hot companion. The large component, which must be of very low density, has been described as an infrared star, but as yet no long-wave radiation has been found. This appears to be the weak point in this otherwise satisfactory model by STRUVE. Later STRUVE suggested that the cool component has an extended circumstellar shell, which consists of a number of distinct "clouds". Other models have been developed by HACK, SAHADE, HUANG and KOPAL. A model by HANDBURY and WILLIAMS (1976) suggests that the eclipsing object has a disk of dust and gas, with opaque inner regions, but which becomes transparent towards the edge. The whole system may be surrounded by an expanding circumstellar shell with a cloud-like structure. The source of the opacity could be electron scattering. The model has certain similarities, as least qualitatively, with that of β Lyr. A detailed description of our knowledge of ε Aur can be found in SAHADE and WOOD (1978, p. 152). The recent minimum will probably throw more light on the nature of ε Aur.

ζ **Aurigae**

This system, whose eclipsing changes have been known since 1931, also proves to be a very unequal pair, but of a completely different nature from

ε Aur. This time the larger star is a supergiant with a K4 spectrum, and the smaller is a main-sequence star of spectral class B7. However, what is remarkable is the ratio of the diameters. The diameter of the B star amounts to only about 1/40 of that of the K star; the radius of the K star is about 200, and that of the B star 5 solar radii (CHAPMAN 1981). The masses are high: about 22 solar masses for the K star, and 10 for the B star. The period is $972^{d}16$ and the range $5^{m}0 - 5^{m}6$ pg. The orbit is quite eccentric ($e = 0.4$). Detailed descriptions can be found in the review by SAHADE and WOOD (1978, p. 121) and in CHAPMAN (1981).

The great significance of ζ Aur for astrophysics lies in the fact that, during the eclipse decline and rise, the B star, which is very small relative to the K star, shines through its very extended atmosphere. The additional lines that then appear in the spectrum of the B star give a means of closely probing the structure of the K supergiant's atmosphere. In this case, however, the partial phases only last $0^{d}8$. According to a model by CHAPMAN (1981), the K star loses about 2×10^{-8} solar masses per year through a stellar wind, part of which is captured by the B star.

In contrast to ε Aur, the larger star is also optically identifiable, and indeed appears alone during the total phase of the eclipse. Both stars are of approximately the same brightness.

VV Cephei

In this eclipsing system, discovered by Miss CANNON in 1908, once again a B star is orbiting a supergiant, which this time has an M2 spectrum. The period amounts to 7430 days = 20.4 years and the range is $6^{m}6 - 7^{m}4$ pg. So far everything is similar to ζ Aur, but here the radii must be still larger. The masses of the M and the B star have been estimated at about 18 and 20 solar masses; the radii at about 1600 and 13 solar radii; and the absolute magnitudes at about $-4^{M}0$ and $-2^{M}3$. If the M star were to be at the Sun's position, it would reach far beyond the orbit of Mars. It is the largest star yet known. Both stars are surrounded by a common gaseous shell, in which forbidden lines occur. In addition, the B star has a ring-like shell (MÖLLENHOFF and SCHAIFERS 1978). The duration of the whole eclipse D is 1.3 years, and of the total phase d 1.2 years. Research into the properties of the system is made more difficult by physical variability of the M component. It is an SRc variable, with a cycle-length of 118 days and an amplitude of 0.3 mag (after McCOOK and GUINAN). Further information about this object can be found in SAHADE and WOOD (1978, p. 126). VAN DE KAMP (1978) has published a light-curve and a distance determination. The (wrongly) rather neglected object AZ Cas ($11^{m}0 - 11^{m}8$ pg; $P = 3404^{d}$, $Sp = $ M0eIb + B0V, $e = 0.55$) appears to be somewhat related.

4.8 Concluding Remarks on the Evolution of Close Binary Systems

At the beginning of this chapter, it was mentioned that the earlier distinction between changes in magnitude that have physical and geometrical causes, has lost some of its justification. The foregoing discussion shows that mass-streaming, extended shells and photospheric disturbances occur not only in contact and semi-detached systems, but even in detached systems. In all these systems the question arises: How do two stars of completely different evolutionary states come to form a pair? A well-known, and extreme, case is Sirius. The question does not really come within the province of this book, although it has already been broached in discussing the eruptive binaries (Sect. 3.1), in which it is particularly difficult to answer. However, two factors will be mentioned. First, the evolutionary lifetime of a star – for example, its duration on the main sequence – is strongly dependent upon its mass. The greater the mass, the faster the evolution. In the most frequently encountered range of masses, the timescale varies by a factor of the order of 100. Second, as soon one component begins to undergo evolutionary expansion (climbing the giant branch), and reaches its stability boundary, mass-exchange plays a big part. Since the pioneer work in this field by KIPPENHAHN et al. (KIPPENHAHN, KOHL and WEIGERT 1967, KIPPENHAHN and WEIGERT 1967, and other works) we have gained a very important insight into the evolutionary processes, and there is now a very considerable literature on the subject. More recent summaries that may be mentioned are: BATTEN (1973, Chap. 10), PACZYNSKI (1971), various contributions in the book *IAU Symposium 73* (1976) and SAHADE and WOOD (1978, Chap. 7).

An easily understood description of the probable development of U Cep is given by BATTEN and PLAVEC (1971).

5. Supplement to the Classification

5.1 Variables in Clusters

5.1.1 Open Clusters

Some time ago it was thought that open clusters (otherwise known as galactic clusters) typically had few variables. Nowadays, in particular since the fundamental work by KHOLOPOV (1956), this idea has been revised. In any case there was no support for it on the basis of stellar evolution, as it is plausible to assume that, as a first approximation, a cluster will contain variables that correspond to its age and to the evolutionary development of its members. In fact, we find numerous *T Tauri* variables and related types in extremely young clusters, and among the low-mass stars in "normal" open clusters, e.g. the Pleiades, there are innumerable *flare stars*, which are difficult to discover. GÖTZ has devoted a great deal of work to this problem (1973 with references to earlier publications; 1980a, 1981). His statistical research into extremely young variables (Sect. 3.3) and other variables has led to various conclusions about the origin, structure and evolution of open clusters (for example NGC 2264, the Pleiades and Praesepe).

In what follows we disregard variables that have a very low evolutionary age. The total number of variables found in an open cluster simply corresponds to the number of stars. If we assume that in a normal region of the Galaxy there is about one variable for every 400 constant stars (Sect. 6.2), then we find the results from open clusters are in satisfactory agreement. POPOVA (1975) has carried out a more recent compilation. Her list, which has not yet been published in detail, contains 2253 variables in 362 open clusters, always within a radius of 5 cluster radii. Table 46 gives the relative numbers of variables with respect to distance from the centre.

For the innermost region one obtains an average number of 2.5 variables per cluster. This is about double the number shown by the earlier work by KHOLOPOV (1956), from which Table 47 has also been compiled, showing how the various classes are distributed.

As the distance of variables from the centre of a cluster is used as a criterion for inclusion, the material in this table naturally contains some field stars that do not physically belong to the clusters. This is obviously the case with the three RR Lyrae stars, for example. Novae, U Geminorum stars and related types are not found in the region of clusters. On the other hand, short-period *eclipsing stars* are present in the old clusters M67 and NGC 188 (for example KUROCHKIN 1960, HOFFMEISTER 1964), as well as numerous

Table 46. Relative number of variables in open clusters

Zone	0–1	1–2	2–3	3–4	4–5 cluster radii
Relative number	1	0.47	0.39	0.32	0.29

Table 47. Variables in open clusters

Class	Number of variables	Number of clusters
RR Lyrae	3	2
δ Cephei	5	5
Algol	13	11
β Lyrae	2	2
Unclassified eclipsing stars	8	6
Mira	3	2
Semi-regular, slow irregular	6	6
Irregular	13	7
Unknown, rapid variation	14	13
RV Tauri	1	1

δ Scuti variables (Sect. 2.1.4). The significance of open clusters for the determination of the zero-point of the Period-Luminosity relationship in δ Cephei stars, because of the variables of this type which they contain, has been explained in Sect. 2.1.2 (see Table 7).

5.1.2 Globular Clusters

In the discussion of *RR Lyrae* stars (Sect. 2.1.3), it was noted that such variables are found in many globular clusters, so that they are also known as *cluster-type variables* (Fig. 143). However, the number of these variables in the clusters depends upon many factors. It reflects, in a way that is by no means fully explained, the evolutionary state of the clusters – like the relationships found in the open clusters. Other classes of variable are found in very much smaller numbers.

The description which follows is largely based on the *Third Catalogue of Variable Stars in Globular Clusters* by SAWYER-HOGG (1973) and on the fundamental discussion of this material by ROSINO (1978).

The *statistics* show (see Fig. 144) that (quoting from SAWYER-HOGG loc. cit.) in 1972 "... a recorded search for variables in 108 of the approximately 130 globular clusters belonging to our galaxy has been made. This search has yielded 2119 variables. Certainly variables do not abound in most globular clusters. Of the 108 clusters that have been examined, only 11 contain more than 50 variables each, and 81 contain fewer than 20 variables each. ... from the data in the Third Catalogue, the most frequent number is zero. There are effectively 13 clusters with no variables, if one includes NGC 6397, whose three variables are considered field stars. One variable alone is found in each

Fig. 143. The globular cluster M3. The positions of RR Lyrae variables according to the catalogue by SAWYER-HOGG (1973) are indicated

of 10 clusters." The view that globular clusters are actually rich in variables – that is in RR Lyrae stars – can also be countered by noting the large number of constant stars in such clusters, 50 thousand – 50 million.

The first part of Table 48 gives the clusters with more than 35 RR Lyrae stars, and the second part, those densely-populated clusters that have few or

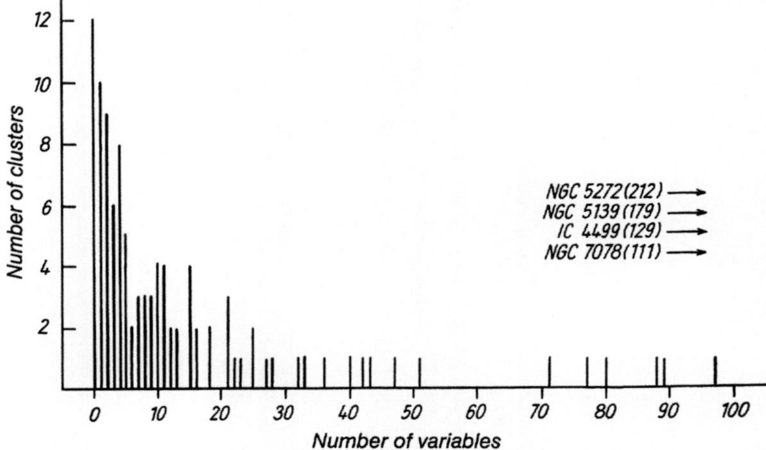

Fig. 144. Frequency distribution of variables in globular clusters (1972 data) (after Sawyer-Hogg 1973)

Table 48. Variables in globular clusters

Cluster NGC	Total no. of variables	Numbers				
		RR	RR [%]	CW	SR + L	Other important objects
5272 (M3)	212	182	86	1	3	1 EW
5139 (ω Cen)	179	142	79	6	8	3 E
IC 4499	129	112	87			
7078 (M15)	111	74	67	3		
5904 (M5)	97	90	93			1 UG
6266 (M62)	89	74	83			
3201	88	83	94			
6715 (M54)	80	63	79	1	2	2 E
6402 (M14)	77	40	52	5		1 N
7006	71	67	94		2	
6934	51	44	86			
5024 (M53)	47	33	70	1	2	
6121 (M4)	43	41	95		2	
4590 (M68)	42	37	88			
2419	41	36	88	1	4	
6981 (M72)	40	39	98			
104 (47 Tuc)	28	2	7		7	3 Mira
6205 (M13)	11	3	27	3	3	
6218 (M12)	1			1		
6254 (M10)	4			2		
6356	10					1 Mira
6637 (M69)	8					
6838 (M71)	4				1	1 E

no RR Lyrae variables (after ROSINO 1978, with additions from SAWYER-HOGG 1973). Only definite cases are included in the totals.

The richest cluster appears to be M3 (212 variables), followed by ω Cen (179) and IC 4499 (129 definite and 41 suspected objects; it is close to the southern celestial pole). The relative frequency, however, calculated from the total number of stars in each cluster, is very different. On the assumption that the total number of stars in a globular cluster is proportional to its luminosity given by M_v, it is found, for example, that NGC 5053 ($M_v = -6.1$) is 44 times poorer in stars than ω Cen ($M_v = -10.2$) as $10^{0.4(10.2 - 6.1)} \approx 44$. The first contains 11 known variables, and the second 179. Dividing by 44, we find that the relative frequency of variables in the cluster ω Cen, well-known for its richness in variables, is about 2.7 times less than in the poorly known NGC 5053 (after KUKARKIN 1972)!

The following classes of variables are present in globular clusters:

RR Lyrae, $P < 1^d$ (RR)
W Virginis, $1^d < P < 20^d$ (CW)
RV Tauri
Yellow and red semiregulars (SR)
Mira (M)
Slow irregular red giants (L)
U Geminorum and Novae (UG, N)
Eclipsing stars (E)

1202 definite RR Lyrae periods have been determined in 46 clusters (Fig. 145), and 26 variables with cycle-lengths between 100 and 219 days (13 clusters), among them 9 definite Mira stars with periods close to 200^d, like those also present in the general galactic halo. In addition, the catalogued material contains 28 definite, Population II pulsating stars with $P > 1^d$ (described in this book as W Virginis stars). Among these there is a group of 14

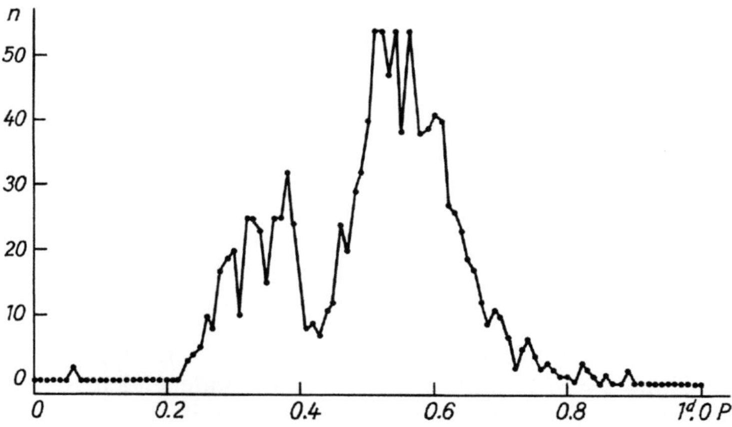

Fig. 145. Number of RR Lyrae stars in 46 globular clusters at 0.01-day period intervals (after SAWYER-HOGG 1973)

stars with $P = 1^{d}.1 - 3^{d}$, whose mean absolute magnitude ($\bar{M}_v = - 0^{M}.39$, $\bar{M}_B = - 0^{M}.01$) agrees with the Period-Luminosity relationship of the remaining W Virginis stars, whereas similar objects (known as abnormal BL Herculis stars) in dwarf galaxies in the Local Group have a luminosity that is about 0.8 mag greater (Sect. 5.2.2).

Three novae found in globular clusters should be mentioned. T Sco, which reached magnitude 7, was discovered in 1860 in M80. WEHLAU (1964) found a faint nova in M14 on plates taken in 1938, and MAYALL (1949) gave details of Nova Sgr 1943 (8^m) at the edge of NGC 6553, but which is possibly a field star.

Variables in globular clusters, with a few casual exceptions, are not given individual designations, and are thus not included in the GCVS. They are generally identified by the name of the cluster (NGC, M, etc.) and the catalogue number after SAWYER-HOGG (1973).

For photographic observation of variables in globular clusters a large instrument with a sufficiently long focal length is essential, especially if variables near the centre of the cluster are to be investigated. On the other hand, globular clusters have the advantage that a large number of variables can be covered with a single exposure. This helps research into the variability of periods in RR Lyrae stars, which has been discussed in Sect. 2.1.3. For example, WILKENS (1964) found that 30 of the 43 RR Lyrae stars investigated in M4 had variable periods. A number of investigations have been carried out concerning the globular cluster M3. SZEIDL (1965) announced that of 112 confirmed RR Lyrae stars, the periods of 22 were increasing with an average rate of 5×10^{-10} days per day; 25 variables showed a decrease at the same rate; there were 7 constant periods and the remainder showed irregular fluctuations in their periods. In the same cluster, BELSERENE (1952) had already found 27 variable periods in 202 stars that she had considered, and OSVÁTH (1957) recognized 37 variable periods in this same material (according to WILKENS 1964). M5 has also been repeatedly examined for the same purpose, for example by OOSTERHOFF (1941) and by COUTTS and SAWYER-HOGG (1969). The examples given should suffice to illustrate the amount of work that has been done; detailed references are given in the catalogue by SAWYER-HOGG (1973), and a comprehensive discussion of period changes of RR Lyrae stars in globular clusters in SZEIDL (1975).

5.2 Variables in Extragalactic Systems

5.2.1 Magellanic Clouds

The two Magellanic Clouds, the Large Magellanic Cloud ($=$ LMC) and the Small Magellanic Cloud ($=$ SMC), are galaxies, that is independent stellar systems. They may be regarded as companions to the Milky Way system, just as the Andromeda Galaxy has companions, such as M32 and NGC 205, although these are, of course, of a different type. The Magellanic Clouds

Fig. 146. The Large Magellanic Cloud. Photograph by HOFFMEISTER (Boyden Station)

belong to the class of irregular galaxies, although some have tried to see the characteristics of a barred spiral in the LMC. Lying in the region of the southern celestial pole, they are a fantastic sight in moderate and high southern latitudes. It is as if two bright clouds of the Milky Way have been moved into regions poor in stars. The northern part of the LMC (Fig. 146) lies in the constellation of Dorado, and the southern portion in Mensa. The SMC (Fig. 147) is totally within Tucana, its southern area reaching the border of Octans. The distances of the Clouds have been determined as 46 kpc (LMC) and 53 kpc (SMC). The apparent diameters amount to about 8° for the LMC and about 3° for the SMC, but both are surrounded by extended haloes, not easily detectable visually or on photographs.

The apparent visual distance modulus is about 18.6 mag for the LMC and 18.8 mag for the SMC (DE VAUCOULEURS 1978). From these figures we may

Fig. 147. The Small Magellanic Cloud and globular cluster NGC 104 (47 Tuc). Photograph by HOFFMEISTER (Boyden Station)

expect an average apparent visual magnitude of 15^m for δ Cephei stars ($P = 10^d$), and $19^m\!.5$ for RR Lyrae stars.

The Magellanic Clouds became very important with the discovery of the *Period-Luminosity relationship* for δ Cephei stars. Nowhere else in the universe can we observe so many δ Cephei stars all at about the same distance from us. It is particularly fortunate that the magnitudes are essentially free from the effects of interstellar extinction in our Galaxy, and that the internal absorption in the Clouds is only moderate. Harvard Observatory conducted pioneer work in investigating the Magellanic Clouds and their variables, as it was able to obtain photographic plates at Boyden Station and at its branch at Arequipa in Peru. (Nowadays the new large telescopes in the Southern Hemisphere, in Chile and Australia, are employed for detailed studies of the Clouds.) Some hundreds of variables were found by LEAVITT, who, as early as 1908, published a list of 1777 variables (LEAVITT 1908), of which 969 were in the SMC and 808 in the LMC.

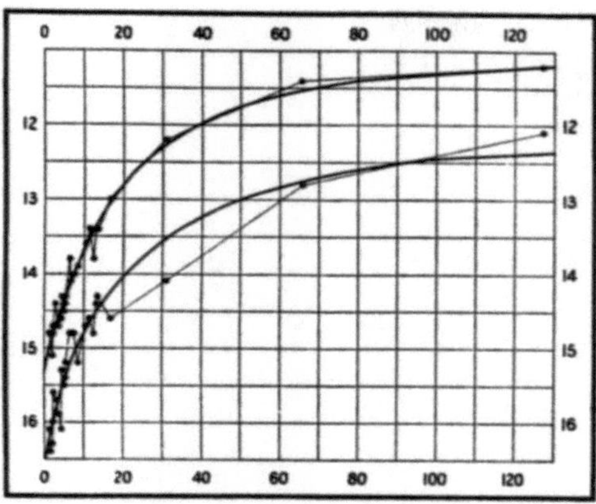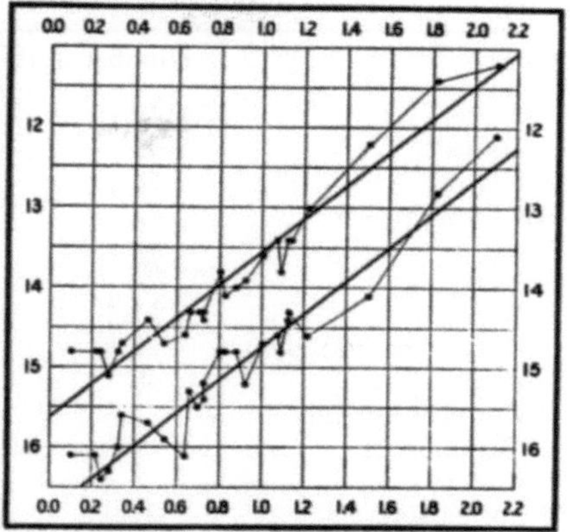

Fig. 148. The first Period-Luminosity diagrams (LEAVITT 1912). *Left*: period in days; *right*: logarithm of the period; *ordinates* apparent magnitude

Relatively short periods were determined for 25 stars, and it was these stars which gave the authoress the first indication of the dependence of the periods on the luminosity (LEAVITT 1912). The paper bears the simple title "Periods of 25 Variable Stars in the Small Magellanic Cloud". The decisive result is contained in two tables; the corresponding graphical representations of this is shown in Fig. 148. The ordinates are magnitudes; the arguments are *left* period in days, *right* the logarithm of this period. In both cases some stars with long periods are included. With the logarithmic argument there is a linear relationship. At that time the true significance of the discovery could not be known; but it did stimulate Harvard Observatory to devote a great deal of attention to the Magellanic Clouds in later work.

There is now a very considerable specialist literature about variables in the Magellanic Clouds. Considerable attention is given, for example, to the problem of the differences that exist between the statistical properties of pulsating stars in the two Clouds, and also between the Clouds and the Galaxy. The causes of these *differences* are to be sought in the *different evolutionary states* and ages of the stellar systems and groups of stars. These points were also touched upon in Sect. 2.1.2 in the discussion of the modern Period-Luminosity relationship.

The position of research on variable stars in the LMC, before the new large telescopes were introduced, can be found in PAYNE-GAPOSCHKIN (1971); she and S. GAPOSCHKIN were greatly involved in this work. Table 49 gives the numbers of variables, according to class, as published by these authors. The question marks show uncertain, but probable, classifications. Strong selection effects naturally distort the distribution, as, at $V \approx 19^{\mathrm{m}}5$, the RR Lyrae stars were difficult objects.

Since then GRAHAM (1972, 1974, 1975) has carried out detailed work on the statistics of RR Lyrae stars in the Magellanic Clouds, using the 1.5 m Ritchey-Chrétien telescope at Cerro Tololo. In just one field $1°3$ square around the globular cluster NGC 121 in the SMC, investigation of 10 inde-

Table 49. Variables in the Large Magellanic Cloud

Class	Number in the LMC
δ Cephei	1110
δ Cephei?	49
RR Lyrae	28
RR Lyrae?	2
Mira	46
Semi-regular	23
W Virginis	17
Irregular (red and blue)	321
Irregular?	48
R Coronae Borealis	5
U Geminorum	1
Novae	3
Eclipsing stars	79
Eclipsing stars?	17
Uncertain	81
	1830

pendent plate pairs yielded 92 variables. Of these, 75 are probably RR Lyrae stars in the SMC's halo, and one is an RR Lyrae star belonging to the globular cluster, in which THACKERAY (1958) had already discovered 3 similar objects. Incidentally, the centre of the field mentioned lies 2° beyond the edge of the bright portion of the Cloud, close to the globular cluster 47 Tuc (Fig. 147), which does not belong to the LMC and whose RR Lyrae stars are much brighter.

A similar programme in the LMC around the globular cluster NGC 1783 gave 63 RR Lyrae stars. The absolute magnitude of these variables seems to be the same in both Clouds ($+ 0^M5 \pm 0.2$ mag). However, there are obvious differences in the period distribution and in the relative numbers of the ab and c subclasses.

The quality of research into the rarer classes of variables in the Magellanic Clouds also continues to increase. In these cases as well it is considerably easier to determine an absolute magnitude than it is in our Galaxy, quite apart from the ease of carrying out simultaneous and comparative studies of evolutionary state, chemical composition and age. Examples are the R Coronae Borealis star *W Men* and the variable hot supergiant *S Dor*, as well as newly discovered objects of these types. *VV Cephei-like stars* (Sect. 4.7) and *symbiotic variables* (Sect. 3.1.6) are also found. FEAST (1974) has given a summary of programmes on the subject. Finally, mention must be made of the (very rare) occurrence of *novae* in the Clouds. A summary by GRAHAM and ARAYA (1971) lists 4 novae for the SMC and 10 novae for the LMC. Light-curves and spectral development appear to be similar to those in galactic objects. A preliminary estimate of the frequency gives the figure of 1 or 2 novae every 1–2 years in each of the Magellanic Clouds.

The discovery of *supernova remnants* in the Magellanic Clouds has already been mentioned in Sect. 3.2.

5.2.2 Other Extragalactic Systems

If supernovae (Sect. 3.2) are excluded, only those systems in the Local Group can be considered. This includes, apart from the Magellanic Clouds, the Andromeda Galaxy M31 and its two companions M32 and NGC 205, the relatively low-mass spiral galaxy M33 in Triangulum, and a series of dwarf galaxies. The distance modulus lies between about 24 mag (M31 and M33) and 19 mag (a dwarf spheroidal galaxy in Sculptor). There is thus a continuous gradation in distance between the Magellanic Clouds and the well-known galaxies in Andromeda and Triangulum. Although, in these latter objects, only δ Cephei stars, novae and certain irregular supergiants are of any real importance, RR Lyrae stars have been successfully studied in the dwarf spheroidal galaxies, just as they have in the Magellanic Clouds.

The number of variables that have been discovered in extragalactic objects (apart from the two Magellanic Clouds) is well over a thousand. The earlier practice of examining restricted areas down to the lowest possible limiting magnitude, using large parabolic telescopes, has now been supplemented by the introduction of large Schmidt telescopes. With these, deep, wide-angle exposures of the whole of a galaxy and its surrounding halo can be obtained.

Examples of this second method are the investigation of the Sculptor galaxy with the Michigan-Curtis Schmidt Telescope (60 cm clear aperture, field 5° by 5°), which produced more than 500 new variables from 10 plate pairs (VAN AGT 1973, 1978); or the search for novae and other variables in M31 and its immediate surroundings by L. MEINUNGER (1971) from plates taken with the Tautenburg Schmidt Telescope (134 cm clear aperture, field 3°4 by 3°4). This last author, in comparing 13 plate pairs taken between 1962 and 1970, discovered, among other objects, 4 novae in the halo of M31 (Fig. 149). It can be estimated that a similar, random coverage of the halo of our Galaxy would require about 6000 plates. This shows the advantages of studying other stellar systems. The present assumption is that a total of about 30 novae per year occur in the Andromeda Galaxy. For our Milky Way system the figure is at least 100, but for an observer within the Galaxy the possibility of observing these events is considerably reduced by the effects of interstellar dark clouds, which prevent the centre of the Galaxy, and all beyond it, from being seen.

As already mentioned, the δ Cephei stars were of particular importance for *distance determinations*. Nowadays, the procedure is generally the reverse: from the known distance of a galaxy, which is essentially the same for all the objects that can be observed within it, it is possible to try to compare the luminosity of variable stars in various regions of the system, or in several galaxies. From the agreement or differences that are found, one can gain information about the theory of stellar evolution, and in particular, the effects of different starting conditions.

A notable example of this is the class of objects known as "abnormal BL Herculis stars", which are pulsating stars with periods of 1–3 days. Mention has already been made of them in the discussion of globular clusters

Fig. 149. Novae in the halo of the Andromeda Galaxy. *Circles*: 4 objects after L. MEINUNGER (1971); *diagonal crosses*: 2 objects after SHAROV and ALKSNIS (e.g. 1975); *vertical cross*: after VAN DEN BERGH et al. (1973)

(Sect. 5.1.2). It appears that their luminosities are significantly different in the Galaxy's globular clusters, in the dwarf galaxies examined so far, and in the Magellanic Clouds. Table 50 gives the average values after ROSINO (1978).

Table 50. Luminosities of the abnormal BL Herculis stars

Stellar system	$\bar{M}_{\mathrm{B}}$
Galactic globular clusters	$-0\overset{\mathrm{M}}{.}01$
Dwarf galaxies	-0.78
LMC	-1.63

If this is really a homogeneous group it should belong to Population II because of the presence of such objects in globular clusters; their luminosity in extragalactic systems, however, is comparable to normal δ Cephei stars. The causes of this are assumed to be differences in helium or metal content (see for example the models calculated by CARSON and STOTHERS 1982), or else differences in age. However, the problem remains unsolved. Moreover, the long-period W Virginis stars ($\bar{P} \approx 21\overset{d}{.}3$) in M31, with $\bar{M}_{\mathrm{B}} = -2\overset{M}{.}3$, appear to be about 1 magnitude brighter than in galactic globular clusters (BAADE and SWOPE 1963, 1965). (As yet this type of variable has not been found in dwarf galaxies.) The frequency distribution of the periods of RR Lyrae stars also shows differences between various systems.

The examples that have been given concerning the presence and observation of variable stars in extragalactic systems touch upon some important individual areas of research, but have been taken somewhat arbitrarily from the mass of relevant publications, to which the reader is referred. Concise summaries about variable stars in dwarf spheroidal galaxies (VAN AGT 1980) and dwarf galaxies like the SMC (LAUSTSEN 1980) with references to the literature are given in *First ESO/ESA Workshop on the Need for Co-ordinated Space and Ground-based Observations*, Geneva 1980.

5.3 Active Galaxies

5.3.1 General

Among the nearly 30 000 objects in the General Catalogue of Variable Stars (GCVS) there are many that are largely neglected as, despite the large amplitudes that at least some of them show, they do not appear to be of particular significance. However, the interest in seven of these "stars" was considerably awakened when it was recognized, at the end of the sixties and in the seventies, that they were not stars! These objects follow.

X Com ($13\overset{m}{.}5 - 17\overset{m}{.}5$) and *W Com* ($13^{m} - 16\overset{m}{.}2$) were discovered at the beginning of the century by the Heidelberg astronomer WOLF (1914 and 1916). Even in the 1969 GCVS the type of variation was marked as "?" for both objects. W Com is identical with the compact radio source ON 231. A recent light-curve is given in Fig. 150.

BL Lac was described by HOFFMEISTER (1929) as probably being a short-period object, but he did not suspect that nearly 50 years later this "star" would become the prototype for a class of extraordinary celestial objects! The 1969 GCVS gives the class as "Ia?". Its magnitude varies between $12\overset{m}{.}5$ and $17\overset{m}{.}0$ pg. Figure 151 shows the light-curve of this interesting object. At the end of the sixties its identity with the radio source VRO 42.22.01 was recognized.

AP Lib was described by ASHBROOK (1942) as a "non-periodic variable". The variations take place within the limits $14\overset{m}{.}0$ and $16\overset{m}{.}4$. It is the optical counterpart to the radio source PKS 1514 − 24.

BW Tau was recognized by HANLEY and SHAPLEY (1940) as varying irregularly within the range $13\overset{m}{.}7 - 14\overset{m}{.}6$ pg. In 1968 PENSTON noted its identity

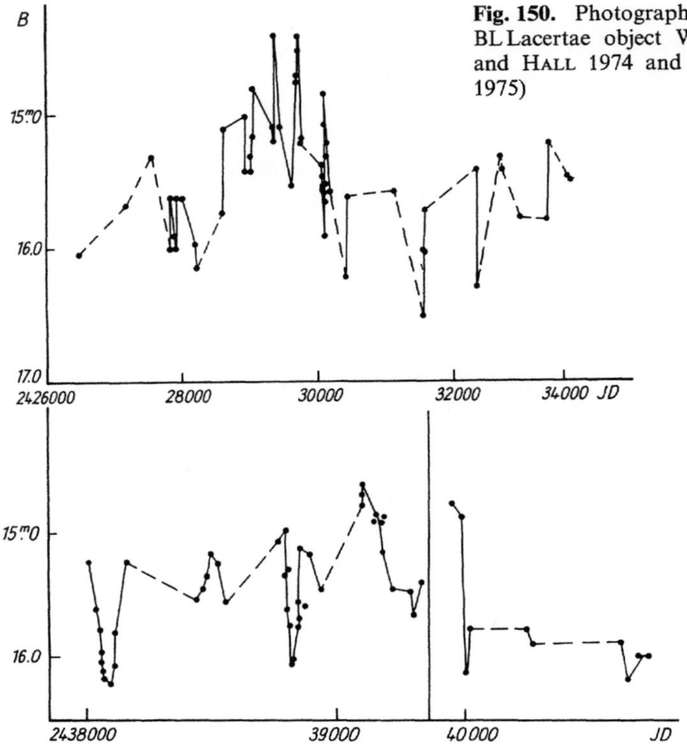

Fig. 150. Photographic light-curve of the BL Lacertae object W Com (after POLLOCK and HALL 1974 and MARKOVA and FOMIN 1975)

with the radio source 3C120. Figure 152 shows the light-curve of the object.

V 396 Her was discovered by HOFFMEISTER in 1959. In the Third Supplement (1976) to the GCVS it was still given as "RR?".

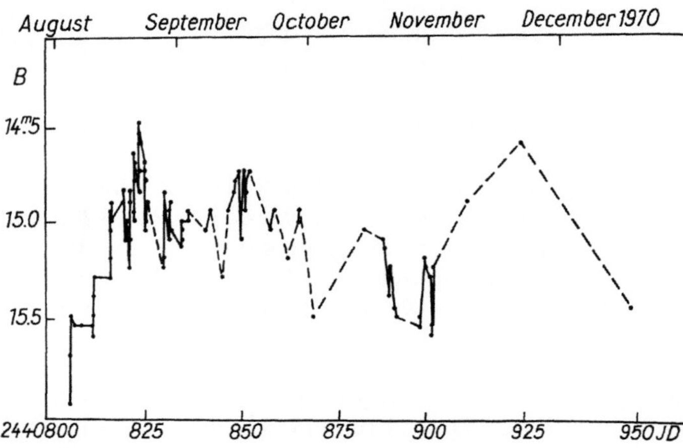

Fig. 151. Light-curve of BL Lac in the *B* region (after LÜ 1977)

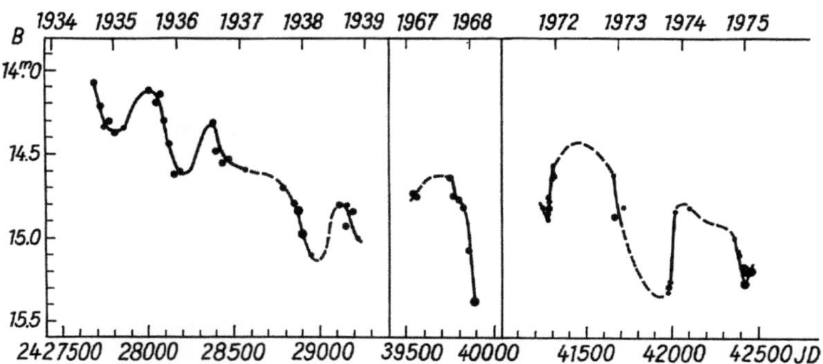

Fig. 152. Light-curve of the Seyfert galaxy BW Tau ($=3$C 120); *small points*: individual measurements; *larger points*: averages of $2-16$ measurements (after USHER 1972; with additional measurements from BERTAUD et al. 1972, 1975, and from OURASSINE and OURASSINA 1975)

GQ Com was discovered by PINTO and ROMANO in 1973. The Third Supplement (1976) to the GCVS gives the class as "L".

With the discovery of their radiation at radio wavelengths, or with investigation of their spectra, it was later realized that these objects were not variable stars, but were variable galaxies! (X Com and BW Tau are Seyfert galaxies; W Com, BL Lac and AP Lib are BL Lacertae objects; V 396 Her and GQ Com are quasars, see BOND et al. 1977.) The seven objects are thus, strictly speaking, outside the theme of this book. Optically variable galaxies are not normally given variable star designations (the seven objects mentioned above are exceptions, as they were named before their extragalactic nature was recognized).

However, we shall briefly discuss variable galaxies because the methods of discovery and analysis of the variations cannot be distinctly differentiated from those applied to variable stars. In addition, it is still not known whether some variable objects are stars or galaxies. Irregularly varying objects with UV excesses and continuous spectra (at low dispersions) could be BL Lacertae objects or else eruptive binaries (Sect. 3.1) where the absorption lines are swamped by emission.

A detailed *catalogue* of all quasi-stellar and BL Lacertae objects known in 1977 (including finder charts, comparison star magnitudes, details of the known variations, and further references) has been given by CRAINE (1977). The latest catalogue of quasi-stellar objects, also with information about optical variability and references to the literature, is that of HEWITT and BURBIDGE (1980).

If we examine the earlier chapters about variable stars, we will see that the time-scale of the variations increases considerably with the increasing spatial dimensions of the objects concerned. For the most compact objects (neutron stars, pulsars) the time-scale is seconds or fractions of seconds; in white dwarfs (ZZ Cet) it is seconds to minutes; in RR Lyrae stars, hours; in red giants, months; and in supergiants, years. At first sight, it therefore seems completely incomprehensible that at even larger, galactic dimensions, the time-scale should decline once more. BL Lac, for example, shows noticeable alterations in magnitude within a day. From this fact alone we can conclude that it is not the whole galaxy, but a small portion of it, an "active nucleus", that is variable. This active nucleus must be brighter than the whole of the surrounding galaxy, which would otherwise overpower it, and prevent its variations from being seen. The time-scale of the variations can only be so short if the dimensions of the active nucleus are no greater than roughly the size of the Solar System, as the signal giving the "command" for the system to become brighter (or fainter) cannot be transmitted from one spatial element to another faster than the speed of light. (The diameter of the Solar System amounts to about 12×10^9 km ≈ 11 light-hours.) So the active nucleus is very small and very bright (not only in visible light, but all the more in the radio, infrared, ultraviolet and X-ray regions). It has a very high energy density corresponding to a radiation temperature of $\geq 10^{12}$ K in visible light.

We shall not discuss attempted explanations of the phenomenon of active nuclei, as this is beyond the scope of this book. Moreover, it all remains in a state of flux. The following references are drawn from the very extensive literature: OZERNOY and USOV (1977), VAN DEN BERGH (1978), HAZARD and MITTON (1979), KELLERMAN (1980), KELLERMAN and PAULINY-TOTH (1981). Popular explanations can be found in KUNDT (1982) and the *Cambridge Encyclopedia of Astronomy*, Jonathan Cape, 1977, p. 363 ff.

The phenomenon of active galaxies is very complex and not fully investigated. At present three types of active galaxies are recognized that may show variability: Seyfert galaxies, quasars and BL Lacertae objects.

5.3.2 Seyfert Galaxies

These are predominantly spiral galaxies that have a very bright, hot, star-like nucleus (diameter considerably smaller than 1 second of arc), which emits non-thermal radiation (known as synchrotron radiation) and is brighter than the galaxy itself. In practically all Seyfert galaxies intense radio emission is observed. The spectrum is continuous, with both permitted and forbidden emission lines. Analysis of the spectrum indicates high temperatures and high expansion velocities for the material close to the nucleus. Two types of Seyfert galaxies are recognized.

Type 2 Seyfert Galaxies

Both permitted and forbidden emission lines are comparatively narrow, corresponding to a range of velocities of about 600 km/s. No variability has yet been noted in these objects. As far as the subject of this book is concerned, they are of no interest to us.

Type 1 Seyfert Galaxies

The *permitted emission lines* are *very wide*, corresponding to a scatter in the velocities of several thousand km/s. The *forbidden lines* on the other hand are *relatively narrow*, on average $\simeq 600$ km/s (Fig. 153).

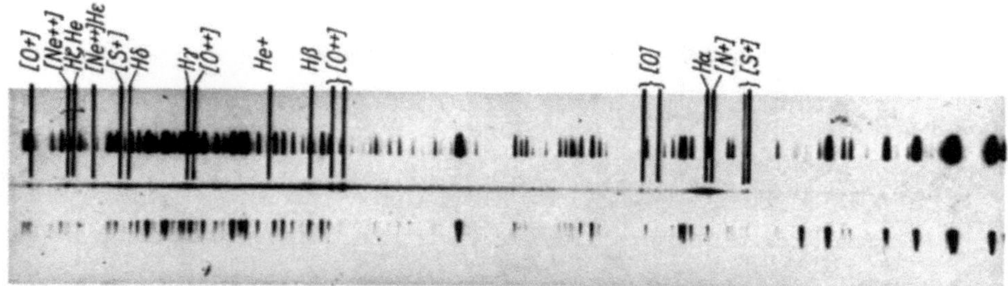

Fig. 153. Spectrum of the variable Seyfert galaxy NGC 4151 (photograph by Notni with the 2-m reflector at Tautenburg)

Optical variability has been established for some of these objects. The variation in luminosity is generally slow and irregular, with low or moderate amplitudes (with a maximum range of about 3 mag). The time-scale of the changes amounts to a few months or years, but there may be faster changes with waves having an amplitude of a few tenths of a magnitude occurring within a few days (Fig. 154). Up to 1976, 19 variable Seyfert galaxies were known.

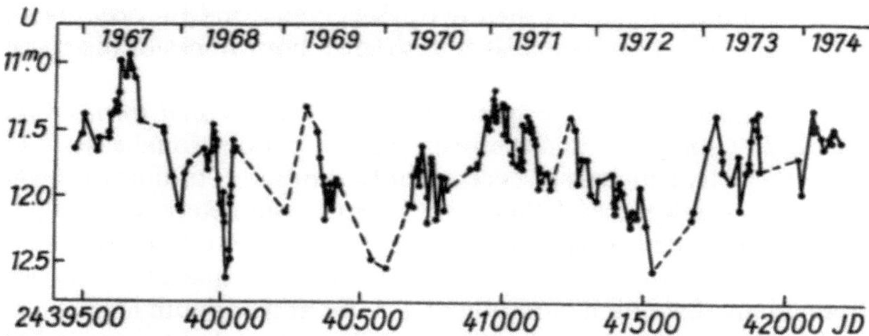

Fig. 154. Ultraviolet light-curve of the Seyfert galaxy NGC 4151 (from observations given by several authors, as compiled by Lyuty and Pronik 1975)

Most Seyfert galaxies are found when spectroscopically suspected nuclei of blue galaxies (known as Makarian galaxies) are examined for variability.

As far as we know, only three Seyfert galaxies have been discovered by the blink procedure (Chap. 6), and these were X Com and BW Tau mentioned above, and S 10838, recently discovered by Gessner (1981 b), the light-curve of which is shown in Fig. 155.

Even in normal spiral galaxies, including the Milky Way system, there are indications of "active" nuclei with synchrotron radiation and high expansion velocities in the inner spiral arms. "Active galaxies" are thus only regarded as being those where the active nucleus makes a significant contribution to the total luminosity. There is presumably a gradual transition between "normal" galaxies and the Seyfert galaxies.

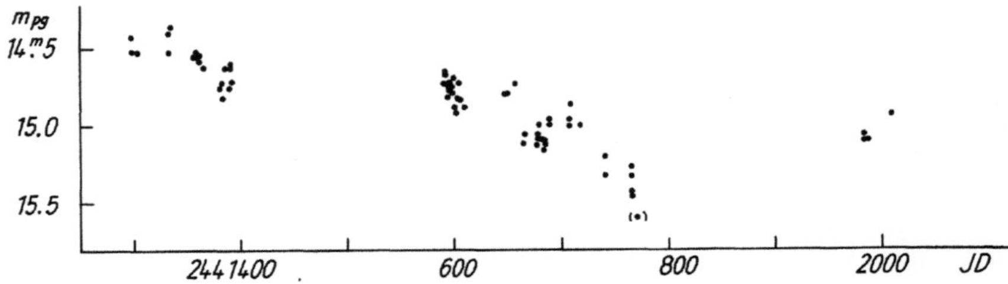

Fig. 155. Light-curve of the Seyfert galaxy S 10838 (after GESSNER 1981)

For more detailed information about Seyfert galaxies see primarily LYUTY and PRONIK (1975) and WEEDMAN (1977). HAMILTON et al. (1978) have published some finder charts and light-curves.

5.3.3 Quasars

Both photometrically and spectroscopically these objects are similar to Seyfert galaxies, but their *optical appearance is stellar*, whence their name *quasi-stellar objects* (QSO). If radio emission is observed, they are also known as *quasi-stellar (radio) sources* (QSS). The (often very considerable) *red-shift in the spectral lines*, however, shows their extragalactic nature. In a very few of the nearer quasars, it has become possible to detect faint traces of the associated galaxies in recent years.

The quasar phenomenon is probably just an *exaggerated form of that found in the Seyfert galaxies* – and there are probably transitional cases – but in the quasars the active nucleus is several tens to thousands of times as bright as the surrounding galaxy, and the latter is thus completely swamped. It should be noted that the quasars have the highest absolute luminosities for any of the objects known in the universe. The brightest quasar yet known, 3C 279, attained an absolute magnitude of $-31^{M}4$ at maximum ($11^{m}3$) – see EACHUS and LILLER 1975, where there is also an excellent light-curve. It was thus 50 000 times brighter than our Milky Way system!

Up to 1976, 36 variable quasars were known, but the list by HEWITT and BURBIDGE (1980) contained nearly 100 such objects.

The greatest luminosity amplitudes reach values of about 3 mag (in 3C 279 even > 6 mag). Changes of $0.1–0.3$ mag can occur within a week, and in one case even 2.2 mag in 13 days (3C 279, see EACHUS and LILLER 1975). The regions that emit most of the light must thus be less than 0.1 pc in extent. Examples of light-curves are given in Figs. 156 and 157.

5.3.4 BL Lacertae Objects

The BL Lacertae objects, for which the ugly and undesirable name "Lacertids" is occasionally used – undesirable because of possible confusion with meteor showers – differ from typical quasars principally in their *continuous spectrum* (see STEIN et al. 1976); emission lines are completely lacking, or at

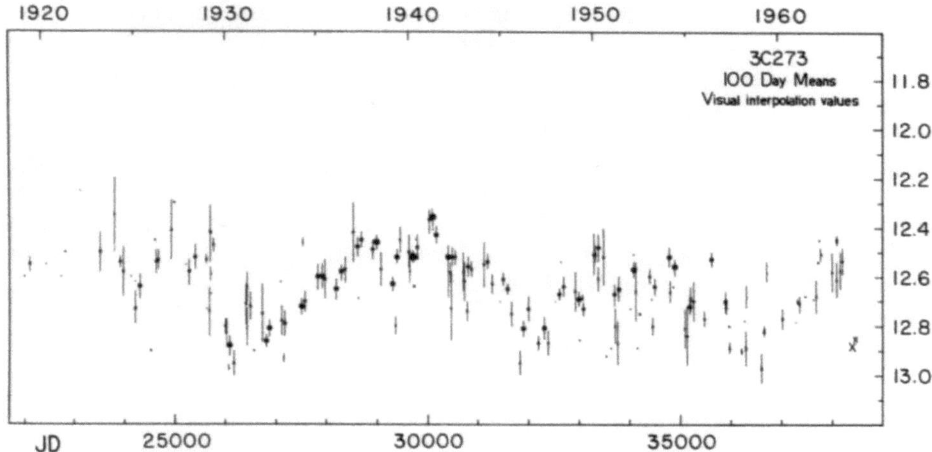

Fig. 156. The first complete light-curve of a quasar (after H.J. SMITH 1965). This light-curve of 3C273 was mainly compiled from observations made at Cambridge (U.S.A.) and Sonneberg

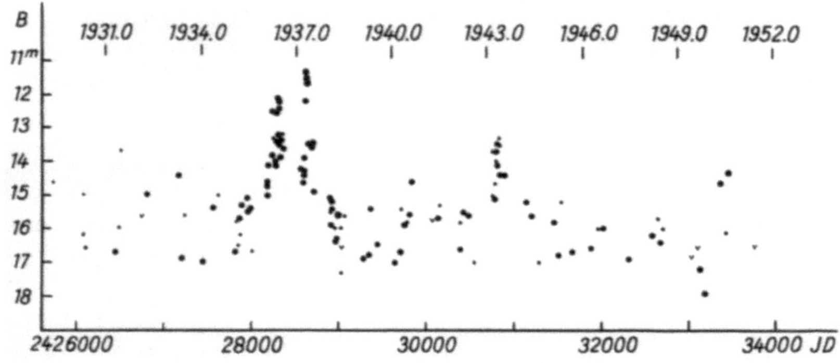

Fig. 157. Light-curve of the quasar 3C279 (Harvard observations, see EACHUS and LILLER 1975); the size of the points indicates the number of individual observations; V indicates observations fainter than the magnitude shown

best, are only very weakly indicated. For this reason, it is difficult to detect the extragalactic nature of these objects, or to determine their distances. In the few cases that have yet been detected, the BL Lacertae objects prove, in contrast to the Seyfert galaxies, to be the extrememly bright nuclei of elliptical galaxies. On average, their luminosities are a little less than those of quasars. Owing to the lack of spectral lines, the absolute luminosity of BL Lacertae objects can only be determined in a very few cases. For the brightest BL Lacertae objects it lies at about $-27^{\rm M}5$; BL Lac itself has an absolute magnitude of $-23^{\rm M}7$. As yet it is unknown whether there are any transitional cases between quasars and BL Lacertae objects (see e.g. VAN DEN BERGH 1978). WEILER and JOHNSON (1980) have discussed the relationship between quasars, BL Lacertae objects and ordinary radio galaxies. The variation in BL Lacertae objects is often very lively and is characterized by *large amplitudes* (Figs. 150 and 151). Variations in magnitude of up to 2 mag may occur within a day. In 1976, 14 optically variable radio sources were known, which are probably BL Lacertae objects (Third Supplement to the GCVS, 1976). STEIN et al. (1976) give a list of 32 BL Lacertae objects; PUSTIL'NIK (1976) discusses 34 BL Lacertae objects and 23 quasars. WEILER and JOHNSON (1980) gave 59 probable BL Lacertae objects. The list by HEWITT and BURBIDGE (1980) contains 58 BL Lacertae objects. For further discussion of variable quasars and BL Lacertae objects see KINMAN (1975) and STRITTMATTER (1976).

6. The Discovery of Variable Stars

6.1 Basic Considerations

Accidental Discoveries

As shown in Sect. 1.1, only 18 variables were known in the middle of the nineteenth century. These were mostly bright stars and long-period variables with large amplitudes that had been accidentally found in the course of other investigations – for example, in the production of star catalogues with meridian circles. In the decades following 1850 the large "Durchmusterungen" appeared, stellar catalogues with more precise positional and magnitude details: the Bonner Durchmusterung (BD), the Cordoba Durchmusterung (CoD) and the Cape Photographic Durchmusterung (CPD). Other variables were discovered as a result of this work. However, as already described, the majority have been found by photography of large areas of the sky. Photography also enabled the systematic search for variables to be carried out, in contrast to the earlier discoveries which were practically all fortuitous. At least the situation became such that planned searches on a visual basis became completely unproductive.

As might be expected, quite a few stars were falsely suspected of variability, either as a result of human fallibility or of faults in the methods used. Personal differences in estimates, particularly in the case of red stars, and many other errors could be the causes of this. Nevertheless, it is not possible to be certain that a real change in brightness did not occur in some individual cases. It is well-known that there are eclipsing stars that have very low discovery probabilities, and we can expect surprises in other types as well, not to mention the stars with very small amplitudes. A special group are the "missing BD stars", listed in the BD, that appear to be confirmed by many observations, but which are no longer present in the sky. Here again, real variability may exist but many types of error are possible.

Nowadays, if one discovers a new variable that is not included in the official catalogues, one must first of all establish if it is really new, or whether it has already been listed elsewhere but, for whatever reasons, has not yet been definitely recognized and named. Extensive literature exists concerning stars suspected of variability, as well as about those where variation has been confirmed, but which await final designations. The most recent and most extensive catalogue of this sort is that published by KHOLOPOV et al. (1982). It contains information of about 14908 objects that had not been named by 1980. It is the successor to the two catalogues prepared by KUKARKIN et al. (1951 and 1965).

Some institutions that specialize in this field of work keep files on both unnamed and named variables, so that it is easy to establish whether a newly-discovered object is already known from work elsewhere. The Sonneberg variable star index may be mentioned in particular (Sect. 9.1).

Purely visual discoveries of variable stars are rare (if one excludes bright novae), but they are possible and are mostly made when a comparison star used for another variable proves not to be constant. Such discoveries are more frequent on photographic plates. In the areas around known variables, such as the brighter Mira stars, there is less chance of discovering a new object, but in the areas around stars that have only recently been recognized as variable the chance of finding a new variable is markedly greater.

Systematic Searches

Most new discoveries come from systematic work, in which a chosen star field is investigated by means of the repeated *comparison of pairs of plates*

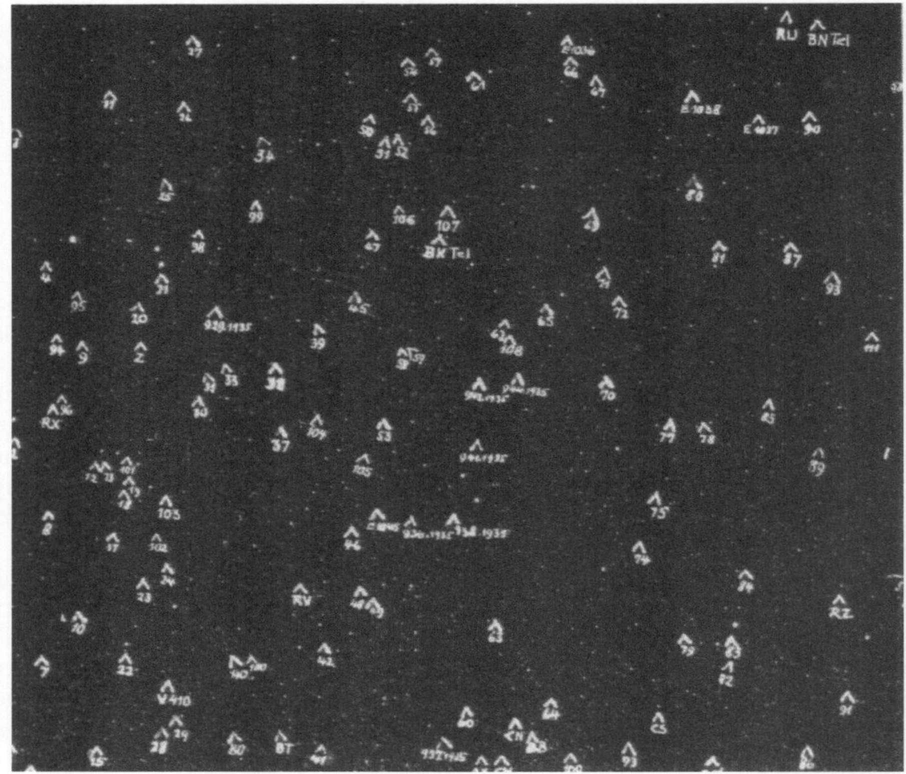

Fig. 158. Section of a discovery plate (the field of η^1 CrA, on which HOFFMEISTER has marked the variables found)

(Fig. 158). Note that in this work it is not sufficient just to compare two plates. It is more a question of having to confirm the objects that they show to be variable or suspect, before the details are announced. For this a series of plates is necessary. A Mira star can be confirmed easily by a few plates spread over a long period of time. But with Algol stars it is very much more difficult. It is not sufficient to announce just the bare fact that an object varies; the discoverer is expected to give the range and class. The determination of the elements of the variation comes at a later stage, of course. In order to meet this expectation, many plates are required. These should be distributed over a period of time in such a way that various types of variability can be distinguished. In many cases 30 plates, taken over 6 months, will permit a decision to be reached. A few of the plates should have panchromatic emulsions, so that it can be established whether the variable is coloured. (This is assuming that the 2-colour Palomar Sky Survey Atlas is not available.) However, for the most difficult Algol stars 30 plates will not be sufficient. Frequently U Geminorum stars will also need many more plates, if, as is often the case, the discovery maximum is only shown on one plate and confirmation has to be obtained. Care must always be taken in such a case as there are insidious plate flaws that cannot be distinguished from real stars. A certain insurance against such errors is given by the fact that such flaws often occur in loose clusters – but not always! Individual plates in a series may thus be affected by this scourge.

The *determination of the class* of a variable follows easily from a knowledge of their characteristics and the resulting statistical distribution of various magnitude values. Some notes will be of use.

Mira stars: no distinct changes over periods of a few days, but strong alterations over months; large amplitudes; red or reddish.

Red irregulars: the same characteristics as Mira stars, but with a smaller overall variation in luminosity, mostly less than one magnitude.

Semi-regulars: frequently with faster changes in luminosity than the previous types; moderate amplitudes; yellow to reddish.

Algol stars: mostly seen at a constant bright level; only rarely faint; amplitudes may be of any size; mostly uncoloured.

β Lyrae stars: continuous changes in light, but are more frequently seen bright than faint; deep minima are rare; uncoloured.

W Ursae Majoris stars: continuous variations; amplitude about 0.7 magnitude or less; uncoloured.

δ Cephei stars: changes mostly recognizable from one day to the next, but no rapid variation; overall as often bright as faint; slightly coloured.

RR Lyrae stars: rapid changes and markedly more frequently faint than bright; average amplitude about one magnitude; uncoloured.

U Geminorum stars: sharp rise, mostly from one day to the next; generally remain bright for some days; large amplitude; maxima infrequent; blue.

RW Aurigae stars (T Tauri stars): rapid changes on occasions; amplitudes generally small, but can amount to 4 magnitudes; mostly occur in connection with gaseous and dark nebulae.

Nova-like stars are not included; here, as with the RW Aurigae stars, positive identification is frequently only possible after an extensive series of observations.

The determination of class is also difficult with rapidly-varying periodic objects. However, the RR Lyrae stars of subclasses a and b are easy to differentiate. But generally it is not possible to distinguish with certainty between the rarer RRc stars and W Ursae Majoris stars. This is only feasible with good light-curves, and then chiefly because the minima in the eclipsing stars are sharper, and the maxima wider, than in the pulsating stars. There remain cases where this criterion also fails, and classification is only possible from the radial-velocity curve or the spectrum.

With regard to the rapidly varying stars, the ideal is to have a series of several plates taken on the same night.

6.2 Methods and Instruments

Photographs

The previous section has shown that the most sensible method of discovering variable stars is by comparing photographic plates. In doing so the scientific aim should not be the discovery of the greatest possible number of variables, but rather the acquisition of homogeneous statistical material, which will further the ultimate objectives that have been described in the Introduction. One has to use instruments that have a large field, that means refractors with relatively short focal lengths. These may be divided into two groups: small objectives up to about 100 mm in aperture, with very large fields (approximately 30×30 degrees), and mostly of the Tessar type; and larger objectives with fields of about 10×10 degrees. Objectives of the first sort permit the whole of the visible sky to be monitored without very great expense, and also under only moderately favourable climatic conditions (Fig. 159). The larger instruments, of which SONNEFELD's four-lens objective with an aperture of 400 mm and 1600 or 2000 mm focal length is probably representative, must be reserved for the observation of selected fields (Fig. 160). Because of their small fields, reflecting telescopes are generally unsuitable, except for special tasks such as research on globular clusters. Schmidt telescopes are more suitable, although again they cannot compete with lens-type objectives for size of field. Additionally, the stellar images given by Schmidt telescopes are so small that the recognition of possible variability is made more difficult.

What are the prospects for success? Let us assume that two plates are to be compared, which reach magnitudes 16 or 17 and are separated by at least a few months. According to previous experience (see RICHTER 1967a), one variable may be expected to about every 400 normal stars. One plate, which contains about 100 000 stars, thus has about 250 variables. This figure would naturally only be attained as a result of the comparison of a number of plate pairs.

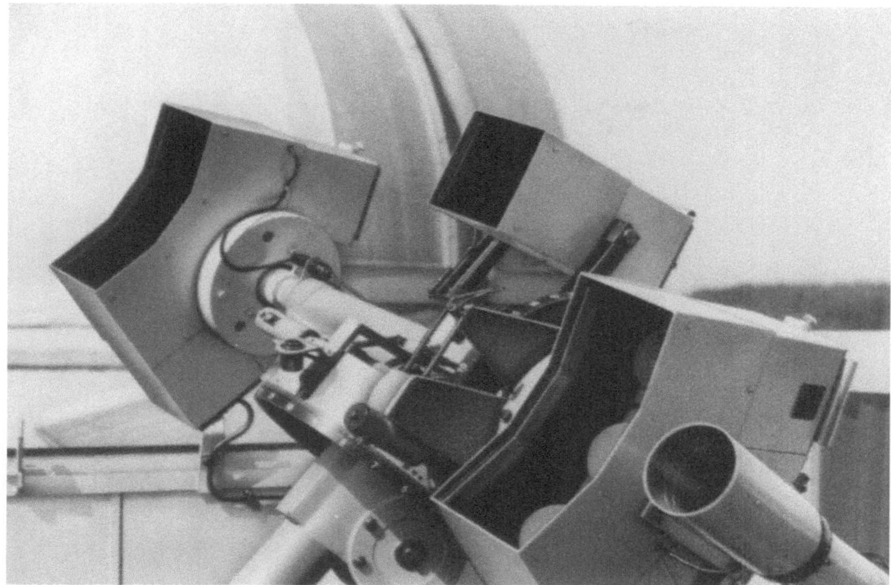

Fig. 159. Seven cameras of the Sonneberg Observatory Sky Patrol on a single mounting

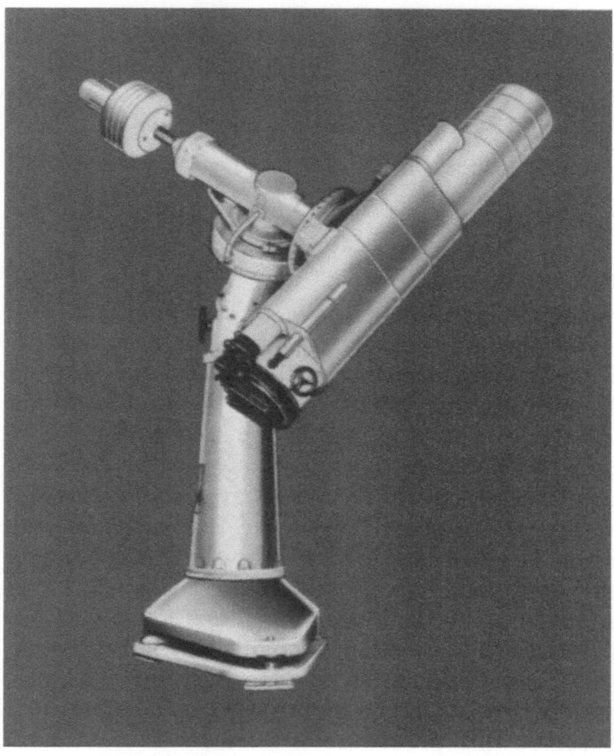

Fig. 160. 400-mm astrograph by VEB Carl Zeiss Jena

Fig. 161. Plate comparator by VEB Carl Zeiss Jena (with the late C. HOFFMEISTER at the machine)

The Comparator

Comparisons are made by using an instrument known as a comparator (Fig. 161). In the *stereoscopic method* the observer sees the left-hand plate with the left eye and the right-hand plate with the right eye. The plates are adjusted so that the images coincide, and the observer has the impression of looking at a single plate. However, this is only the case if the two plates are of comparable quality. Motion of an object approximately in the direction joining the two eyes produces a stereo effect. The object appears to be either in front of, or behind, the plane of the plate. If an object is present on only one of the plates then its optical appearance is altered and it is immediately recognized as being unusual. This also happens with all plate flaws. If a star image is large on one plate and small on the other, or if it has equal sizes but unequal densities, then an experienced observer recognizes it immediately. The impression on the observer is difficult to describe precisely, but variables, as long as the plates show a sufficient difference, simply appear unlike normal stars. With an suitable observer this procedure is very productive, once sufficient experience has been gained. A prerequisite is that the pair of plates should be well matched. Success largely depends upon their being suitably paired.

The *blink method* is somewhat easier to use. For this the binocular head is removed and replaced by a monocular. A blink mechanism, powered either by hand or by a motor, alternately presents the images of the left-hand and right-hand plates to the ocular in moderately rapid succession. The image of a variable "pulsates" and can be easily recognized. The procedure is immediate and hardly needs any practice on the part of the observer.

The Positive-Negative Method

A completely different method was tried many decades ago at Harvard College Observatory and applied with success. From one of the two plates to be compared a contact positive is made on glass. If this positive and the other negative are superimposed then normally the stellar disks will cancel one another. If a variable is bright on the positive plate, however, and faint on the negative, then when they are combined a bright disk with a dark core will appear, or, in the case of only a small difference in brightness, a bright ring. Another variant has been developed in Holland and described by BORGMAN (1956). In this version, the combined picture appears on a television screen and can be examined by the observer with both eyes, while sitting in a comfortable position. Attempts have also been made to use *colour filters* to assist the eye in the recognition of differences. WACHMANN (1961) has reported very good results with such a method, recommended earlier by PLAUT and BORGMAN (1954). But the proficiency of the observer still plays a part.

All these procedures are the same in that the plates are searched in strips. All variables or suspected objects that are found are marked, preferably on the glass side of one of the plates. The results of the comparison are then checked in order to decide what is real and what is new.

Automation

In the electronic age the idea of a fully automatic machine, capable of discovering variables, naturally suggests itself. The principle is relatively simple: the two light paths in a stereo comparator are fitted with electron multipliers and the equipment is constructed so that it reacts to differences in the two photo-currents. It is not difficult to build a machine that will print or punch onto paper tape the co-ordinates of the position that it is examining. This sounds a good idea, but the technical effort needed to construct such a machine, which has to be controlled by a computer, is very considerable. Apart from this the machine records all plate flaws, and as these are far more frequent than the variables, the idea is impractical in this form. A great improvement can be made by choosing small measuring diaphragms as, with suitable programming, the machine can examine every suspected object and determine whether it is a plate flaw or not from the way in which the density is distributed within the supposedly stellar image. The expenditure is naturally still greater, and even so it is still not possible to say with certainty that a particular case is not a plate flaw. Another method of excluding plate flaws would be to take simultaneous pairs of plates with a double astrograph, and to construct a quadruple comparator that would only register a signal arising from *both* plates of the pair that were simultaneously exposed.

Machines already exist that can evaluate plates fully automatically for surface photometry and spectroscopic work, and after modification might be suitable as a "discovery machine". At present, however, few machines seem to operate in a truly satisfactory manner.

6.3 The Theory of Discovery Probability

Let us assume that a star field is to be examined by one of the methods described in the preceding section, with the object of finding as many as possible of the existing, but as yet unknown, variables in the field. Based on this examination, we want to gather *statistics about the variables* according to number and class. As well as any new objects, one will naturally find those that are already known, and these must also be recorded. This is important, as otherwise the rules of probability theory cannot be applied in the appropriate manner to obtain the number of variables that remain undiscovered. In the second and later comparisons, the known and new groups of objects will be joined by a third, that of re-discovered new objects, that is those that were found in a previous comparison. For statistical purposes they are to be treated with the known objects.

The concept of the *discovery probability* w is of great importance for the theory. If it requires an average of n comparisons to discover a variable of a particular class and brightness, then $w = 1/n$. The idea is most simply illustrated for an Algol star. If we have a plate that shows the star at its bright normal light and want to know the probability that another plate, to be compared with it, will show the star fainter and will thus lead to discovery, then the value is obviously given by the ratio D/P where P is the period and D the duration of the eclipse. To be precise, this probability value is too high, because discovery is only possible when the decline has passed some threshold value. If one compares any two plates in a series, then the probability of observing an eclipse minimum is

$$w = 2\frac{D}{P}\left(1 - \frac{D}{P}\right),$$

where the second factor takes account of the possibility that the star could be faint on both plates, resulting in a smaller value of w. In general it may be assumed that the value of w in Algol stars lies between 0.05 and 0.15. We shall return later to the difference between the theoretical and actual discovery probabilities.

We make the simplifying assumptions that there are N undiscovered variables in a field to be examined. and that these variables have a uniform discovery probability w. This simplification is not necessarily unrealistic. It approaches the conditions that actually apply if we do not consider all the variables in the field, but only homogeneous groups, RR Lyrae stars of subclasses a and b, for example. In many globular clusters we would include nearly all of the variables with such an approach. According to theory we may expect to find Nw objects in the first comparison, and $(N - Nw)w$ in the second, as the number of undiscovered variables is reduced to $N - Nw$ after the first comparison. We thus obtain the scheme:

$$
\begin{aligned}
\text{1st comparison} \quad A_1 &= Nw & &= Nw \\
\text{2nd comparison} \quad A_2 &= (N - Nw)w & &= Nw(1 - w) \\
\text{3rd comparison} \quad A_3 &= (N - 2Nw + Nw^2)w &&= Nw(1 - 2w + w^2) \\
\text{4th comparison} \quad A_4 &= \dots\dots\dots\dots\dots\dots &&= Nw(1 - 3w + 3w^2 - w^3)
\end{aligned}
$$

and at the n-th comparison

$$
\begin{aligned}
A_n &= Nw\left[1 - \binom{n-1}{1}w + \binom{n-1}{2}w^2 - \binom{n-1}{3}w^3 + \dots \mp \binom{n-1}{n-1}w^{n-1}\right] \\
&= Nw(1 - w)^{n-1}.
\end{aligned}
$$

This last formula can be rearranged, so that the unknowns N and w may be calculated, as the values A_i (where $i = 1, 2, 3$, etc.) are obtained from the plate comparisons. If A_i and $A_{i'}$ are two such values, where $i > i'$, then by division we obtain the corresponding equations

$$
\frac{A_i}{A_{i'}} = \frac{(1 - w)^{i-1}}{(1 - w)^{i'-1}} = (1 - w)^{i-i'}.
$$

Fundamentally one can count i from any values within the series of comparisons. If the zero point is set at the beginning of the series then

$$
A_i/A_1 = (1 - w)^{i-1}
$$

can be used to determine w. The total number N of variables is deduced from

$$
A_1 = Nw.
$$

This relationship is valid for the beginning of the series, but it can be similarly calculated for other points.

The question of how many variables can be expected in a larger number of comparisons is of practical significance. So too is how great an expenditure of time and effort must be made to discover a certain percentage of the variables that are still unknown. The total number of new discoveries to be expected theoretically from the first n comparisons is given by the summation of the n part series

$$
\sum_1^n A_i = N\left[\binom{n}{1}w - \binom{n}{2}w^2 + \binom{n}{3}w^3 - \dots \mp \binom{n}{n}w^n\right] = N[1 - (1 - w)^n].
$$

If we assign an index i to the value N, we can express the total after the i-th comparison by the value N_i. Between N_i and the value N_0, applicable at the beginning of the series, we have the relationship.

$$N_0 = N_i + N_0[1 - (1 - w)^i] = N_i(1 - w)^{-i}.$$

As is well-known the binomial series is convergent for $w < 1$. This relationship is always fulfilled here as the probability w is always a proper fraction. However, the rate of convergence is concerned in the question of how many plate-pairs must be compared – for a given value of w – in order to find a certain percentage z of the undiscovered variables. In this case, the equation

$$1 - (1 - w)^n = z/100$$

·must be solved for n (HOFFMEISTER 1933).

Table 51. Number of comparisons required

	$z = 70\%$	80%	90%
$w = 0.4$	2	3	5
0.2	5	7	10
0.1	11	15	22
0.05	23	31	45
0.03	40	53	76
0.02	60	80	114
0.01	120	160	229

Table 51 shows the relationship between w and the total number of comparisons that will be required, on average, to find 70, 80 or 90% of the variables in a given field.

It should be noted that these values are unrealistic, as the variables in a given field have a whole range of values of w. The figures only have any real significance when a strictly limited group of variables with approximately equivalent discovery probabilities is considered. Nevertheless, they do allow one to estimate the amount of effort that will be required to achieve a certain target.

VAN GENT (1933) considered the problem about the same time as HOFFMEISTER, in a practical application to a field in Corona Austrina that is rich in variables. He showed that the probability that a variable will be discovered k times in n comparisons ($k = 1, 2, 3 \ldots n$) is

$$a_k = N\binom{n}{k} w^k (1 - w)^{n-k}.$$

In addition

$$G = nw\,\frac{N}{N - a_0}$$

is the mean frequency with which a variable, already discovered, was found in n plate comparisons.

Comprehensive work by Kvíz (1956a) was chiefly concerned with the probability of meteors being observed by several observers, but he later dealt with its application to the discovery of variable stars (Kvíz 1959). Here Kvíz investigated certain assumptions that are made in the theory just described, and which are not fulfilled in practice. We shall return to this point later.

RICHTER (1967a) has published a re-examination of the problem and an extension of the theory relating to discovery probability. He starts from the fact that in practice variables never have the same discovery probability w not even when homogeneous groups of variables are involved. This is because of the differences in the shape of the light-curves, amplitudes, magnitudes, etc. Instead the discovery probabilities scatter around a mean value $\bar{w}$. As RICHTER has shown (1967a) a_k is no longer given by VAN GENT's formula mentioned above but by

$$a_k = N \binom{n}{k} \prod_{i=1}^{k} (h + i) \prod_{i=1}^{n-k} (g - h + i) \bigg/ \prod_{i=1}^{n} (g + 1 + i)$$

where

$$g = \frac{\bar{w}(1 - \bar{w})}{\sigma^2} - 3$$

$$h = \frac{\bar{w}^2(1 - \bar{w})}{\sigma^2} + 1 + \bar{w}$$

($\bar{w}$ = mean of the discovery probabilities, σ = standard deviation in the discovery probabilities). If the value of a_0 from the equation mentioned earlier

$$G = nw \frac{N}{N - a_0}$$

is specially entered, then an arbitrary number of pairs of values $\bar{w}$ and σ can be found to satisfy this equation. The most likely pair of values are those that best explain all the observed values a_k. For this a computer is helpful. However, one can obtain the right pair of values for $\bar{w}$ and σ, without a great deal of effort, with the help of suitable nomograms, published by RICHTER (1967a). Naturally, the same considerations and method of calculation of the values of $\bar{w}$ and σ can also be applied to HOFFMEISTER's technique.

The mean discovery probabilities calculated with this later method are smaller than those calculated by VAN GENT's technique, by about the following factors (for 5 plate comparisons):

Irregulars, RW Aur, U Gem	0.65
Algol	0.70
β Lyrae	0.76
Long-period, semiregular,	
RR Lyr, δ Cep	0.84 .

Kvíz notes that the discovery probability of any given variable also differs from one pair of plates to another, owing to the varying quality of the plates. This problem can, however, be minimized if particular care is taken in choosing the plates to be compared. Then the effect is small. Let us assume that in the various plate comparisons the discovery probability w varies by a factor F. Then after n comparisons the mean will be:

$$\bar{w} = \frac{(n-1)\,w}{n-2+4\,F/(1+F)^2}$$

For $F = 2$ we obtain:

$$n = 2 \qquad 4 \qquad \infty$$

$$\frac{\bar{w}}{w} = 1.125 \quad 1.030 \quad 1.000\,.$$

The error can thus be neglected. KIANG (1962) came to the same conclusion in a more general discussion.

In a later work, RICHTER and I. MEINUNGER (1972) have published formulae that simultaneously take into account both effects (variation in discovery probability from one object to another, and from plate comparison to plate comparison).

6.4 Determination of Discovery Probability from Given Light-Curves

Kvíz (1956b) has described a *geometrical method* of determining the discovery probability from light-curves.

In Fig. 162, which is taken from Kvíz's discussion and only slightly altered, a light-curve with a sine-wave form is shown above. Δm is the minimum difference in brightness that must be attained for discovery to be possible. The strip of width $2\Delta m$, which one must imagine as lying at any arbitrary position on the light-curve, always defines the portion of the curve within which the difference in magnitude is too small. If we project this to give the phase scheme shown below, as illustrated for phase zero, we can see in which portion of the light-curve discovery is possible, for each of the 24 phase points. In other words, we find the phases that the star must have on the second plate for it to be discovered. The ratio of the shaded portion to the total area of the square is then the discovery probability. Kvíz also gives a curve of the relation between discovery probability and amplitude, expressed in units of Δm. We shall return to this point.

HOFFMEISTER (1962a) has analysed a series of light-curves by a method similar to that of Kvíz. He confined his investigations to the RR Lyrae class, which is the most homogeneous. Three typical curves were taken as a basis,

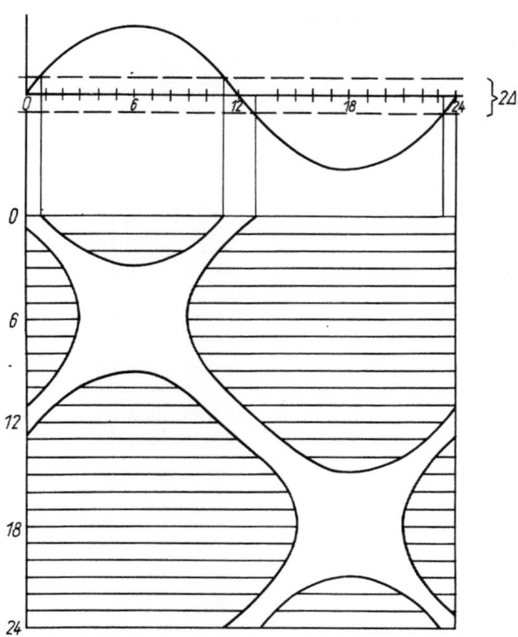

Fig. 162. Graphical determination of discovery probability for a variable star, after Kvíz (see text)

corresponding to the subclasses RRa, RRb and RRc. Table 52 gives the *theoretical discovery probability* thus calculated, relative to the critical difference in brightness Δm and amplitude A.

A simple method of deriving *empirical discovery probabilities* is as follows: if a variable is found k times in n comparisons of independent pairs of plates, then $w = k/n$. Naturally, in any individual case this value is strongly determined by chance, but usable mean values can be obtained from many stars of the same class. In a larger study of this sort, based on plates taken by HOFFMEISTER at Boyden Observatory near Bloemfontein (Republic of South Africa) with a 250/1250 mm objective, and again restricted to RR Lyrae stars, the data given in Table 53 were obtained. Here $\bar{m}$ indicates the mean value of the average magnitude $0.5(m_{\max} + m_{\min})$ and n the total number of stars.

Table 52. Theoretical discovery probabilities

I. Sub-class RRa				II. Sub-class RRb			III. Sub-class RRc (Sine-wave)	
	A				A			A
Δm	1.5 mag	1.0 mag	0.5 mag	Δm	1.0 mag	0.5 mag	Δm	0.5 mag
$0^{m}10$	0.79	0.65	0.55	$0^{m}10$	0.79	0.62	$0^{m}10$	0.67
0.15	0.68	0.59	0.45	0.15	0.72	0.50	0.15	0.56
0.20	0.64	0.54	0.35	0.20	0.64	0.40	0.20	0.47
0.25	0.60	0.50	0.27	0.25	0.57	0.31	0.25	0.37
0.30	0.57	0.46	0.19	0.30	0.51	0.23	0.30	0.32

Table 53. Empirical discovery probability (RR Lyrae stars)

Amplitude 1.5 mag			1.0 mag			0.5 mag		
$\bar{m}$	w	n	$\bar{m}$	w	n	$\bar{m}$	w	n
13.58	0.29	6	11.80	0.20	5	11.25	0.18	3
14.25	0.23	9	13.00	0.21	5	12.62	0.30	4
14.75	0.28	8	13.50	0.24	23	13.25	0.16	11
15.25	0.22	13	14.00	0.26	39	13.75	0.17	29
15.75	0.16	15	14.50	0.24	72	14.25	0.16	25
16.25	0.13	8	15.00	0.24	62	14.75	0.14	40
16.75	0.14	2	15.50	0.16	119	15.25	0.13	44
			16.00	0.13	52	15.75	0.10	25
			16.50	0.10	74	16.25	0.09	14
			17.00	0.07	6	16.75	0.06	4

The field of 67 Oph, observed with the 400/1600-mm Astrograph, was used as a control. Ten comparisons were available, and 24 RR Lyrae stars gave $\bar{w}=0.20$. Stars were found as follows:

		$\bar{m}$
10 stars	once	15.65
7 stars	twice	15.36
3 stars	3 times	15.05
2 stars	4 times	15.50

Most are faint objects. Naturally more plates than the 10 comparison pairs were used to establish the cases that were only discovered once. Obviously the whole of the available material should be used to determine discovery probabilities. Note, however, that then the material will not be as homogeneous as with a restriction to the comparison pairs.

BORGMAN (1956) has employed a similar procedure, which will be used in the next section.

Table 53 shows that the discovery probability lessens with both increasing and decreasing mean brightness $\bar{m}$. This is primarily because the photographic plate's density curve is steepest over a certain optimum range of brightness. Both at greater and lesser magnitudes the curve is flatter; a greater difference in brightness Δm is necessary for the variability of a star to be recognized.

6.5 Comparison of Theoretical and Empirical Discovery Probabilities – The Influence of Observer and Method

The tables in the previous section show that the values of discovery probabilities found statistically are *barely half* those expected from theory. At first sight this result is surprising. The cause is to be sought in the fact that theory assumes ideal plates and an ideal observer at the comparator. In the real world these conditions are not fulfilled. This must be the correct interpretation, and is also supported by experiences described in the literature.

Anyone who has worked extensively with sky photographs knows that the quality is very variable and that it is often very difficult to pick out good pairs of plates. The faults arise from focussing and guiding, and from atmospheric effects on the sharpness and steadiness of the images. Even at focal lengths of 1 metre, the latter effects are strongly in evidence.

The observer's influence is also doubtless very considerable, as the visual methods – the stereoscopic procedure probably more than the blink method – require a long period of practice if they are to be fully effective. HOFFMEISTER (1933) deduced "coefficients of efficiency" from 121 plate comparisons undertaken for the photographic sky patrol. He found a rise to about three times the initial value over the course of about 4 years. Even after those 4 years (in 1933) there were still no signs that a stable value had been reached. The cause of this is certainly to be found in the stereoscopic procedure, where recognition of the smaller differences in brightness requires a much longer period of practice than with the blink method. It is also significant that new discoveries are frequently made at the very faintest magnitudes reached by the plates.

RICHTER (1967 a) has investigated HOFFMEISTER's numerous comparisons of pairs of plates taken with a 170/1200 mm triplet, establishing an increase in performance from 1927–1930, but with a practically stable value from then until 1945. These observations were made by the stereoscopic method, as were those described next.

A good measure of performance, and its changes with time, is the total number of variables found in one comparison, that is the sum of the previously known, re-discovered, and new objects. HOFFMEISTER compared the results found for 181 plate pairs, taken with four-lens objectives, which were spread over 30 fields, each having an area of about 103 square degrees. The comparisons fall into two groups, well separated in time: 1945–1946 and 1962–1967. The objective and comparator were not identical in both series, but the technical data approximately agree. The objective with which the first series of plates were taken had a somewhat shorter focal length than that put into service in 1961. The difference in the field sizes is taken into account in the following discussion. In addition, the type of plates had to be changed. The ratio between the mean success rate for the 57 plate pairs compared in 1945–1946, and that for the 124 plate pairs, covering the same fields, compared in 1962–1967, is 1.000 : 0.966. It is thus possible to say that the efficiency of the objective + plate + comparator + observer system remained practically constant over a period of 20 years. There could be reasons for the decrease of 3.4%, if it is real. However, as 11 of the 30 fields showed an increase in efficiency, the final result must be strongly governed by chance.

BORGMAN (1956) defined a "quality function" $Q(\Delta m, m)$ by the integral equation

$$w(A, m) = \int_0^1 Q(\Delta m, m)\, \Theta\left(\frac{\Delta m}{A}\right) d\left(\frac{\Delta m}{A}\right)$$

which he uses to compare various procedures, concluding that the "electronic scanning" method which he employed is several times as effective as the blink

Table 54. Empirical discovery probabilities

Class m_Max \ A	δ Cep, RRab		RRc, W UMa	β Lyr
	0.3 – 1.0	1.1 – 1.8	0.3 – 1.0	0.3 – 1.0 mag
] 12^m	0.15	0.20	0.14	0.06
12 – 13	0.18	0.23	0.14	0.06
13 – 14	0.14	0.17	0.11	0.04
14 – 15	0.08	0.10	0.06	0.03

Class m_Max \ A	Algol	$D/P = 0.03 - 0.12$		$D/P \geqq 0.13$		
	0.3 – 1.0	1.1 – 1.8	1.9 – 3.0	0.3 – 1.0	1.1 – 1.8	1.9 – 3.0 mag
] 12^m	0.03	0.05	0.09	0.05	0.09	0.15
12 – 13	0.03	0.06	0.10	0.06	0.11	0.17
13 – 14	0.04	0.06	0.05	0.05	0.10	0.15
14 – 15	0.02	0.04	0.05	0.04	0.06	0.08
15 – 16	0.02	0.02		0.03	0.03	

Class m_Max \ A	Long period, semi-regular, RV Tau, moderate periods				
	0.5 – 1.0	1.1 – 1.8	1.9 – 3.0	3.1 – 4.9	$\geqq$ 5.0 mag
] 12^m	0.09	0.21	0.39	0.52	0.56
12 – 13	0.10	0.23	0.39	0.44	
13 – 14	0.10	0.20	0.29	0.32	
14 – 15	0.06	0.15	0.21		
15 – 16	0.04	0.07			

Class m_Max \ A	Irregular			
	0.3 – 0.6	0.7 – 1.0	1.1 – 1.8	1.9 – 3.0 mag
] 12^m	0.04	0.07	0.10	0.20
12 – 13	0.04	0.08	0.12	0.19
13 – 14	0.03	0.07	0.11	0.19
14 – 15	0.02	0.05	0.07	
15 – 16	0.02	0.04		

procedure of van Gent and Ferwerda. $Q(\Delta m, m)$ is the probability that an observer at the comparator will recognize the difference in magnitude Δm of a star (close to the apparent magnitude m) on a particular pair of plates. $\Theta(\Delta m/A)$ is the probability that a particular type of variable star of amplitude A has a difference in magnitude Δm on this pair of plates.

In the work by Richter, already frequently mentioned, empirical values are also deduced for the discovery probabilities of various classes of variables, which come from the comparisons made by Hoffmeister of pairs of plates taken with the 170/1200 mm triplet. The limiting magnitude of these plates is approximately 16^{m}5. A summary of the results is given in Table 54.

To avoid possible misinterpretation the following should be noted. When the table indicates that, for example, a long-period star with a maximum magnitude between 12^m and 13^m, and an amplitude of 4 mag, has a discovery probability of 0.44, this means that on an arbitrary pair of plates of the field, taken with a suitable time interval, the chance of it being found is 44%. Two factors determine this value. First, the probability that with a maximum amplitude of about 4 mag, the brightness of the star will differ sufficiently on the two plates for it to be recognized. Second, the probability that the observer will really note it, and that accidental effects on the image, such as the normal variation in density and image diameter, or other disturbing effects in the emulsion, will not prevent recognition. (It should also be noted that the experiments described here were carried out with more than one instrument, that is with the 170/1200 mm triplet with a limiting magnitude of about $16^{m}5$, and with the four-lens objective 400/1600 mm having a limiting magnitude of about 18^{m}. It should be possible to make a comparison, introducing limiting magnitude as an argument.)

It can be shown, particularly in the case of the RR Lyrae stars, which can be treated most easily, that the empirical probabilities are found to be considerably smaller than those obtained from given light-curves or from theory. In a purely formal sense, one could obtain approximate agreement if the minimum difference for recognition were raised to over 0.5 mag. But this is not generally admissible, as variables are not uncommonly found with a total amplitude of less than 0.5 mag. As already described, the primary cause is that theory has to assume ideal plates and an ideal observer, but that in the complex process of discovery there are various deviations, both of a technical nature and also arising from various mental and physical effects. In addition, in the case of rapidly varying stars, including RR Lyrae stars, it is well-known that the relatively long exposure times tend to flatten the photographic light-curves.

6.6 Discovery Statistics

As described in the Introduction, prior to the application of photography the discovery of variable stars was practically only by chance. Photographs of the same field at various times first made it possible to systematically search for unknown objects by comparing the plates. For some decades the leader in this field was Harvard College Observatory, Cambridge, Massachusetts, under the direction of PICKERING and SHAPLEY. The investigation of the two Magellanic Clouds, mainly by Miss LEAVITT, took place in the period around and after the turn of the century.

Apart from ARGELANDER, early observers yet to be mentioned are POGSON with 14 new discoveries, and HIND with 20. The first catalogue of variable stars was published by PIGOTT; it contains 12 recognized cases and 14 doubtful ones. Then ARGELANDER's list, already mentioned, followed in 1844 with 18 stars, and in 1868, in the *Vierteljahresschrift* of the Astronomische Gesellschaft, SCHÖNFELD and WINNECKE's catalogue with 126 stars. In the period between 1884 and 1904 there were several publications by GORE and CHANDLER. The great increase was then shown by the publications of the

Table 55. Distribution of the various classes of variables

Class	Number			
	1948	1958	1969	1976
δ Cephei and W Virginis	497	610	705	773
RR Lyrae	1720	2426	4470	5817
δ Scuti and RRs	0	5	64	157
Mira	3025	3659	4568	5212
Semiregular and irregular (L)	2019	3045	3916	5151
RV Tauri	72	92	104	105
β Cephei	6	11	20	50
α_2 Canum Venaticorum	0	9	28	73
Novae	114	146	166	194
U Geminorum	77	112	215	253
Z Camolopardalis	15	15	20	32
Nova-like, γ Cas, Z And and S Doradus	25	35	48	107
UV Ceti and related classes	0	15	99	838
T Tauri	173	590	979	1117
R Coronae Borealis	35	39	32	40
ZZ Ceti	0	0	0	7
BY Draconis	0	0	0	9
Eclipsing stars (all types)	1913	2763	4051	4714
Ellipsoidal variables	3	5	8	10
Pulsars	0	0	55	147
Optically variable quasars and galaxies	0	0	66	102
Unknown or unclassified	1071	992	836	813
Probably constant	152	142	148	148

Harvard College Observatory: PICKERING 701 stars in 1903; CANNON (in 1907) 1425 stars, of which 551 were in clusters. About this time the Astronomische Gesellschaft (AG) established a commission for variable stars, whose task was the publication of the *Geschichte und Literatur* and of the yearly summary *Katalog und Ephemeriden*. The total number of stars for certain years is as follows:

1910	677 stars
1923	2233 stars
1933	5826 stars
1943	9476 stars

The Harvard Catalogue of 1907 contained 1425 variables, the difference basically lying in the fact that objects discovered in stellar clusters and the Magellanic Clouds were excluded from the publications of the AG. Stars were only accepted as being new that had been previously recognized and given their final designations in the *Benennungslisten* ("Name Lists"). The title *Katalog und Ephemeriden* ("Catalogue and Ephemerides") should not be taken to mean that it was a mere list, as it also gave predictions of the maxima of long-period stars, and minima of eclipsing binaries. The work was carried out at Bamberg by HARTWIG and after his death in 1923, by PRAGER and then SCHNELLER at Babelsberg. The last year of publication was 1943.

Table 56. Variable star designations

Prefix to number	Meaning	Year in which started	Country
HV	Harvard Variable	1890	USA
Innes	Name of discoverer	1914	South Africa
Ross	Name of discoverer	1924	USA
S	Sonneberg	1926	DDR
Zi	Zinner	1929	Germany
SVS	Soviet Variable Star	1933	USSR
P	R. Prager	1934	Germany
VV	Vatican Variable	1950	Vatican
BV	Bamberg Variable	1955	BRD
GR	G. Romano	1958	Italy
Wr	R. Weber	1958	France
HBV	Hamburg-Bergedorf Variable	1961	BRD
A	Asiago	1966	Italy
T	Tonantzintla	1968	Mexico
B	Byurakan	1970	USSR

By 1945 the Astronomische Gesellschaft had lost its international character, so the International Astronomical Union requested the astronomical institution in Moscow to continue the compilation, with the aim of publishing a catalogue every 10 years and supplementary lists in the interim periods. This was done, but without giving ephemerides. The editions which have appeared to date contain:

1948	10912 stars
1958	14711 stars
1969	20437 stars
Supplements to 1976	25842 stars

Table 55 gives the subdivision of these figures into the various classes of variability. (For pulsars, optically variable quasars and galaxies, unnamed objects are also included in the table.)

6.7 Preliminary Designations of Variable Stars

As indicated in Sect. 6.1, with the introduction of photography, the greatest number of new discoveries came from systematic searches. This resulted in individual observers making whole series of discoveries. However, as final designation of variable stars only takes place after basic investigation of the variations has been made, many authors, or their institutions, introduced preliminary designations, mostly letters followed by a sequential number. The most important series of discoveries (number of new discoveries > 100) are given in Table 56. The greatest number of variables were found by Harvard Observatory (13023); several ladies must be particularly mentioned: CANNON, FLEMING, HOFFLEIT, LEAVITT and SWOPE. Sonneberg, with 10848 variables, comes second; principally owing to HOFFMEISTER, with about

10 000 objects, and MORGENROTH with 500-odd. Apart from the series given in Table 56, considerable lists have been published by: Mrs HARWOOD at the Maria Mitchell Observatory (U.S.A.), M. and G. WOLF (Heidelberg), BRUN (France), LUYTEN (Holland and U.S.A.), PLAUT and OOSTERHOFF (Holland), BAADE (U.S.A.), and PIGATTO and ROSINO (Italy). Other successful authors are STROHMEIER (Bamberg) and WACHMANN (Bergedorf near Hamburg). At present nearly 16 000 stars still await confirmation and designation.

A number of variable objects received some form of designation before their variability was discovered. They have already been catalogued on the basis of their peculiarities; whether photometric (X-ray, ultraviolet, infrared, or radio radiation), spectroscopic (emission-line objects and spectrum variables), or morphological (planetary nebulae). These designations may also be preliminary or final; and consist of abbreviations of the names of the institutions, satellites, or discoverers, together with either a sequential number or an abbreviated form of the equatorial – or for planetary nebulae, the galactic – co-ordinates.

A summary of the most important astronomical catalogues can be found in ZOMBECK (1980) and in the *Information Bulletins* of the Stellar Data Centre in Strasbourg. Some of the most important compilations are: the catalogue of X-ray sources by AMNUEL et al. (1979); the catalogues of radio sources by DIXON (1970), and by FINLEY and JONES (1977); the catalogue of quasars by BURBIDGE et al. (1977); and the catalogue of planetary nebulae by PEREK and KOHOUTEK (1967). Examples:

1) X-ray source XRS 04305 + 052 (XRS = catalogue by AMNUEL et al. 1979; the figures are the abbreviation for $\alpha = 4^{\mathrm{h}}30^{\mathrm{m}}5$ and $\delta = +5°2$) $\equiv$ 4U 0432 + 05 (4U = Fourth Uhuru Catalogue) $\equiv$ radio source 3C 120 (Third Cambridge Catalogue) $\equiv$ 4C + 05.20 (Fourth Cambridge Catalogue) $\equiv$ NRAO 182 (Green Bank) $\equiv$ OA 140 (Ohio) $\equiv$ OF + 052 (Ohio) $\equiv$ PKS 0430 + 05 (Parkes) $\equiv$ BW Tauri (variable Seyfert galaxy)
2) XRS 16560 + 354 $\equiv$ 4U 1656 + 35 $\equiv$ 2A 1655 + 353 (Ariel catalogue) $\equiv$ Her X-1 $\equiv$ HZ Herculis (variable X-ray star)
3) Planetary nebula 60−7°1 (catalogue by PEREK et al. 1967) $\equiv$ He 1−5 (catalogue by HENIZE) $\equiv$ FG Sagittae (unusual variable star).

7. The Significance of Variable Stars for Research on the Structure of the Galaxy and Stellar Evolution

7.1 Methods

General Procedure

An important aim of stellar statistics is to obtain clues to the true distribution of objects in space from their apparent distribution, thus leading to a knowledge of the structure of the Galaxy. *Three procedures* are available for this purpose.

1) From the celestial co-ordinates and the distance one can derive the spatial co-ordinates of an object, usually in rectangular co-ordinates x, y and z, relative to the galactic plane and the centre of the system. A knowledge of the distance is essential. It is here that we encounter the limitations of this method. From a large number of individual determinations of spatial co-ordinates, it is possible to obtain a picture of the sub-system concerned. The procedure has been applied to globular clusters with the great success. It is also suitable for variable stars, if their distances can be determined from the absolute and apparent magnitudes. Its application has generally been restricted to relatively rare objects.

2) The second method is of a purely statistical nature. From the apparent area density of objects on the celestial sphere, that is their number per square degree, and the distribution of their apparent magnitudes, the space density at various distances is derived. For this, it is necessary to know the frequency with which the various absolute magnitudes occur (the luminosity function) for the class of stars involved. One obtains an integral equation, from which the space density at various distances can be calculated.

3) The third procedure is completely different. It makes use of the emission lines from gaseous clouds, especially neutral hydrogen (21-cm line), and also various lines of CO, OH, CH, H_2CO, etc. Measurement of the intensity and Doppler shift of these lines enables conclusions to be drawn about the rotation and spiral structure of the Galaxy; the variation in the chemical composition of the gas in the z and R directions; and about expansion of the gas in the nuclear region of the Galaxy and the existence of a ring of higher density material that surrounds the centre of the Galaxy at a distance of 5 kpc. More information about this subject may be found in the recent book *The large-scale characteristics of the Galaxy* (IAU Symposium No. 84, 1978).

As well as the first procedure, the second may also be applied to variable stars, in fact with a great deal of success, as for some groups of objects, at least, the luminosity function may be replaced by a constant. This is approximately the case, for example, with the RR Lyrae stars, but it also applies to the δ Cephei stars if just restricted period ranges are considered. It will be seen that it is important to discover as many variables as possible as the more objects that one has at one's disposal, the closer one can come to fulfilling this requirement. The total number of δ Cephei stars may be rather too small for a strict application of this principle, but it is essential for the work to be carried out with individual absolute magnitudes. The procedure then takes place in the following manner, given a field with a certain area in square degrees, in which a certain number of variables are known to exist. The solid angle determined by the field is divided by equidistant transverse planes into sections, whose volumes may be easily calculated. The distance of the variables is determined from the absolute magnitudes, thus finding how many objects fall within each of the sections. This gives the space density per cubic kiloparsec (kpc^3). It only remains to calculate the galactic co-ordinates of the centres of the sections, and to repeat the process in other fields for the same type of stars. If many fields are available, then a picture of the "subsystem" of the chosen class of stars can be built up from their extent and density gradients (a measure of the increase or decrease in space density with distance). The *population type* may thus be determined; the degree of concentration towards the galactic plane being the crucial factor here.

The principal difficulty in the application of these methods comes from *interstellar extinction* in dark clouds. Only the radio-astronomy method is little affected by this, as the dielectric dust particles do not absorb radio waves. But these observations do not relate to stars, and it is just their distribution that we wish to establish. Consequently, the extinction has to be investigated in every individual portion of the field. The amount of the total absorption is given by a count of the number of extragalactic systems that are visible in the field, but this is also subject to uncertainties, as it is well-known that galaxies have a tendency to occur in clusters. However, a basic indication may still be obtained.

Unfortunately the dark clouds are not evenly distributed along the line of sight, and we can only approximately succeed in eliminating their effect. Use is made, for example, of the reddening of starlight from scattering by very small dust particles, very many investigations having been carried out on this subject. Use may also be made of star counts that take apparent magnitudes into consideration, which may be interpreted by a method suggested in principle by M. WOLF, and later greatly improved, which permits the extent of individual dust clouds in depth to be determined. It is not necessary for us to discuss these problems in detail, but we should note that as many fields as possible should be investigated, in order to minimize the effect of systematic errors that are unavoidably connected with individual fields. Important monographs about this question in relation to variable star research have been published by KUKARKIN (1949), PAYNE-GAPOSCHKIN (1954) and PLAUT (1965a).

Questions of Stellar Evolution

Before we discuss results, it should be noted that the distribution of objects in the Galaxy, and thus of variable stars as well, is closely linked with stellar evolution. The various Populations have often been mentioned and also the fact that, according to our present knowledge, the formation of stars (and therefore the existence of the T Tauri stars and related types, for example), is only possible in those parts of the Galaxy that are rich in *interstellar gas and dust* (extreme Population I). It has also been noted that pulsational instability is associated with a late evolutionary phase, related to the transition from hydrogen "burning" to the helium-carbon process; and, finally, that eruptive binaries such as novae, U Geminorum stars and, above all, X-ray binaries, probably correspond to an advanced stage of evolution, and supernovae to a final stage.

It seems essential to restrict our description concerning stellar evolution. Advances in recent years have been so great that many facts are in danger of becoming out-of-date in a very short time. It therefore seems advisable to consider only those results which appear to have been confirmed and which are of major significance.

The most important fact is that the *evolution of a star* is essentially determined by two properties: the *initial chemical composition* and the *initial mass*. The influence of the primary chemical composition has been estimated in many studies, it appears that it is fairly considerable. Calculation is easier with regard to the initial mass: the greater the mass, the faster evolution takes place, and indeed by a very considerable factor (see e.g. the corresponding short table in Sect. 3.3.2). An increase in the mass by several solar masses can cause a reduction in the main-sequence lifetime by a factor of 100. This means that stars of similar age can be at very differing stages of evolution, which significantly complicates the interpretation of observed phenomena. Reference may be made to the book *Variable Stars and Stellar Evolution*, edited by SHERWOOD and PLAUT (1975) on this subject. All in all, it is not surprising that there are many difficulties and contradictions if one tries to fit variable stars into the overall pattern of the Galaxy according to their age and distribution in space.

7.2 Results

Distance of the Galactic Centre

In order to arrive at the distribution of variable stars in the Galaxy (galactocentric distribution) from the distribution in the neighbourhood of the Sun (heliocentric distribution), a knowledge of the position and the distance of the centre of the Galaxy is essential. As the centre is a strong radio, infrared and X-ray source, and as these wavelengths are relatively little affected by the intervening interstellar material, the co-ordinates of the centre are exceptionally well-known. It is more difficult to determine the distance $R_\odot$ between the Sun and the galactic centre, because of the large amount of interstellar extinction in this direction. The most precise method is by determining the density distribution (the number of objects per cubic kiloparsec, kpc^{-3}) with respect to distance from the Sun, for particular classes of object. This is done in selected fields with low interstellar extinction (in so-called "windows") at small, but different, angular distances from the galactic centre. With increasing distance from the Sun the space density v for the objects concerned first increases, tends to a maximum, and then declines again.

By extrapolating the systematic trend of the distance of this density maximum with reducing angular distance from the galactic centre, it is possible to determine the distance of the galactic centre $R_\odot$.

Objects that are suitable for determining $R_\odot$ are Population II objects, and in particular some classes of variable stars. Table 57 gives a summary of the most important recent values. The main errors arise from insufficient knowledge of residual extinction and uncertainties in the absolute magnitudes of the sample objects. All investigations show, however, that $R_\odot$ lies between 7 kpc and 9 kpc.

Table 57. Determinations of the distance $R_\odot$ of the galactic centre

Objects employed	$R_\odot$ [kpc]	Source
46 globular clusters	8	BARKHATOVA et al. (1973)
980 RR Lyrae stars	8.7 (± 0.6)	OORT and PLAUT (1975)
111 globular clusters	8.5 (± 1.6)	HARRIS (1976)
76 globular clusters	6.8 (± 0.8)	FRENK and WHITE (1982)
70 Mira stars	8.8	GLASS and FEAST (1982)

A detailed discussion of the $R_\odot$ problem is given by GRAHAM (1979).

Distribution Within the Galaxy

The work by RICHTER (1967a) on this subject has the advantage that it is based upon fairly homogeneous material. However, its results should only be seen as preliminary as since then much more, and better, material has been gathered; so a revision would be worthwhile.

RICHTER begins with the integral equation for stellar statistics:

$$A(m) = \omega \int_0^\infty r^2 v(r) \phi(M) \, dr$$

were $A(m)$ is the total number of objects per square degree in the brightness interval $m \pm 0.5$; $v(r)$ the desired space density as a function of the distance r from the observer; $\phi(M)$ the luminosity function and ω the solid angle. Average luminosities and spectral classes have been brought together in Fig. 163 (a Hertzsprung-Russell diagram). For other aspects of this work, especially in connection with interstellar extinction, reference should be made to the original paper, which has also appeared in a shortened form (RICHTER 1967b).

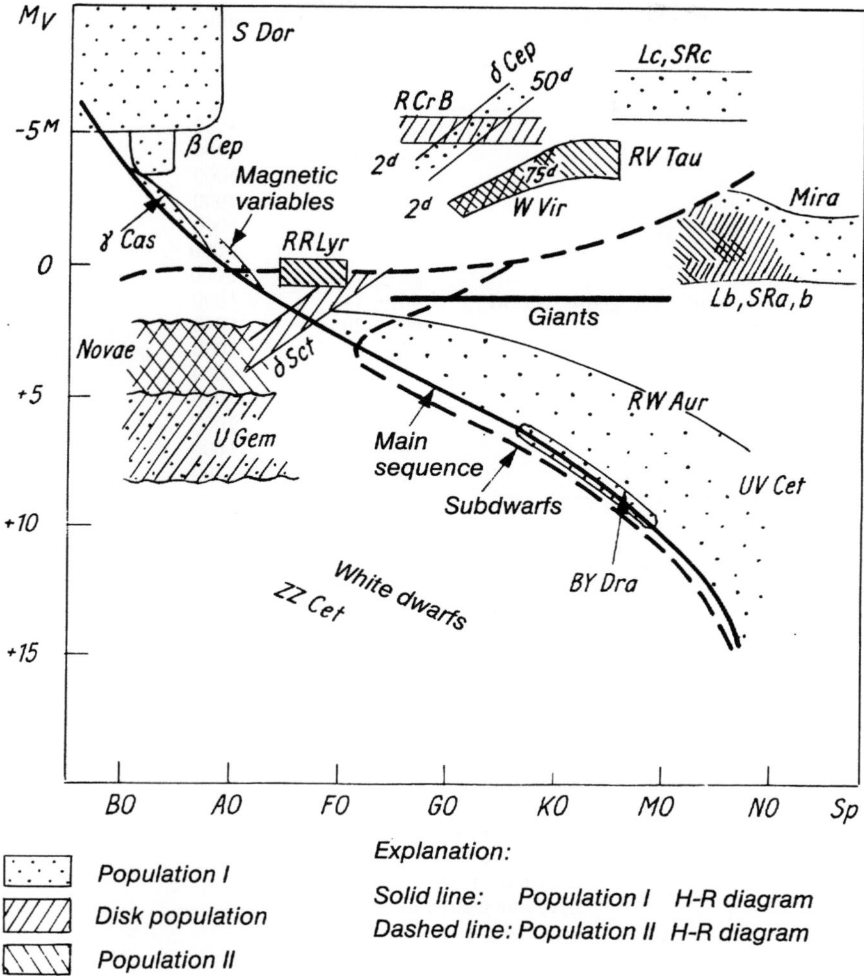

Fig. 163. The position of various classes of variable stars on the Hertzsprung-Russell diagram

PLAUT (1965 a) has published a similar comprehensive paper on the problem of the distribution of variables in the Galaxy. He used the statistical results from many observatories and organizations, which by its very nature cannot be homogeneous. But he also took dynamics into account, both using OORT's theory, which relates to the whole system, and also collating proper motion and radial-velocity determinations for individual objects.

Two tables (58 and 59) may be given from RICHTER's results. The first shows under $v_\odot$ the total number established for that particular type of variable per cubic kiloparsec in the vicinity of the Sun; and under N the total number estimated to exist in the Galaxy. In the original paper the logarithmic density gradients were also calculated in galactocentric cylindrical co-

Table 58. Space density $v_\odot$ and total N of variables in the Galaxy – preliminary results from the Sonneberg Field Survey

Class	$v_\odot$ [kpc^{-3}]	N
U Gem and Z Cam	$5\,000 \pm 1\,000$	$5\,600\,000$
Mira, $P = \quad 90^d - 150^d$	2.5 ± 0.5	$8\,000$
$\qquad\quad 150^d - 200^d$	2.5 ± 1	$10\,000$
$\qquad\quad 200^d - 300^d$	13 ± 5	$60\,000$
$\qquad\quad 300^d - 400^d$	30 ± 10	$111\,000$
$\qquad\qquad > 400^d$	13 ± 8	$9\,000$
Mira, spectra S, C, R, N	25 ± 15	$6\,000$
Semiregular and irregular red giants, *Sp.* M	146 ± 14	$268\,000$
Similar objects, *Sp.* S, C	160 ± 20	$21\,000$
RR Lyr ab, $P < 0^d\!.45$	6 ± 2	$30\,000$
$\qquad\qquad P > 0^d\!.45$	8 ± 2	$88\,000$
RR Lyr c	3 ± 1	$40\,000$
W Vir	25 ± 1	$9\,000$
δ Cephei, fainter than $-3^M\!.4$	16 ± 8	$3\,600$
δ Cephei, brighter than $-3^M\!.4$	10 ± 4	$1\,700$

Table 59. Variables and Population classes – Sonneberg Field Survey results

Population	Classes of variables
Halo Population II	Yellow semiregulars RR Lyr ab, $P > 0^d\!.45$ Mira, $P = 150^d - 200^d$, *Sp.* M
Intermediate Population II	RR Lyr ab, $P < 0^d\!.45$ RR Lyr c Mira, $P = 90^d - 150^d$, *Sp.* M Mira, $P = 200^d - 300^d$, *Sp.* M SR giants, *Sp.* M Irregular, little coloured Irregular giants, mostly *Sp.* M
Disk Population	Mira, $P = 300^d - 350^d$, *Sp.* M Mira, $P > 350^d$, *Sp.* M W Vir U Gem + Z Cam Eclipsing variables
Older Population I	RV Tau Mira, *Sp.* generally C, R, N, S Irregular and SR giants, *Sp.* C, R, N, S δ Cep, fainter than $-3^M\!.4$
Extreme Population I	Red irregulars and SR supergiants δ Cep, brighter than $-3^M\!.4$ RW Aur

ordinates, that is the distance R from the axis perpendicular to the galactic plane and passing through the centre of the Galaxy, and the distance z from the equatorial plane. These values also serve to differentiate between the Populations to which the various classes belong.

According to OORT and PLAUT the space density of the Mira, semi-regular and irregular stars taken together amounts to $v_\odot \approx 200\,\mathrm{kpc}^{-3}$, whereas WOOD and CAHN (1977) find $v_\odot \approx 245\,\mathrm{kpc}^{-3}$ for the Mira stars alone. For RR Lyrae stars OORT and PLAUT (1975) and I. MEINUNGER (1977) find $v_\odot = 12\,\mathrm{kpc}^{-3}$, and KINMAN et al. (1966) $v_\odot = 9\,\mathrm{kpc}^{-3}$. BEYER gives $v_\odot = 30\,\mathrm{kpc}^{-3}$ for RR Lyrae stars; $2.4\,\mathrm{kpc}^{-3}$ for W Virginis stars, and $89\,\mathrm{kpc}^{-3}$ for δ Cephei stars. According to WARNER (1974a), for U Geminorum stars $v_\odot = 500\,\mathrm{kpc}^{-3}$, and GORBATSKIJ (1975) finds that the total number in the Galaxy is $10^7 \leq N \leq 10^8$.

The most surprising result, at least for those who are not familiar with this field, is that the number of U Geminorum and Z Camelopardalis stars in the Galaxy is greater than that of all the other variables taken together, even when we use WARNER's density ($v_\odot = 500\,\mathrm{kpc}^{-3}$), which is lower by a factor of 10. The reasons for this are first, that these stars have faint absolute magnitudes, so they can only be seen out to relatively small distances and thus appear rarer than they actually are; and second, that because of their light-curves, their discovery probability is very low.

In general the data given in Table 58 should be assessed in a qualitative rather than a quantitative manner, not only on account of the incomplete nature of the material, but also because of various natural factors: in many classes, especially with extreme Population II, there is a great deal of uncertainty, as to the total number in the galactic centre; it is also very difficult to estimate with any degree of accuracy the influence of dark clouds. The table does not include RW Aurigae stars, which belong to extreme Population I, and mostly lie within dark clouds. According to WENZEL's (1961) statistics, their value for $v_\odot$ is about $10000\,\mathrm{kpc}^{-3}$, but N cannot be estimated in view of their tendency to occur in clouds. Locally these stars may be the most numerous ones, which is understandable, considering their great density in the Orion Nebula and in the Taurus Cloud. Also omitted from the table are R Coronae Borealis stars as their total number is so small that a statistical discussion is not possible; the same applies to some sub-classes of irregulars and semiregulars.

Regarding *membership of the various Populations*, RICHTER arrived at the summary given in Table 59, which is based on preliminary Sonneberg Field Survey material.

As may be seen from the table, frequently there is no close association between the types of Population and of variable stars. In addition, here and there within the groups given in Table 59 a scatter in the Population classes can be seen. There is, however, general agreement with earlier investigations. The scatter in the RR Lyrae stars, for example, had already been noted by PRESTON (1959). RICHTER has himself tried to clarify some of the contradictions. KOPYLOV (1957) puts the U Geminorum stars in extreme Population II, but it should be noted that in this case both statements are very uncertain. According to PARENAGO (1957) the W Virginis class is Population II, but PLAUT (1963) maintains that the low systematic velocity and proper motion indicate membership of the disk Population. A summary of modern ideas on the position of W Virginis stars in the Galaxy is given by KUKARKIN (1975).

According to this the W Virginis stars are not a homogeneous group, but some are members of the disk Population and others of Population II. The latter include W Vir itself, which has the greatest distance from the galactic plane, 2.845 kpc. Despite this, most of these stars are closer to the classical δ Cephei stars than to the RR Lyrae stars.

Mainly on account of their low luminosity, the U Geminorum stars' population remains very uncertain. If we assume that they occur in a thin layer extending on both sides of the galactic equator, thus belonging to the disk population, then the smaller the radius of space that is considered, the more spherical their distribution will appear. The observational material so far evaluated has not been suitable for deciding such questions, owing to the lack of coverage at high galactic latitudes. Moreover, little is known about the scatter in the absolute magnitudes of U Geminorum stars at minimum.

Novae are not included in these tables. Their attribution is by no means unequivocal; they show a strong concentration towards the galactic plane, but also a strong increase in numbers towards the galactic centre. In a detailed discussion of modern work concerning novae in M31, ARKHIPOVA and MUSTEL (1975) conclude that the novae do not form a single sub-system. Some belong to the spherical, and others to the flattened sub-systems (see also WENZEL and I. MEINUNGER 1978 and RICHTER and BÖRNGEN 1981).

BOYARCHUK (1975) gives a general review of the situation regarding the symbiotic stars. They belong to the old disk population; the space density in the solar neighbourhood and the total number in the Galaxy amount to $v_\odot \approx 0.4 \, \mathrm{kpc}^{-3}$ and $N \approx 10^3$.

One of the problems in galactic research that has received a great deal of attention is that of recognizing and mapping the *spiral arms*. This is a difficult task as we are situated within the system, and also on account of interstellar extinction. We can, of course, only use high-luminosity objects for which we can determine distances. Star clusters, H II regions, OB associations and δ Cephei stars all contribute towards solving the problem. Reference may be made to work by KRAFT and SCHMIDT (1963), BECKER (1964), OORT (1965) and HUMPHREYS (1978). These have shown that the δ Cephei stars brighter than $-4\overset{M}{.}3$, which appear to be relatively young, show a distinct correlation, but that the ones with fainter absolute magnitudes do not (Fig. 164). Older objects have moved too far from the sites of their formation, owing to their peculiar motions and the blurring (the shearing effect) caused by the differential rotation of the Galaxy. In addition, KRAFT and SCHMIDT (loc. cit.) investigated the relationship between galactic rotation and the radial velocity of δ Cephei stars. For determining the positions of the spiral arms, radio observations of H I regions and interstellar molecular clouds play an important part.

Stellar Lifetimes

The most fundamental advance in variable star research in recent decades has been the recognition that variability is not an abnormal state, but corresponds to phases of normal stellar evolution. A normal star passes through

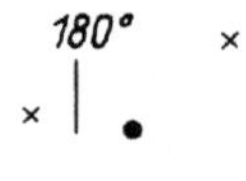

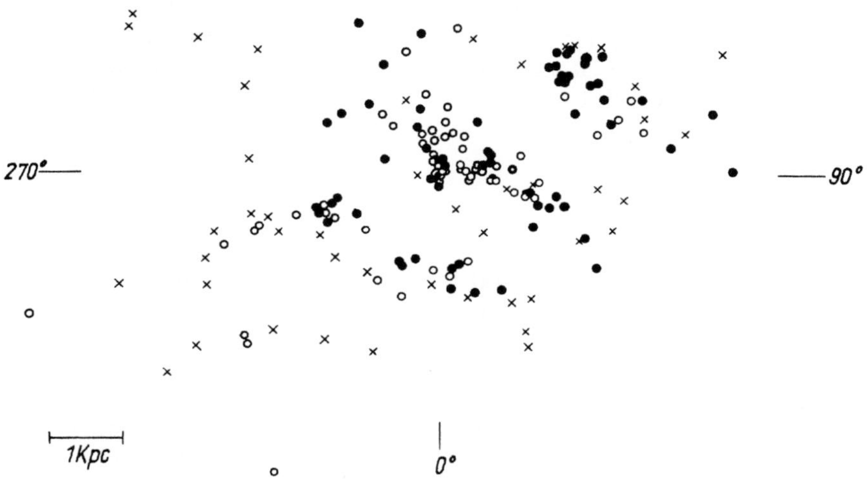

Fig. 164. The spiral structure of the Galaxy as shown by young clusters and H II regions (*points* and *circles*) and δ Cephei stars of $M_v = -4.3$ and brighter (*crosses*)

regions of instability, such as the contraction phase at the beginning of its evolution (RW Aurigae stars) and the pulsation phases after it leaves the main sequence (δ Cephei stars, RR Lyrae stars, slowly varying red giants, for example). The question immediately arises whether statistical methods can be used to estimate the lifetimes. This is undoubtedly an important task for future research. From general statistical data on stars, RICHTER has established that, in the vicinity of the Sun, out of 4010 supergiants, 610 are δ Cephei stars, and concludes from this that the duration of the pulsating stage amounts to 15.2% of that of the supergiant stage. EFREMOV and KOPYLOV (1967) reach practically the same result, finding 10–20% of the duration of the supergiant stage, based upon material from open clusters and associations.

From the theoretical side, HOFMEISTER et al. (1964, 1965) arrived at the figures given in Table 60 for models of 5 and 7 solar masses. According to theory, the duration of the pulsating stage is 7.6% of the supergiant stage at

Table 60. Length of evolutionary stages

Mass	Period	Duration of the supergiant stage	Duration of the pulsating stage
5 solar masses	$4^{d}.4$	2.1×10^7 years	16.1×10^5 years
7 solar masses	$11^{d}.4$	1.0×10^7 years	7.3×10^5 years

5 solar masses, and 7.3% at 7 solar masses. In view of the great uncertainties in the basic assumptions, both in the statistics and in the theory, the agreement is not bad. In such a situation we must be grateful if we obtain the correct order of magnitude. It is worth noting that a star may pass through the instability zone as many as five times.

Mention may also be made of the lecture given by KIPPENHAHN, *Stellar Evolution and Variability* (1965), and to the thesis by HOFMEISTER (1965), *Delta-Cephei-Sterne vom Standpunkt der Sternentwicklung* (δ Cephei stars from an evolutionary point of view).

7.3 The Galactic Halo

It is still not known precisely how far the spherical component of the Galaxy (the halo) – including RR Lyrae stars and faint blue stars – extends into space, how massive it is, or what structure it possesses. An attempt by PEREK (1951) suffers from the incomplete nature of the material used, and errors in the magnitude scale. PRESTON (1959) and KINMAN (1959) later found that with respect to periods RR Lyrae stars in the halo are not a homogeneous group, and that there is also a relationship between the periods and the kinematic properties. KINMAN (1964), LAFLER and KINMAN (1964) and KINMAN et al. (1964) carried out comprehensive investigations with the 20-inch Lick astrographs, but at different galactic latitudes. Some of the RR Lyrae stars, particularly those with short periods, must belong to the disk population.

KINMAN et al. (1966) discuss 3 fields close to the North Galactic Pole, and find some surprising results. The apparent density is low here: 54 RR Lyrae stars in 74 square degrees, down to the limiting magnitude of $18\overset{m}{.}4$. The density gradient was determined for distances of 5–15 kpc from the galactic plane. From this it was calculated that between magnitudes 18^m and 20^m only about 0.05–0.12 RR Lyrae stars would be found per square degree. This is in contrast to the more than 1000 RR Lyrae stars per square degree (according to an older determination by BAADE) found in the direction towards the galactic centre, in the bright star clouds in Sagittarius.

The 3 authors advanced an idea linked with this, which still requires proof. They surmise that the halo extends beyond the Magellanic Clouds, that is more than 50 kpc, and that the closer dwarf systems included in the Local Group are not independent systems, but concentrations in the galactic halo, especially as their stars belong to Population II.

Further contributions have been made by the Palomar-Groningen project, which examined fields close to the galactic centre to determine the number of variables that they contained, using the 122-cm (48-inch) Schmidt Telescope at Palomar. PLAUT (1966, 1968a, 1968b, 1970) has established the density values from the total number of 1180 RR Lyrae stars and 1012 long-period stars observed. Tables 61 and 62 have been compiled from these data. The initial increase in the density of stars with increasing distance from

Table 61. Space density of RR Lyrae stars (number per kpc^3)

Distance from the Solar System	Direction				
	$l = 0°$ $b = +29°$	$l = 4°$ $b = +14°$	$l = 4°$ $b = +10°$	$l = 0°$ $b = -12°$	$l = 0°$ $b = -8°$
0 kpc	13	13	13	13	13
5	18	130	170	260	420
10	3.5	30	60	34	33
15	0.4	1.5	4.5	1.0	0.1
20	0.1		0.2:		

Table 62. Space density of long-period variables (number per kpc^3)

Distance from the Solar System	Direction			
	$l = 4°$ $b = +14°$	$l = 4°$ $b = +10°$	$l = 0°$ $b = -12°$	$l = 0°$ $b = -8°$
0 kpc	40	40	40	40
5	50	100	100	150
10	30	50	50	90
15	10	16	16	30
20	3	5.5	4	9
25	0.8	1.2	0.9	3.3

the Solar System is primarily due to the fact that at first our line of sight passes somewhat closer to the galactic centre. According to all the investigations yet undertaken, although they are very incomplete, it would seem that the density of the halo decreases outwards, at first very rapidly, but then very slowly. How far it extends into space is undecided.

HOFFMEISTER (1963) made an attempt to determine the extent of the halo at low galactic latitudes. He used plates that had been taken with the 2-m reflector at the Karl Schwarzschild Observatory, of "windows" in the region of the galactic anticentre. The "windows" are areas where a view is obtained between dark clouds; they may be recognized by the fact that extragalactic systems are visible within them. The faintest RR Lyrae star discovered has the following data:

$$\text{KN Aur} \quad l = 163°38' \quad b = +13°12' \quad \text{Range} = 18^m5 - 20^m0$$
$$\text{Period} = 0^d58246$$

The extinction cannot be very great as $3'6$ south of the variable there is a moderately darkened 17^m galaxy. Trial values of interstellar extinction from $1-3$ mag give the following distances for the star from the galactic centre:

Extinction	Distance
1 mag	33 kpc
2	23
3	16

The first value is more probable than the last, but even the latter, at 52 000 light-years, lies far outside the conventionally accepted boundaries of the Galaxy. Even greater distances from the galactic centre (namely 40–50 kpc) have been recently announced by SAHA (1982) for 17 RR Lyrae stars in the direction of $l = 110°$, $b = +30°$. These were found during a survey of the galactic halo with the Palomar 122-cm Schmidt Telescope. These results confirm the great extent of the halo.

As discussed elsewhere, the galactic distribution of novae also suggests that the halo is very extensive.

Mention may also be made of a future possiblity. The structure of the galactic halo may, of course, be determined in fields with the least amount of interstellar extinction.

If it is thus estabished that the halo is rotationally symmetrical and free from major inhomogeneities, then it will be possible to obtain the amount of the total absorption in given fields by comparing the statistics of RR Lyrae stars with the calculated unaffected values. The use of galaxies is known to have a disadvantage in that they have a strong tendency to occur in clusters. However, it should still be remembered that in one case the discrepancy between the total number of RR Lyrae stars and that of galaxies has led to the probable discovery of an intergalactic absorption cloud belonging to the Local Group (HOFFMEISTER 1962b, see also I. MEINUNGER 1976).

In principle, it is possible to determine the acceleration K_z perpendicular to the galactic plane, which is very important for understanding the structure of the Galaxy. This is feasible if both the density gradient in the z-direction and also the range of peculiar velocities in the z-direction σ_z, are known for a homogeneous group of objects (the RR Lyrae stars are particularly suitable). Then for a cylinder through the Sun's neighbourhood with its axis in the z direction, we have the equation

$$- \partial \log v / \partial z = 134 \times 10^6 \, K_z / \sigma_z^2 \, .$$

On the basis of the preliminary Sonneberg material already mentioned, RICHTER (1967a) obtained

$$K_z = -(6.2 \pm 1.1) \times 10^{-9} \text{ cm/sec}^2 \quad \text{(for } 0 < z < 2 \text{ kpc).}$$

KINMAN et al. (1966) obtained

$$K_z = -3 \times 10^{-9} \text{ cm/sec}^2 \text{ (for } z \text{ between 5 and 10 kpc).}$$

SCHMIDT (1956) gives theoretical values to be expected from a particular model:

$$K_z = -6.5 \times 10^{-9} \text{ cm/sec}^2 \text{ (for } z = 5 \text{ kpc)}$$

and

$$K_z = -2.8 \times 10^{-9} \text{ cm/sec}^2 \text{ (for } z = 15 \text{ kpc).}$$

For further discussion of this problem see KING (1977).

7.4 Remarks on the Evolution of the Galaxy

The still unsolved problem of the evolution of the Galaxy will only be discussed here in so far as it concerns variable stars. New stars are formed from interstellar material. The greater portion of this consists of hydrogen; among other constituents are dust particles and heavy elements such as Ca, Na, Fe and others. For a long time, the origin of the nuclei of the heavy elements posed a problem, and the general opinion was that they were formed at the time of the origin of the universe. Nowadays we know that conditions suitable for their formation exist within stellar interiors, and we believe that we may safely assume that these fractions of the interstellar material have already passed through a stellar stage at least once. We thus observe both the formation of new stars from interstellar material and also the loss of stellar material into interstellar space. This immediately raises the question of whether there may be some quasi-stable state where, within the spiral arms of the Galaxy, as much material is made available through the destruction of stars as is taken up by the stellar formation that takes place at the same time. This question was discussed by NERNST in a Berlin Academy lecture "The Universe in the Light of Modern Research" as long ago as 1921. Even then the novae already appeared to be the source of stellar material. Since then the outbursts of supernovae have been recognized as being of greatest importance. However, the normal corpuscular radiation from stars, the "stellar wind", also seems to be an effective mechanism, both at the initial stages of evolution (in the T Tauri stars), and also after the evolution of stars away from the main sequence, particularly in the red giants and supergiants that lie close to the stability boundary. RR Lyrae stars in globular clusters appear to have only half the mass of their precursors on the main sequence, so they have apparently already shed half of their material back into the interstellar medium! For further discussion of mass-loss in stars see the book *Mass Loss and Evolution of O-Type Stars, IAU Symposium No.83 (1979)*.

8. Observational Methods and Organizations

8.1 Observations

The discussion in this chapter should serve as a summary principally for those not familiar with the subject, and for beginners. Reference must be made to more specialized literature for details of methods of observation, reduction or analysis that require greater technical expertise. In what follows we cover fairly comprehensively those procedures that enable the brightness of variables to be determined in a simple manner, generally in magnitudes.

8.1.1 Photometric Observations

General Remarks

As in other photometric fields there are visual and non-visual methods, the non-visual being represented primarily by photography and photoelectric photometry. A similar distinction applies to the analysis of photographic observations because the determination of magnitudes may also be made visually or by means of impersonal photometers. In contrast to other fields of study, direct visual procedures, either on the sky or on photographic plates, are not only economical in effort, but also capable of a high degree of accuracy, with suitable experience and aptitude on the part of the observer. However, direct photoelectric observations of stars are generally very much more accurate, whereas on photographic plates they cannot avoid the errors that these contain.

Visual Observations

All visual observations rely upon the variable being estimated against a sequence of invariable neighbouring stars. At least two stars should always be used, one of which is brighter, and the other fainter than the variable. In the following description we primarily consider ARGELANDER's Step Method, which is very frequently used. If the variable is only slightly fainter than star a, then ARGELANDER describes the difference as being one photometric step. It is written a1v; if it is markedly brighter than b, by about three times the step, then the estimate is given as a1v3b. As the variable becomes brighter or fainter, then further comparison stars are added. The difference in their magnitudes should not, in general, be greater than about half a magnitude, but one is naturally constrained to use whatever the sky permits, as large separations from the variable are also to be avoided. Stars with large amplitudes require a whole sequence of comparison stars. It is not expedient to give too many instructions. Observers can best determine what a step is by their own experiments, as we are not dealing with an objective amount, but with what is, to a certain extent, a quantity that depends upon the individual. The step difference between two invariable stars, furthermore, will not always be found to be the same, but this does not have a detrimental influence on the observations. Two of the method's peculiarities should be borne in mind: first, one should be careful not to subdivide small magnitude differences into a large number of steps; it is detrimental to try to ascribe a step value of 2 or 3 to the smallest differences, in the hope of detecting still smaller differences. Second, experienced observers occasionally use up to 10 steps or even more for correspondingly large magnitude differences. However, it must be clearly understood that this is where the method's limits lie, owing to the danger of systematic errors (large differences are generally underestimated). It is better to employ another comparison star, even if it is not particularly well placed.

Some problems arise in direct visual observations of red variables. First, the colour generally distorts the estimation of steps; second, the perceived magnitude depends upon the aperture of the telescope and the brightness of the sky background. In a large instrument, at twilight, or in a moonlit sky most observers will see a red star as too bright. The cause of this (the Purkinje Effect) is the strong contribution that is made by the colour receptors (the cones) in the retina of the eye. With a dark sky background and faint stars essentially only the rods are sensitive. The perception of anything that is red can vary greatly from observer to observer. It is not uncommon for an observer, even in a dark sky, to see a red star as fully half a magnitude brighter or fainter than someone else. As a result, if a series of observations from various sources have to be combined, they must be reduced relative to one another, or else to an ideal mean observer. The treatment of the red variable α Herculis by VAN SCHEWICK (1937) may be mentioned in this connection.

The success of the Argelander method lies partly in the fact that the size of the step can be shown to be a personal constant. This does not apply at the beginning, as it may well be a year before an assiduous observer's step reaches its final value. Generally it is about 0.1 mag initially, reducing to a final value of 0.06 or 0.07 mag. However, these are only approximate figures, from which individual values may deviate. The amount of the step has hardly any influence on the accuracy of observations, especially when the brightness of the variable lies between those of two comparison stars. Such a use of two comparison stars really produces a combination of the Argelander and fractional methods.

The use of visual point or surface photometers is hardly ever employed nowadays; they are still used now and again for the determination of the magnitudes of comparison stars.

There are some variants of the Argelander method, which have not, however, become of great importance. The fractional method recommended by PICKERING sets the difference between two comparison stars equal to 10, whatever its actual amount, the magnitude of the variable then being estimated in these tenths. One drawback with this method is that it is not possible to obtain individual comparison star magnitude scales (see below).

The AAVSO has produced a large number of charts for variables (based on Harvard Observatory material), on which the magnitudes of nearby stars are marked, correct to 0.1 mag. The direct estimation of the magnitude of the variables to an accuracy of a tenth of a magnitude is recommended. The method is sufficiently economical of effort for the continuous surveillance of variables that is carried out by the American Association of Variable Star Observers (AAVSO).

Photographic Observations

The step method may be used just as well on photographic plates as on the sky. Owing to the differences in the plates caused by exposure times and the state of the atmosphere, a greater scatter is to be expected in the step differences of the comparison stars. This is harmless, however, if the variable is estimated from two comparison stars. The final accuracy will be somewhat less than with good visual observations on the sky, because in addition to the errors in the estimates there is the inherent inaccuracy of the photographic plate, which may be estimated as about $\pm\,0.05$ mag. With Schmidt Telescopes it is even greater, as the stellar images are "too good" optically, being too small for step estimation. Short-focus Tessar lenses have the same failing. This disadvantage can be partially compensated by using higher magnifications in the measuring microscope.

A photoelectric photometer is often used to evaluate plates, and it is always recommended when detailed research on individual objects is concerned. However, when many variables have to be evaluated for statistical purposes, then the time-saving step method is preferable. It should be noted that no photometer is capable of eliminating errors that occur on the plate. Nowadays, measurements are almost exclusively on focal plane images, for which the iris-diaphragm photometer has proved to be best. Some decades ago it was thought that plate errors could be minimized by allowing the star to darken a greater area, either by extrafocal setting or by a special Schraffierkassette technique (moving the images in a zig-zag pattern on the plate during exposure). These methods are no longer used; they have been superseded by direct photoelectric photometry with its much greater accuracy, especially as it has considerably reduced the times required. The limits of photoelectric photometry are set by the magnitude of the stars: the majority of faint objects will remain the province of photography.

Reduction of Step Estimates

In the step method we are essentially estimating differences. However, if, as is usually the case, two comparison stars are used, then an observation of the form a s_1 v s_2 b (where s_1 and s_2 indicate step differences) establishes a ratio. For example, a 3 v 5 b indicates that the variable is fainter than a by 3/8 of the difference between a and b. Consequently the result is independent of the amount of the individual steps. Let the magnitude of the brighter comparison star be m_1, and m_2 that of the fainter, and s_1 and s_2 again the step differences. Then in stellar magnitudes $m_1 < m_v < m_2$. The magnitude m_v of the variable can be calculated from:

$$m_v = m_1 + \frac{s_1}{s_1 + s_2}(m_2 - m_1) = m_2 - \frac{s_2}{s_1 + s_2}(m_2 - m_1).$$

Normally, when many observations of the same star are available, a graphical procedure is preferable. Let us assume that 5 comparison stars a, b, c, d, and e have been used. If the approximate stellar magnitudes have be obtained by some method, then the first task is to adjust these magnitudes to agree with the observed series of steps. The mean step differences a − b, b − c, etc. are calculated from all the observations. The faintest comparison star is assigned the value of 0.0, and by addition of the step differences the values of the other stars are obtained (in steps). If these values and the stellar magnitudes were precise, the points representing the stars would lie on a straight line in a diagram such as Fig. 165. Because of the various inaccuracies this is not the case; however we can fit a straight line to these points, and shift them parallel to the x-axis so that they lie on the line. The mean step differences will then remain unaffected. Using millimetre paper one can then read off the corrected magnitudes of the comparison stars. Table 63 contains the numerical values for the example given here. The mean amount of a single step is 1.83 mag: 27.4 steps = 0.067 mag/step. This is the figure that is used if, exceptionally, only one comparison star has been employed.

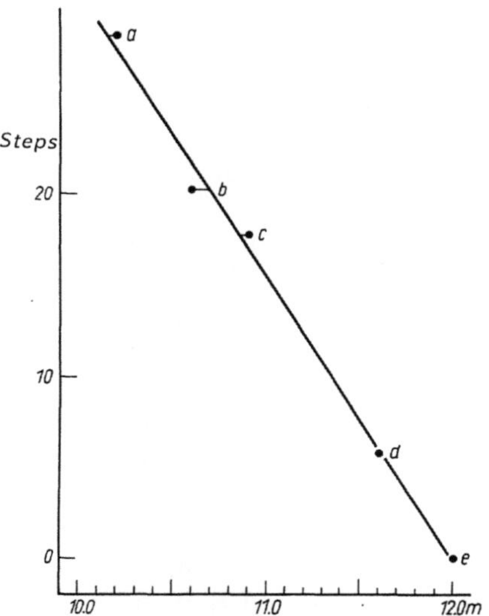

Fig. 165. Relationship between step intervals and magnitude (data given in Table 63)

Table 63. Step intervals and stellar magnitudes (example)

Star	Δs	m	s	m_c
a		$10^{\mathrm{m}}2$	27.4 st.	$10^{\mathrm{m}}16$
b	7.2 st.	10.6	20.2	10.70
c	2.4	10.9	17.8	10.85
d	12.0	11.6	5.8	11.62
e	5.8	12.0	0.0	11.99

(Δs) average step difference, (m) assumed stellar magnitude, (s) stellar
magnitude in step intervals, (m_c) corrected magnitude

The linear relationship between the step difference Δs and the magnitude
interval in photographic observations can only be guaranteed over a restricted photographic range. When approaching a plate's limiting magnitude,
for example with Mira stars which become invisible at minimum, then the
size of the step expressed in magnitudes becomes greater, and the accuracy
less. The trend line is no longer a straight line, and its slope decreases towards
the x-axis. This is because the diameter of star images cannot be reduced
below a certain limiting value, so that just above the threshold value a
decrease in brightness only results in a smaller degree of darkening.

A similar effect is also found with very bright stars. Here the situation is
such that darkening cannot increase beyond a certain maximum level, so that
towards greater intensities the number of steps only increases slowly. The
energy in the starlight is partially expended in "solarization", which may be
recognized by the fact that the images of very bright stars show a small,
almost glass-clear, central point.

In general, however, a linear relationship between step differences and
magnitude intervals may be expected over a range of about 4 to 5 magnitudes. In the main, this will be between approximately 1.5 mag and 6 mag
above the plate's threshold.

With direct visual observations of very bright and very faint stars the
accuracy of step estimates is also less; the size of the step is larger.

Determination of Comparison Star Magnitudes

The usual way is to relate the comparison star, the magnitude of which one
wishes to determine, to stars of known magnitude. This is often difficult to
do. Standard sequences, that is series of stars with accurately determined
magnitudes, are available in the *North Polar Sequence* (NPS) and in the
Selected Areas (SA). The former is given in many handbooks in the form of
charts. The Selected Areas are fields distributed over the whole sky, 206 in
total, generally at intervals of 1^h in right ascension and $15°$ in declination,
following a plan devised by KAPTEYN at Groningen. In photographing wide
fields one can therefore expect to have an SA on the plate. If this is not the
case, then an intermediate plate is taken, with a suitably displaced centre.
Linking the comparison stars is easily carried out with an iris-diaphragm
photometer. However, an auxiliary apparatus can also be built, where, by
means of two adjustable arms, the parts to be compared may be brought

together in a divided field of view, so that the procedure becomes purely visual. Care must be taken regarding the position on the plate and differential atmospheric extinction at differing zenith distances.

Special reference may be made to the *Atlas of Selected Areas* by BRUN and VEHRENBERG (1965). The brighter stars down to 12^m5 are shown on fields 40' square; the fainter on fields 15' square near the centre of the larger fields. The limiting magnitude lies at about 16^m-17^m. It should be noted that the atlas gives Harvard magnitudes. For Areas 1–139 a re-determination of the magnitudes has been carried out at Mt Wilson, on the basis of the International System, which has a considerable colour equation and a systematic scale deviation when compared with the Harvard scale. As a rule, when comparing Harvard magnitudes with the International System, they will be found to be too bright by up to about 0.6 mag, but differing from Area to Area, and dependent upon the stellar magnitude within a given field. Tables exist for this correction (SEARES 1925). Observers who do not have access to these sources should make it clear that their magnitudes relate to the Harvard system.

Special lists of comparison star magnitudes and the relationship between the photometric systems are given in the very comprehensive discussion by LAMLA (1965), for example.

The *Pleiades* are particularly suitable for obtaining a standard sequence of magnitudes as here, better than in the North Polar Sequence, stars from magnitude 3 down to the very faintest magnitudes are found within a restricted area, and are thus easily located. The North Polar Sequence has the advantage that it is available at any time of the year, and is always at the same zenith distance.

For older variables it is usual to find that comparison star magnitudes have been published somewhere, for example in the *Atlas Stellarum Variabilium* by HAGEN, or for 350 long-period variables by MITCHELL (1935) as well as on the AAVSO charts already mentioned. In the case of a bright variable (and occassionally even for a faint one) it can be worthwhile to check lists of photoelectric magnitudes (for example those issued by the International Stellar Data Centre in Strasbourg, see Sect. 9.5), for suitable neighbouring stars of known brightness. An amateur who does not have these aids available can, in the case of short-period variables and eclipsing stars, make use of the maximum and minimum magnitudes given in the literature, as a basis for a system of comparison stars. With faint stars, one can also use the limiting magnitude of the instrument as determined from the North Polar Sequence or the Pleiades, as a link, if no other possibility exists.

If it is not possible to use a Selected Area directly, then, at a pinch, a special plate of the SA or of the North Polar Sequence can be taken, naturally under exactly the same conditions as the observation plate itself. Both plates must be from the same package. According to the circumstances, it may be possible to cover part of the plate during the first exposure, and then to use this for the comparison field. It is not advisable to superimpose the two exposures upon one another, as has been done occasionally in the past, even though it is possible to distinguish between the stars which belong to one field

and those of the other. But a latency effect may occur, causing the emulsion
to react differently to the two exposures.

The determination of comparison star magnitudes is generally a somewhat difficult task. On the other hand, very often a systematic error of half
a magnitude in the comparison star series is without much significance, as the
shape of the light-curve is not distorted. It is only of importance if the aim
is to determine absolute magnitudes or distances. Amateurs have no choice
but to be guided by the organizations that are concerned with the study of
variables stars, or by specialized or professional observatories. We are unable
to discuss the matter in any further detail in this book.

Determination of Maxima and Minima

The observer must take pains to obtain as many points as possible on the
light-curve around maxima and minima, if these are to be used for determining or improving the elements. With a normal RR Lyrae or eclipsing star,
one should try to obtain an observation every 5–10 minutes; with a Mira star
every night. The extreme value can be read off a *mean curve* laid through the
points. It is not advisable to take the time of the brightest or faintest observation as the time of maximum or minimum. Problems can occur if, as is
often the case with Mira stars, the light-curve shows irregularities at the time
of maximum. Secondary waves must not be given too great a weight.

Pogson suggested a procedure that can occasionally be of use. On a plot
of the light-curve, lines are drawn parallel to the time axis, which thus join
points of equal magnitude on the ascending and descending branches. The
chords lying between these two branches of the light-curve are then bisected
and a smooth curve laid through those points. The time of maximum is taken
to be when an extension of this curve cuts the light-curve (*Pogson's method
of bisected chords*). As can be seen from Fig. 166, the time of maximum that
is thus obtained does not necessarily coincide with the highest point of the
curve; however it does better represent the general course of the variation.
The slope of the bisecting curve is an indication of the asymmetry of the
light-curve; in symmetrical curves it is a straight line, perpendicular to the
time axis. Various mathematical methods of calculating the times of the
extremes have been proposed, but they are of hardly any significance as they
have to assume that the curve follows an approximately mathematical form,
such as a parabola. It should be noted that none of these procedures can

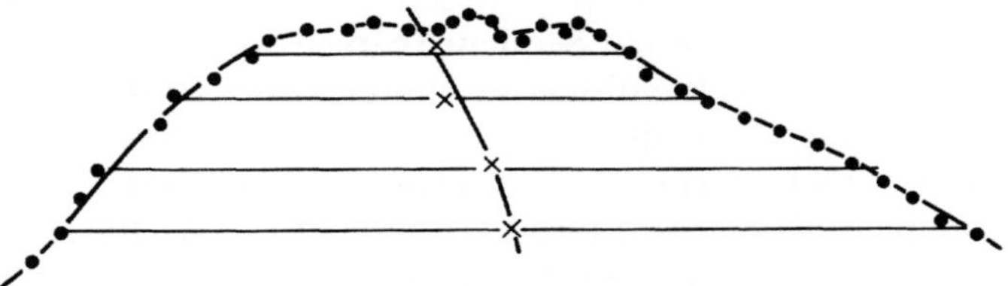

Fig. 166. Pogson's method of bisected chords

attain a greater accuracy than is inherent in the observations. The only purpose of any method of analysis is to bring out the full information content of the observations.

Observational Errors and Accuracy

The disturbing effect of *colour differences* has already been mentioned. Other psychological and physiological effects may also induce errors in step estimates, which can differ very greatly in various observers. They are probably also time-dependent, being connected with the state of the observer when the particular observations are made. The most dangerous is certainly *position-angle error*, which is determined by the relative position of stars that are to be compared in the field of view of the telescope, that is, the angle between the line joining the two stars and the vertical. The result is a systematic error in the observations if a star is followed over a period of several hours or – in stars that are observed once a night – an annual variation. *Separation error* results in an equal magnitude interval being estimated differently according to the separation between the two stars. *Magnitude error* consists of a relationship between the size of the step and brightness, while *interval error* is connected with the magnitude interval between the stars being compared.

It is known that only position angle error can introduce really serious errors, if the observer suffers from it, but even here large values should not occur. With the mean light-curves of short-period stars, however, this error will merely increase the scatter in the observations. The other errors will be more or less compensated if individual comparison star scales are used. The effects of these errors have certainly been overestimated from time to time in the literature. Nevertheless, in choosing comparison stars, one should try to avoid large separations and great differences in brightness.

Technical developments, particularly in electronics, have led to visual methods often being regarded as of lesser value. Certainly the newer methods are a great advance, particularly in the accuracy of measurements. This is also true when visual estimates are compared with photoelectric photometry. However, in any observational method we are not only concerned with the accuracy, but also with the effort; in other words with the amount of time that is required to determine the magnitude of a variable. On this score the Argelander method is far superior to photoelectric photometry. The technical effort must also be kept in mind. In addition, it should be noted that one hardly needs the greatest accuracy for the determination of the class and elements of a new variable, or later on, for checking the period. It is obviously different if one wishes to determine the light-curve in various spectral regions or, in eclipsing stars, to obtain the system constants.

An example will show what the Argelander method can achieve. During his second stay in South Africa in 1952–1953, HOFFMEISTER visually observed a large number of new short-period variables that he had discovered on photographic plates made during his previous visit. His wife took over supervision of the photographic instruments for the greater part of the night, so the programme could be even further expanded. As a result, in about 13

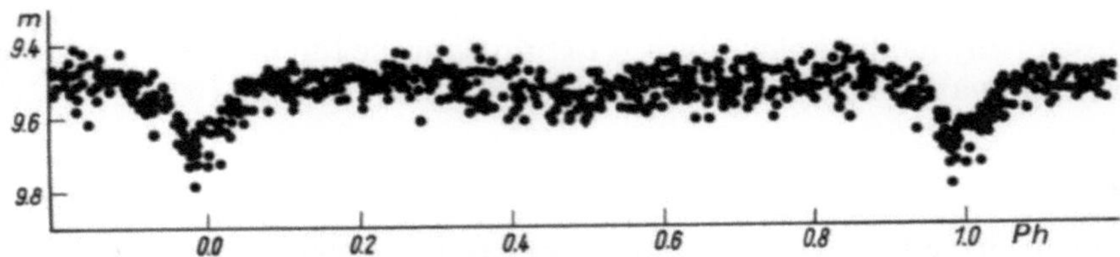

Fig. 167. Mean visually estimated light-curve of V 673 Cen (after HOFFMEISTER)

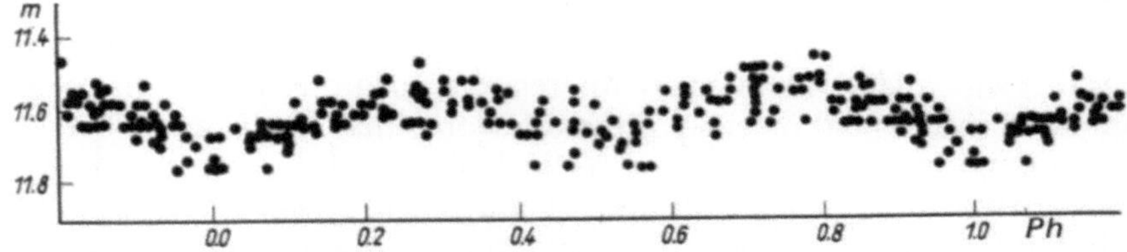

Fig. 168. Mean visually estimated light-curve of V 677 Cen (after HOFFMEISTER)

months he obtained 22 416 observations of 96 stars. The telescope, with an aperture of 130 mm and a focal length of 1160 mm, was very simply mounted as an altazimuth. Figures 167 and 168 show two light-curves of small-amplitude eclipsing stars (A_1 = primary minimum; A_2 = secondary minimum):

V 673 Cen 9^m5-9^m7 $P = 0^d932792$ $A_1 = 0.2$ mag $A_2 = 0.05$ mag
V 677 Cen 11^m55-11^m7 $P = 0^d325067$ $A_1 = 0.15$ mag $A_2 = 0.1$ mag

The two light-curves probably show the most that can be achieved with the Argelander method, and the causes of this lie not only in the observer but also with the object. First, suitable comparison stars must be available, and second, the course of the variation must be very similar in the various cycles. In the close contact systems, to which both V 673 and V 677 Cen belong, this is not always the case, as has been shown in the chapter on eclipsing stars. In addition, the magnitude must lie in a range suitable for the instrument. Photographically, variables with an amplitude of 0.2 mag can hardly be confirmed; they will only be noted as being suspect.

Photoelectric Photometry

The principle of the method consists of changing the energy present in light into electrical energy, it being much easier to measure the latter than directly measuring the intensity of light. Photoelectric photometry was introduced into astronomy in about 1910, in North America by STEBBINS and in Germany by GUTHNICK for direct measurements on the sky and by ROSENBERG for measurements on photographic plates. At that time vacuum or gas-filled cells were used, where light quanta caused the release of electrons from a thin evaporated layer of potassium, caesium or rubidium (among others). These electrons were collected by a (normally ring-shaped) electrode inside the cell,

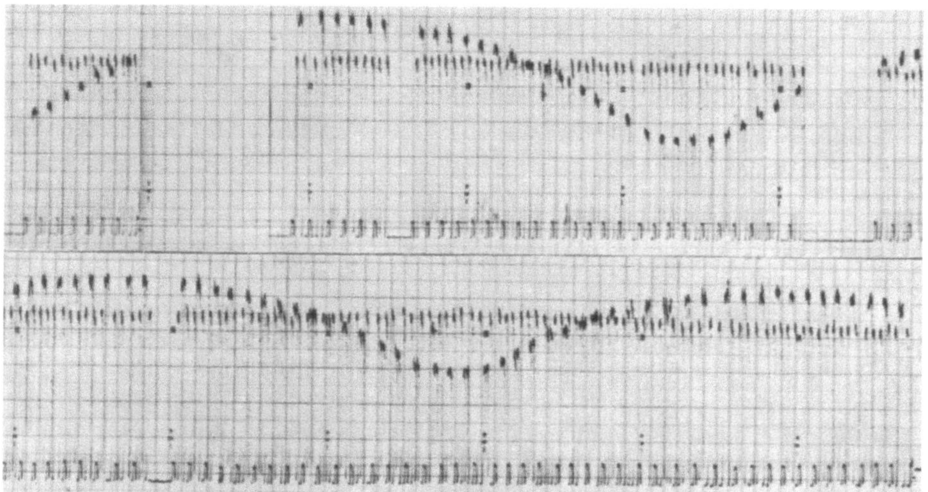

Fig. 169. Example of a photoelectric chart recording. These are measurements of CC Com (after ROSE and WENZEL). The undulating series of marks, each 60 seconds in length, arises from the variable. The amplitude of the star is 0.94 mag at primary minimum (*upper trace*) and 0.76 mag at secondary minimum (*lower trace*). The approximately horizontal series of marks, each 30 seconds long, are those of the comparison star; they show a surprising degree of stability in the equipment and in the atmosphere (the slight decline towards the right is due to atmospheric extinction with increasing zenith distance). Finally, the marks at the lower edge give the brightness of the sky background and the zero point set by the dark current. The rapid fluctuation within the traces, particularly those of the two stars, results partly from scintillation, and partly from "noise" in the equipment (particularly in the PMT). This effect determines the accuracy (standard error) of the measurements. The interval between the centres of the primary and secondary minima is $P/2 \approx 2.6$ hours. The lower trace is a direct continuation of the upper one

by means of a positive potential. The charge or current could then be measured.

Later the cells were replaced by secondary electron multipliers, otherwise known as *photomultiplier tubes* (PMT). These essentially depend upon the same principle, but an amplification process takes place within the vacuum tube. The primary electrons are given additional energy by an applied potential of about 100 V, and encounter a surface from which they, by virtue of their high energy, are able to liberate a greater number of secondary electrons. This process is repeated 10 or more times, according to the design, so that a considerable amplification ("multiplication") is obtained. With 12 steps each of 100 V, a PMT voltage of about 1200 V is required. Voltage supply units are needed for this. They must meet rigid standards as the DC voltage, frequently of more than 1000 V must be held constant to within 1 part in a thousand. The photocurrent delivered by the PMT can be registered by a chart recorder or digitally, if necessary with the use of an intermediate stage and other equipment.

Just as with other methods it is advisable to carry out differential measurements, that is by using comparison stars; indeed it is only by this method that variations in atmospheric transparency can be compensated. Occasionally light-curves are produced, using just the differences in brightness between the variable and a primary comparison star, without the apparent magnitude of the latter having been obtained through a sequence of standard stars and added to the measured differences in magnitude. This book contains some light-curves of this sort, the differential magnitude being indicated by Δm.

To assist reduction of the observations it is necessary to measure occasionally the sky background and the "dark current". The latter results from thermal electrons within the vacuum tube, which always exist, even when the cell is not illuminated. The smaller the dark current in a PMT, the more suitable it is. The dark current can be considerably reduced by cooling the cell with CO_2 snow, and this is frequently used in practice.

The PMT has also made fainter stars available for photometry. As an example, the mean error of a single measurement on a 12^m star with a 600 mm telescope amounts to about $\pm$ 0.02 mag. Apart from the method of measurement mentioned above, there are nowadays many other procedures (integrating and pulse-counting photometers, light-chopping methods, among others) which will not be discussed here.

Figure 169 shows a section of a record, reproducing a series of measurements made by a photoelectric photometer on the Sonneberg 600 mm reflector (see also Fig. 170).

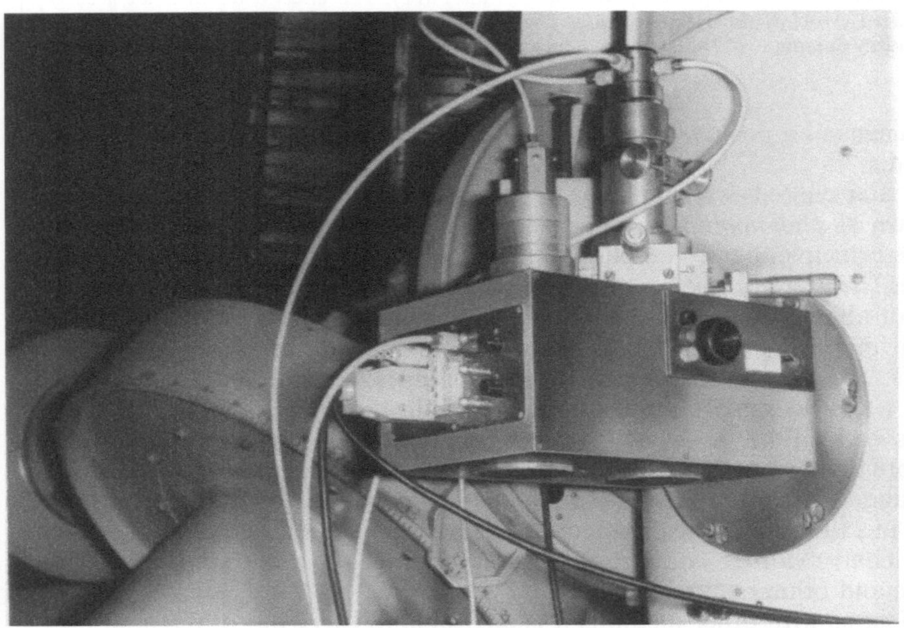

Fig. 170. Photoelectric photometer on the 60-cm Sonneberg No. 1 Reflector

Thanks to the great advances in *semiconductor technology*, the infrared spectral range ($\lambda > 0.8\,\mu\mathrm{m}$) down to sub-millimetre waves is now also available for measurements.

8.1.2 Spectrographic and Other Observations

Knowledge of the spectrum and its changes is of very special significance for research into variables as it is the only means by which the phenomena shown by the light-curve can be explained physically. This is not to say that we have even distantly achieved this aim in most cases. Two phenomena observable spectroscopically are of particular importance: first the Doppler effect, which has (among other things) enabled the pulsation theory to be confirmed; and second, the effects of gaseous shells and disks, which not only cause line-shifting and line-widening, but also the appearance of emission lines.

As is well-known there are several methods by which stellar spectra may be obtained. In one a prism is placed in front of an objective, both having the same diameter, so that the spectra of all the stars in the field are obtained on the photographic plate, provided they are bright enough. This objective-prism method is therefore suitable for the classification of a large number of stars, and also for the discovery of special objects, such as Hα stars, those stars that have the red Hα line in emission. However, for precise physical investigations on individual stars, greater dispersion is required. A slit spectrograph is used for this, which can form the spectrum of only a single star at a time, and has provision for introducing comparison spectra.

This book contains numerous examples of notable spectra of variables. Note that negative reproductions are used (lines appearing dark are emission lines) and that in a number of cases the comparison spectra just mentioned can be seen alongside the stellar spectra.

Observations with radio telescopes at centimetre and metre wavelengths have, as yet, only been successful in a few individual cases of variability. Prediction of future developments is hardly possible at present.

Three other technical fields should be at least mentioned. One is the use of *image intensifier tubes* to increase the light-grasp of photographic exposures, or to decrease exposure times. Another is the field of *electronic data processing*, which is continually increasing in importance for the reduction of astronomical observations. The third technique that we may mention is that of using *charge-coupled devices* (CCD), where a linear or rectangular array of numerous microscopic semiconductor detectors provides high-sensitivity structural resolution. This book's purpose is to describe fairly well-established facts and techniques, so there is little point in giving further references, which would be quickly out-of-date, as development is advancing so rapidly. The reader is therefore referred to the literature for further information about this subject.

The observation of variable stars from Earth satellites has begun to take place, especially in the infrared, ultraviolet and X-ray regions that are difficult or impossible to observe from the ground. A whole series of new discoveries may be expected here. However, these can only be successfully combined with our current state of knowledge if comparative "optical" information can be obtained simultaneously or in any other appropriate way.

8.1.3 Bright Variables

Among the stars visible to the naked eye there are quite a number of variables. A list of these objects is of interest in several respects. Table 64 has been compiled from the General Catalogue and its 3 Supplements (GCVS, KUKARKIN et al. 1969–1976).

Table 64. List of bright variables

Name	Range		Class	Period	Spectrum
λ And	$4\overset{m}{.}9$	$5\overset{m}{.}3$	SR?	54^d	G8
η Aql	4.1	5.4	Cδ	7.177	F6
48 (RT) Aur	5.0	5.8 v	Cδ	3.728	F4
ε Aur	3.5	4.5	EA	9892	F0ep
ζ Aur	5.0	5.6	EA	972	K4 + B7
29 (UW) CMa	4.5	4.8	EB	4.393	O8 + O8
27 (EW) CMa	4.3	4.6	I?		B4e
FW CMa	5.0	5.3	γC	–	B3e
R Car	3.9	10.0 v	M	309	M4e
S Car	4.5	9.9 v	M	149.5	K7e
η Car	−0.8	7.9 v	SD	–	Pec
l Car	3.4	4.1 v	Cδ	35.522	F6
γ Cas	1.6	3.0 v	γC	–	B0e
ϱ Cas	4.1	6.2 v	RCB?	–	F8p
μ Cen	2.9	3.4 v	γC	–	B2
δ Cep	3.5	4.3 v	Cδ	5.366	F5
μ Cep	3.6	5.1 v	SR		M2e
o Cet	2.0	10.1 v	M	331.9	M5e
ε CrA	4.7	5.0 v	EW	0.591	F0
T Cyg	5.0	5.5 v	L?		K3
o^1 (V 695) Cyg	4.9	5.3	EA	3784	K4 + B4
f^1 (V 832) Cyg	4.5	4.9 v	γC	–	B2e
χ Cyg	3.3	14.2 v	M	407	S7e
P Cyg	3	6 v	SD	–	cB1peq
β Dor	3.5	4.1 v	Cδ	9.842	F4
ζ Gem	3.7	4.2 v	Cδ	10.151	F7
η Gem	3.3	3.9 v	SR (E)	233 (2984)	M3
β Gru	2.0	2.3 v	L?		M3
α Her	3.0	4.0 v	SR		M5
u Her	4.6	5.3	EB	2.051	B3 + B5
R Hor	4.7	14.3 v	M	404	M5e
R Hya	4.0	10.0 v	M	390	M6e
EW Lac	5.0	5.3	γC	–	B3ep
R Leo	4.4	11.3 v	M	312	M6e
RX Lep	5.0	7.0 v	SR	150 ±	M4
δ Lib	4.9	5.9 v	EA	2.327	A0
13 (R) Lyr	3.9	5.0 v	SR	46	M5
β Lyr	3.3	4.3 v	EB	12.914	B8p
ε Oct	5.0	5.4 v	SR	55 ±	M6
χ Oph	4.2	5.0	γC	–	B2pe
U Ori	4.8	12.6 v	M	372	M6e
α Ori	0.4	1.3 v	SR	2335	M2e

Table 64. (continued)

Name	Range		Class	Period	Spectrum
$\varkappa$ Pav	$3^{\mathrm{m}}9$	$4^{\mathrm{m}}8$ v	CW	$9^{\mathrm{d}}088$	F5
λ Pav	3.4	4.3 v	γC	–	B2e
β Peg	2.3	2.7 v	L	–	M2e
β Per	2.1	3.4 v	EA	2.867	B8
ϱ Per	3.3	4.0 v	SR	$50 \pm$	M4
ζ Phe	3.9	4.4	EA	1.670	B6 + B8
δ Pic	4.6	4.9 v	EB	1.673	B0
47 (TV) Psc	4.6	5.4 v	SR	$65 \pm$	M3
V Pub	4.7	5.2	EB	1.454	B1 + B3
KQ Pup	4.9	5.2	?		M2e + B2e
MX Pup	4.6	4.9	γC	–	B2e
L_2 Pup	2.6	6.2 v	SR	140	M5
W Sgr	4.3	5.1 v	Cδ	7.595	F4
3 (X) Sgr	4.2	4.8 v	Cδ	7.012	F5
RR Sco	5.0	12.4 v	M	279	M6e
α Sco	0.9	1.8 v	SR	1733	M1
μ^1 Sco	2.8	3.1	EB	1.440	B2 + B7
R Sct	4.4	8.2 v	RV	140	G0e
δ Sct	4.9	5.2	δSc	0.194	F3
d Ser	4.9	5.9 v	?		G0 + A6
28 (BU) Tau	4.8	5.5 v	γC	–	B8ep
λ Tau	3.3	3.8	EA	3.953	B3 + A4

Designations (international abbreviations):

EA	Algol		M	Mira
EB	β Lyrae		RV	RV Tauri
EW	W Ursae Majoris		SR	Semi-regular
E	Eclipsing variable, sub-class unknown		L	Irregular (late spectral classes)
Cδ	δ Cephei		I	Irregular (early spectral classes)
CW	W Virginis		γC	γ Cassiopeiae
δSc	δ Scuti		RCB	R Coronae Borealis
			SD	S Doradus

Variables have been included that have maximum magnitudes of $5^{\mathrm{m}}0$ or greater, and with amplitudes of 0.25 mag or greater (visual or photographic). Alongside the names of the stars the table gives the magnitude ranges: v indicates visual or *V*-band, all others have been determined photographically or in the *B*-band. The other columns give the class designation, the period (to a limited precision), and the spectral class at maximum (disregarding any other spectral characteristics).

Most of the stars listed can be located easily with atlases of naked-eye stars, so co-ordinates do not have to be given here.

Novae and supernovae are not considered. Because of the criteria mentioned, the much-observed star R CrB is not included, although it has given its name to a whole class of variables (Sect. 3.5). Mention should also be made of α UMi, the Pole Star, which is a W Virginis variable with a period of $3^{\mathrm{d}}970$ and the very small amplitude of 0.15 mag. There is a considerable literature about this star. The period is variable, but it has not yet been possible to derive a generally valid formula. Interferometric measurements by

WILSON jr (1937) showed that the primary is double: there is a 4^m companion at a separation of $0\overset{''}{.}24$. This component is possibly responsible for a term with a period of 29.6 years that occurs in the magnitude variations. There is a 9^m companion at a distance of $18\overset{''}{.}3$.

The fact that it is still possible to discover notable variability in bright stars is shown by 1 (V 436) Per. RUFENER (Geneva) found some deviant photoelectric measurements in using it as a standard star. A group of French amateurs then visually observed the primary and secondary minima of the eclipsing star which was thus discovered, determined the period $(25\overset{d}{.}936)$ and established strong ellipticity in the orbit – eccentric secondary minimum – (NORTH and RUFENER 1981). The range of variation is $V = 5\overset{m}{.}51- > 5\overset{m}{.}84$. A similar case is that of 71 (DE) Dra, an eclipsing system, discovered by FÜRTIG (1975) and more closely investigated by L. MEINUNGER (1979), which has a visual magnitude at maximum of $5\overset{m}{.}7$.

8.1.4 Amateur Collaboration

Hardly any other field of activity is as suitable for amateur astronomers as is the study of variable stars. There are many reasons for this: first, the instrumental requirements are minimal; every instrument from binoculars to the largest reflecting telescope can serve as the basis for a programme. Second, the method of observation is easily learnt, even if this does require some care and a little time. Third, if observations are carried out properly one can be sure of success and has an extremely wide range of activity that can be followed.

It would probably be more confusing than explanatory if we were to list all the possibilities. Careful study of the sections dealing with the classification of variables will give numerous suggestions. Yet it may be as well to set out some important considerations.

Period Changes

The very topical subject of period changes should be mentioned first. If we were to say that there are no constant periods among variable stars, it would be somewhat overstating the case, but it is probably not far from the truth. Even many eclipsing variables have variable periods, as is shown by the case of Algol, and the investigations carried out to date. Period changes are also found in δ Cephei stars. In the RR Lyrae stars they occur periodically in relation to changes in the amplitude and the form of the light-curve. The Mira stars are a special case as they probably have sudden irregular changes in their periods.

Although we may recognize a change of period in a star, very frequently we do not know when it took place. A good maximum or minimum would, in that case, be capable of giving information about when the change occurred, or at least of narrowing down the interval when it may have happened, as well as saying something about the mechanism of the alteration. Certainly photographic sky patrols can provide a lot of help here, but with short-period stars they do not provide fully-observed light-curves, quite apart from interruptions caused by the weather.

It is better to obtain closely covered curves of single maxima or minima, than to attempt to observe as many stars as possible. An experienced observer can certainly cope with a large programme, but then he will give precedence in each hour to those stars which are just before minimum or maximum. It is normally sufficient to observe rapidly-varying stars at 30-minute intervals. If we see that an Algol star has become fainter, or an RR Lyrae star that was previously at minimum has begun to brighten, then it should be observed every 5 or 10 minutes, continuing until minimum or maximum light has been definitely passed.

With Mira stars and other classes with slow variations, photographic observation is also recommended, the more so since in the rich fields of the Milky Way there will be a large number of suitable objects on the plate. As these are often coloured stars, care must be taken that the type of plate is always accurately described, and is not altered in a way that cannot be checked. The photographic method is less suitable for rapidly-varying stars as the plates do not give a sharp epoch, but give an average magnitude, unless they are exposed for a very short period of time. This is not possible if we wish to photograph very faint stars and obtain the maximum information. A well-covered visual light-curve of an RR Lyrae star can give the maximum with an error of $\pm$ 5 minutes. With a 30- or 60-minute photographic exposure the time of a maximum can lie anywhere within that period, and we introduce considerable inaccuracy if we give it, as we must, as being the middle of the exposure time. If we have to find a period that is as yet unknown, then in general we have to turn to such observations, and no detriment results. However, if minor details such as period changes are to be investigated, then the situation is different.

Eruptive Variables

Another very worthwhile task is the monitoring of nova-like variables and old novae, where a further outburst may be expected. A list of such novae prepared by the International Astronomical Union is given in the appropriate section (3.1.2). Those old novae were chosen that have a relatively small amplitude, as these may be presumed to have short intervals between outbursts ($\approx$ 10 to $\approx$ 60 years). Here it is a question of recognizing the start of the rise as soon as possible, so that the course of the outburst may be observed spectroscopically. The nova-like stars can also provide surprises. In either case, an observatory should be alerted.

The U Geminorum stars must also be mentioned, many of which require a good deal of patience on the part of the observer, owing to the rarity and short duration of the outbursts. Cases like UV Per could never have been explained without visual monitoring over a period of years.

Constant monitoring of the rare R Coronae Borealis stars is also very important. R CrB itself is 6^m at maximum. The earliest possible notification that the magnitude is beginning to decline to one of its unpredictable minima is of the very greatest importance to professional observatories that are equipped with spectrographs or infrared photometers. It is highly desirable to obtain densely covered light-curves of these objects, as the form, date and depth of the minima, when taken in conjunction with series of physical measurements and observations, enable fundamental conclusions to be drawn about the nature of the circumstellar material causing the minima.

8.2 Organization

The organization of national and international *co-operation* for the investigation of variable stars is especially important on at least three counts.

First, the knowledge of the light-curves of variables with as unbroken a coverage as possible, over suitable periods of time – whether the variation is irregular or cyclic – is often decisive for the correct interpretation of the observational evidence and for the development of models. Owing to the influence of weather and the Earth's rotation, this can generally only be achieved by international co-operation.

Second, for simultaneous observations in several fundamental wavelength regions (radio frequencies, IR, visual region, UV, X-rays), together with spectrographic investigations, it is necessary to use various types of observational equipment, which would not be available to a single observatory or scientist.

Third, a co-operative division of labour between theoreticians, computers and observers is generally also necessary.

The organization of world-wide international co-operation between professional astronomers is the task of the *International Astronomical Union* (IAU). Approximately 40 Commissions in the IAU deal with individual specialized areas covering the whole range of astronomical research. Commission 27 is responsible for variable stars, and Commission 42 for eclipsing stars, but naturally other commissions are also involved, such as those for photometry, spectroscopy, binary stars, galactic structure, stellar structure, and stellar evolution. In the variable star field, the organization of international observational programmes, and the combined use of specialized observational sites and instruments in many countries are of especial practical importance, as is the holding of scientific symposia, colloquia and workshops.

Apart from the IAU there are other forms of organization which are more restricted, for example the "Commission for the study of problems of stellar physics and evolution by multilateral co-operation between Academies of socialist countries". Here, Sub-commission 3 (Non-stationary Stars) is particularly concerned with co-operation in the field of variable stars.

Amateur astronomers are generally combined on a national basis in more or less loosely organized associations. In recent years many new groups have been formed. A complete listing cannot be given here. Some examples of organizations that are concerned with furthering work in the field of variable stars may, however, be given.

AAVSO   American Association of Variable Star Observers
AFOEV   Asociation Française d'Observateurs d'Etoiles Variables
AKV   Arbeitskreis "Veränderliche Sterne" im Kulturbund der
             DDR
BAA-VSS   British Astronomical Association (Variable Star Section)
BAV   Berliner Arbeitsgemeinschaft für Veränderliche Sterne
             (West Berlin)
BBSAG   Beobachter von Bedeckungssternen in der Schweizerischen
             Astronomischen Gesellschaft (Switzerland)
RASNZ-VSS   Royal Astronomical Society of New Zealand (Variable
             Star Section)
SUAA-VSS   Scandinavian Union of Amateur Astronomers and Ursa
             Astronomical Association (Variable Star Section)

In many countries professional observatories are centres for such groups
(Czechoslovakia – Brno, Poland – Krakow), while in other countries they
may be based on school or public observatories. Contact between amateurs
and professional astronomers is essential, and this can also lead to international co-operation.

9. Literature

9.1 Brief Notes About Stellar Catalogues and Atlases

Indispensable for the discovery, observation and analysis of variable stars are the "Obshchii katalog peremennykh zvezd" ("General Catalogue of Variable Stars") and its three Supplements, published by KUKARKIN et al. (1969, 1971, 1974, 1976); the "Name Lists" that appertain to it (e.g. KHOLOPOV et al. 1981); and also the catalogue of stars, still unnamed, but suspected of variability, now in its third edition (KHOLOPOV 1982: "New catalogue of suspected variable stars"). The first work is generally referred to in this book as the "GCVS"; the last is known as the "NSV" (its precursors were the "CSVS 1951" and "CSVS 1965" by KUKARKIN et al., 1951 and 1965).

The GCVS and its Supplements also contain the classification approved by the IAU, to which repeated reference has been made in this book. A recent, revised description of this system is given by KHOLOPOV (1981).

Among general star catalogues, the most important are the *Durchmusterung catalogues* and the charts drawn from them, as long as positions do not have to be very precisely determined. This subject, which comes within the province of astrometry, will not be discussed here. The details of the Durchmusterung catalogues are as follows:

Northern Bonner Durchmusterung, Epoch 1855.0, declination boundaries $+90°$ and $-2°$, abbreviation BD
Southern Bonner Durchmusterung, Epoch 1855.0, declination boundaries $-2°$ and $-23°$, abbreviation BD
Cordoba Durchmusterung, Epoch 1875.0, declination boundaries $-22°$ and $-90°$, abbreviation CoD
Cape Photographic Durchmusterung, Epoch 1875.0, declination boundaries $-18°$ and $-90°$, abbreviation CPD

The most important older *catalogues of stellar spectra* are:

Henry Draper Catalogue of Stellar Spectra, Ann. Harvard Obs. **91–99** (1918–1924)
Henry Draper Extension, Ann. Harvard Obs. **100; 105; 112** (1925–1949)
Potsdam Spectral Durchmusterung, Publ. Astrophys. Obs. Potsdam **88–93** (1929–1938)
Bergedorf Spectral Durchmusterung, 5 Vols. (1935–1953)

Charts have been published for the BD and the CoD, drawn to a scale of $1° = 20\,\text{mm}$. They reach to about 10^{m} visual. The positions of stars not included in the Durchmusterung can be read to an accuracy of about $0\rlap{.}''5$ by interpolation. However, with a new variable it is advisable for a special small chart to be drawn and published, so that the object can be unambiguously identified.

A detailed catalogue of the *positions* and proper motions of almost 260 000 stars is available in the "Star Catalog" of the Smithsonian Astrophysical Observatory (SAO Catalog). There also exists an Atlas based on this, which includes, apart from the catalogue stars (Epoch 1950.0), nonstellar objects, and has a scale of $1° = 8.6$ mm.

In addition there are other atlases with smaller scales, which may be of use from time to time. Some are those by BEYER-GRAFF, BEČVÁŘ, MIKHAILOV and VEHRENBERG. The latter has produced two widely-used *photographic atlases*. Photographic charts are especially important, as they are also a record of the date on which they were taken.

The oldest project of this sort is the set of photographic charts by PALISA and WOLF. A total of 210 sheets were published, each covering an area of $6° \times 7\rlap{.}°5$ at a scale of $1° = 36.7$ mm. Their significance lies in the fact that they reach at least 15^m, and that a highly accurate grid is incorporated, which permits positions of stars to be determined to within an error of about $0\rlap{.}''2$. Unfortunately the charts do not cover the whole of the northern sky. The exposures on which it is based were primarily taken for study of the Milky Way and of minor planets. Note that the epoch is not uniform; one section of the charts has a grid for 1875.0, another for 1900.0.

The photographic charting of the southern sky that was carried out by the Johannesburg Observatory, in the "Map of the Sky South of $-19°$", on the other hand, is completely consistent. The epoch is 1875.0, in agreement with the CoD and the CPD. The area of individual sheets is $5\rlap{.}°4 \times 6\rlap{.}°8$ at a scale of $1° = 35.5$ mm. The grid is as accurate as that on the PALISA and WOLF charts. Unfortunately the limiting magnitude is unsatisfactory on many sheets. These photographic charts can be very useful, particularly in the determination of positions, which are more accurate than those taken from the Durchmusterung charts.

The Palomar Sky Survey Atlas, published by Palomar Observatory and the National Geographic Society, has become specially significant. The charts cover the sky from the north pole to $-33°$. The exposures were made with the large 122-cm Schmidt Telescope. The plates cover a field of $6\rlap{.}°5 \times 6\rlap{.}°5$, length of side 35 cm, scale $1° = 53.5$ mm. The limiting magnitude is about 21^m. Each of the fields has been photographed twice, in the blue and the red. By comparing the two plates the colour of a star may be determined. This is the reason for the Palomar charts being so important for variable-star research.

The Palomar Atlas has since been extended by another two zones down to $-45°$ declination, but only in the red. A survey that will complete the coverage of the southern sky, and be fully equivalent to the Palomar Sky Survey, is at present being carried out with the British 124-cm Schmidt telescope at Siding Spring in Australia and the 100-cm Schmidt telescope at the European Southern Observatory in Chile.

Unfortunately, a large number of the variables discovered photographically since 1900 generally lack the highly-desirable *finder charts*. The only complete series of close-field charts is that for the more than 10000 variables discovered at Sonneberg Observatory. These are available, first in the Observatory's "Mitteilungen", and then a long series in "Mitteilungen über Veränderliche Sterne" (MVS) Nos. 245–330, and then in the journal "Astronomische Nachrichten". Mention should also be made of the collection of close-field charts for RR Lyrae stars and other variables that has been prepared by TSESEVICH and KAZANASMAS, Odessa (1963, 1971); to the numerous series of "Charts of Southern Variable Stars" issued since 1952 by BATESON et al. on behalf of the Variable Star Section of the Royal Astronomical Society of New Zealand; and also to the "AAVSO Variable Star Atlas" by SCOVIL (1980).

The increasing volume of astronomical data has led to the foundation of the *International Stellar Data Centre* in Strasbourg (France). All important catalogues are stored there, in machine-readable form. Unpublished collections of data from various institutions are also entered. The advantage is that research or compilations of desired catalogue information can be carried out automatically, and the material stored can be continuously revised. Such listings, brought completely up-to-date, are, for example:

"A General Catalogue of UBV Photoelectric Photometry" by MERMILLIOD and NICOLET, containing details of about some 53000 stars; "MK Extension Catalogue" by KENNEDY, containing more than 32000 modern spectral data on the MORGAN-KEENAN spectrum/luminosity system.

The Stellar Data Centre also has on file the "Bibliographic Catalogue of Variable Stars" (BCVS) by HUTH and WENZEL (1981), which gives more than 270000 references for variable stars named up to 1976, and which was based on the card index that had been maintained for many years at Sonneberg Observatory (see WENZEL 1981). The BCVS may serve as successor to the "Geschichte und Literatur des Lichtwechsels der Veränderlichen Sterne", a multi-volume work, the first edition of which was prepared by HARTWIG and MÜLLER, and the second edition (1934–1963), still frequently in use today, by PRAGER and SCHNELLER. The BCVS will be continuously revised.

9.2 General Summaries, Compilations and Review Articles

The following is a selection of important reviews and contributions. Most of the older descriptive works given in the first German edition are omitted. No attempt has been made to be comprehensive.

IAU Symposia (Reidel Publishing Company, Dordrecht):
No. 59 Stellar Instability and Evolution – 1974
    67 Variable Stars and Stellar Evolution – 1975

   70 Be and Shell Stars – 1976
   73 Structure and Evolution of Close Binary Systems – 1976
   83 Mass Loss and Evolution of O-type Stars – 1979
   88 Close Binary Stars – 1979
   98 Be Stars – 1981

IAU Colloquia (various publishers):
No.   4 Non-Periodic Phenomena in Variable Stars – 1969
     6 Mass Loss and Evolution in Close Binaries – 1969
  15 New Directions and New Frontiers in Variable Star Research – 1971
  21 Variable Stars in Globular Clusters and in Related Systems – 1972
  29 Multiple Periodic Variable Stars – 1975
  32 Physics of Ap Stars – 1975
  42 The Interaction of Variable Stars with their Environment – 1977
  46 Changing Trends in Variable Star Research – 1978
  53 White Dwarfs and Variable Degenerate Stars – 1979
  59 Effects of Mass Loss on Stellar Evolution – 1981
  69 Binary and Multiple Stars as Tracers of Stellar Evolution – 1981
  70 The Nature of Symbiotic Stars – 1981
  72 Cataclysmic Variables and Related Stars – 1982

Astrophysics and Space Science Library (Reidel Publishing Company, Dordrecht):
Vol.   6 UNDERHILL, The Early Type Stars – 1966
  13 HACK (Ed.), Mass Loss from Stars – 1968
  45 COSMOVICI (Ed.), Supernovae and Supernova Remnants – 1974
  48 GURSKY and RUFFINI (Eds.), Neutron Stars, Black Holes and Binary X-Ray Sources – 1975
  65 FRIEDJUNG (Ed.), Novae and Related Stars – 1977
  66 SCHRAMM (Ed.), Supernovae – 1977
  68 KOPAL, Dynamics of Close Binary Systems – 1978
  77 KOPAL, Language of the Stars – 1979
  89 CHIOSI and STALIO (Eds.) IAU Colloquium 59
  95 FRIEDJUNG and VIOTTI (Eds.), IAU Colloquium 70
 101 LIVIO and SHAVIV (Eds.), IAU Colloquium 72

Stars and Stellar Systems (University of Chicago Press, Chicago and London, 1963 ff.):
Vol.   3 KUKARKIN and PARENAGO, Surveys and Observations of Physical and Eclipsing Variable Stars
       WOOD, Empirical Data on Eclipsing Binaries
       KRAFT, The Absolute Magnitudes of Classical Cepheids
       PAYNE-GAPOSCHKIN, GAPOSCHKIN, The Luminosities of Variable Stars
Vol.   5 KRAFT, Distribution of Classical Cepheids
       PLAUT, Distribution and Motions of Variable Stars
       PLAUT, Distribution of Novae in the Galaxy

Vol. 6 KRAFT, The Spectra of Supergiants and Cepheids of Population I
UNDERHILL, Early Type Stars with Extended Atmospheres
WILSON, Eclipses by Extended Atmospheres
MERRILL, Spectra of Long Period Variables
McLAUGHLIN, The Spectra of Novae
JOY, Spectra of Dwarf Variable Stars
Vol. 7 HARO, Flare Stars
Vol. 8 SCHATZMAN, Theory of Novae and Supernovae
ZWICKY, Supernovae

Handbuch der Physik, Bd. 51 (Springer Berlin, Göttingen, Heidelberg, 1958):
LEDOUX and WALRAVEN, Variable Stars
PAYNE-GAPOSCHKIN, The Novae
ZWICKY, Supernovae

LANDOLT-BÖRNSTEIN, Zahlenwerte und Funktionen, Neue Serie, Gr.VI
(Springer Berlin, Heidelberg, New York, 1965 and 1982):
Bd. 1 HERCZEG, Spectroskopische Doppelsterne (Spectroscopic Binaries)
BEYER, Veränderliche Sterne (Variable Stars)
Bd. 2b DUERBECK and SEITTER, Variable Stars
SEITTER and DUERBECK, Peculiar Stars
APPENZELLER, Protostars, Pre-Main Sequence Objects
WEIDEMANN, White Dwarfs
HERCZEG, Double Stars
Bd. 2c FINK, GREWING and TRÜMPER, Compact Objects, X-Ray and γ-Ray
Sources

Advances in Astronomy and Astrophysics (Academic Press, Chicago, from 1962):
Numerous reviews

Annual Review of Astronomy and Astrophysics (Annual Reviews Inc., Palo Alto, from 1963):
Numerous reviews

Other conference proceedings (with editor, and place and year of the conference):
ARAKELYAN, Nestatsionarnye Zvezdy (Non-Stable Stars), Byurakan 1956
LEDOUX, Problèmes d'Hydrodynamique Stellaire, Liège 1975
MIRZOYAN, Vspykhivayushchie Zvezdy (Flare Stars), Byurakan 1976
ZYTKOW, Nonstationary Evolution of Close Binaries, Warsaw 1977
MIRZOYAN, Vspykhivayushchie Zvezdy (Flare Stars), Byurakan 1979
HILL and DZIEMBOWSKI, Nonradial and Nonlinear Stellar Pulsations. Tucson 1979

WHEELER, Type I Supernovae, Austin 1980
LEDOUX, Variability in Stars and Galaxies, Liège 1980
CARLING and KOPAL, Photometric and Spectroscopic Binary Systems, Maratea (Italy) 1980
TREMKO, Ejection and Accretion of Matter in Binary Systems, Tatranská Lomnica 1980
ANDRESEN, X-Ray Astronomy, Amsterdam 1981
GEVON and STERKEN, Workshop on Pulsating B Stars, Nizza 1981
REES and STONEHAM, Supernovae, Cambridge 1981
MARIK and SZABADOS, Magnetic and Variable Stars, Szombathely 1982

Recent monographs:

ALKSNE and IKAUNIEKS, Uglerodyne Zvezdy (Carbon Stars) (Zinatne, Riga 1971); English version: Astron. Astrophys. Series *11* (Pachart Publ. House, Tucson 1981)
CLARK and STEPHENSON, The Historical Supernovae (Pergamon Press, Oxford 1977)
GLASBY, Variable Stars (Constable, London 1968)
GLASBY, Dwarf Novae (Constable, London 1970)
GLASBY, The Nebular Variables (Pergamon Press, Oxford 1974)
GURZADYAN, Flare Stars (Pergamon Press, Oxford 1980)
IKAUNIEKS, Dolgoperiodicheskie Peremennye Zvezdy (Long-Period Variables) (Zinatne, Riga 1971)
KUKARKIN et al., Nestatsionarnye Zvezdy i Metody ikh Issledovaniya (Non-stable Stars and Methods for their Investigation), 5 Vols. (Nauka, Moscow 1970 ff.)
KUKARKIN (Ed.), Pulsating Stars (Wiley, New York 1973)
MIRZOYAN, Nestatsionarnost' i Ehvolyutsia Zvezd (Stellar Variability and Evolution) (Izd. A.N. Arm. SSR, Erevan 1981)
PAYNE-GAPOSCHKIN, The Galactic Novae (Dover, New York 1964)
PETIT, Les Etoiles Variables (Masson, Paris 1982)
SAHADE and WOOD, Interacting Binary Stars (Pergamon Press, Oxford 1978)
STROHMEIER, Variable Stars (Pergamon Press, Oxford 1972)
STROHMEIER, Veränderliche Sterne I (Treugesell, Düsseldorf 1974)
TSESEVICH, RR Lyrae Stars (Wiley, New York 1969)
TSESEVICH, Peremennye Zvezdy i Sposoby ikh Issledovaniya (Variable Stars and Methods for their Investigation) (Pedagogika, Moscow 1970)
TSESEVICH (Ed.), Eclipsing Variable Stars (Wiley, New York 1973)
ZHILYAEV et al., Zvezdy Tipa R Severnoi Korony (R Coronae Borealis Stars) (Nauka Dumka, Kiev 1978)

9.3 References

The abbreviations for publications and the transliteration of the Russian alphabet follow the system employed in "Astronomy and Astrophysics Abstracts" (Springer Berlin, Heidelberg, New York), which originates in decisions and recommendations by the Abstracting Board of the International Council of Scientific Unions. For simplicity, and where no misunderstanding is likely to occur, only the first page number of the particular paper is cited.

Acker, A., Marcout, J. (1977): Astron. Astrophys., Suppl. Ser. **30,** 221
Ahnert, P. (1939): Astron. Nachr. **269,** 241
Alekseev, G. N. (1973): Astron. Tsirk. No. 788, 3
Alexander, J. B. et al. (1972): Mon. Not. R. Astron. Soc. **158,** 305
Alksne, Z., Ikaunieks, J. (1971): Trans. Riga Obs. **13**
Alksnis, A, Alksne, Z. (1977): Trans. Riga Obs. **16,** 7
Allen, D. A. (1981): Anglo-Aust. Obs., Prepr. No. 146
Allen, D. A. (1983): Anglo-Aust. Obs., Prepr. No. 185
Allen, D. A., Ward, M. J., Wright, A. E. (1981): Astron. Astrophys. **195,** 155
Ambartsumyan, V. A. (1949): Astron. Zh. Akad. Nauk SSSR **26,** 3
Amnuel, P. R., Guseinov, O. H., Rakhaminov, Sh. Yu. (1979): Astrophys. J., Suppl. Ser. **41,** 327
Andersen, C. M., Casinelli, J. P. (1981): Wisconsin Astrophys. No. 127
Andriesse, C. D., Viotti, R. (1979): IAU Symp. **83,** 47
Appenzeller, I., Mundt, R., Wolf, B. (1978): Astron. Astrophys. **63,** 289
Argue, A. N., Sullivan, C. (1982): Observatory **102,** 4
Arkhipova, V. P., Mustel, E. R. (1975): IAU Symp. **67,** 305
Ashbrook, J. (1980): Sky Telesc. **60,** 21

Baade, W., Swope, H. (1963; 1965): Astron. J. **68,** 435; **70,** 212
Baglin, A. et al. (1980): Proc. fifth Europ. Reg. Meet., Liège, p. B. 3, 1
Bailey, J. (1979): Mon. Not. R. Astron. Soc. **189,** 41 P
Balazs, B. (1980): private communication
Balog, N. I., Goncharskij, A. V., Cherepashchuk, A. M. (1981): Astron. Zh. Akad. Nauk SSSR **58,** 67
Balona, L. A. (1977): Mem. R. Astron. Soc. **84,** 101
Balona, L. A. (1983): Observatory **103,** 163
Barbaro, G., et al. (1969): Mitt. Sternw. Ungar. Akad. Wiss. **6,** 41
Barkhatova, K. A. et al. (1973): Astron. Tsirk. No. 743, 4
Barlow, M. J., et al. (1981): Mon. Not. R. Astron. Soc. **195,** 61
Barnes, T. G., Du Puy, D. L. (1975): Astrophys. J. **200,** 364
Barrell, S. L. (1982): Mon. Not. R. Astron. Soc. **200,** 139
Bath, G. T. (1972): Astrophys. J. **173,** 121
Bath, G. T. (1976): IAU Symp. **73,** 173; Publ. Univ. Obs. Oxford No. 163
Bath, G. T. et al. (1974): Mon. Not. R. Astron. Soc. **169,** 447
Bath, G. T., Shaviv, G. (1978): Mon. Not. R. Astron. Soc. **183,** 515
Batten, A. H. (1973): *Binary and Multiple Systems of Stars* (Pergamon Press, Oxford)
Batten, A. H., Plavec, M. (1971): Sky Telesc. **42,** 213
Becker, W. (1964): Z. Astrophys. **58,** 202
Belserene, E. (1952): Astron. J. **57,** 237
Bertaud, Ch., Dumortier, B., Pollas, C. (1975): Inf. Bull. Variable Stars 970
Bertaud, Ch., Véron, M.-P., Pollas, C. (1972): Inf. Bull. Variable Stars 703
Bertout, C. (1980): Preprint fifth Europ. Reg. Meet., Liège
Bessell, M. S. (1969): Astrophys. J., Suppl. Ser. **18,** 195
Beyer, M. (1948): Erg. Astron. Nachr. **11,** Nr. 4

Beyer, M. (1965): Landolt-Börnstein Neue Serie Gruppe 6, **1**, 517 (Springer, Berlin, Heidelberg, New York)
Beyer, M. (1977): Veröff. Remeis-Sternw. Bamberg **12**, Nr. 123
Bianchini, A., Sabbadin, F., Hamzaoglu, E. (1982): Astron. Astrophys. **106**, 176
Bidelman, W. P. (1979): IAU Symp. **83**, 306
Biermann, P., Kippenhahn, R. (1971) Astron. Astrophys. **14**, 32
Binnendijk, L. (1960): *Properties of Double Stars* (Univ. of Pennsylvania Press, Philadelphia) Chap. VI
Blair, W. P. et al. (1981): Astron. Astrophys. **99**, 73
Böhm-Vitense, E. et al. (1974): Astrophys. J. **194**, 125
Bohusz, E., Udalski, A. (1980): Acta Astron. **30**, 359
Bond, H. E. (1976): Publ. Astron. Soc. Pac. **88**, 192
Bond, H. E. (1978): Sky Telesc. **56**, 12
Bond, H. E. (1980): Sky Telesc. **60**, 106
Bond, H. E., Kron, R. G., Spinrad, H. (1977): Astrophys. J. **213**, 1
Borgman, J. (1956): Publ. Groningen No. 58
Borkowski, K. J. (1980): Acta Astron. **30**, 393
Bowers, P. F., Cornett, R. H. (1974): Astrophys. Lett. **15**, 181
Boyarchuk, A. A. (1969): Mitt. Sternw. Ungar. Akad. Wiss. **6**, 395
Boyarchuk, A. A. (1975): IAU Symp. **67**, 377
Brandt, R. (1967): Sterne **43**, 4
Brecher, K., Morrison, P., Sadun, A. (1977): Astrophys. J. **217**, L139; see also Sky Telesc. **54**, 364
Breger, M. (1979): Publ. Astron. Soc. Pac. **91**, 5
Breger, M. (1980): *The Nature of Dwarf Cepheids V* (Preprint)
Breger, M. (1981): Preprint
Brown, D. A., Huang, S.-S. (1977): Astrophys. J. **218**, 461
Bruch, A., Duerbeck, H. W., Seitter, W. C. (1981): Mitt. Astron. Ges. **52**, 34
Brun, A., Vehrenberg, H. (1965): *Atlas der Kapteynschen Eichfelder* (Atlas of Kapteyn Selected Areas) (Treugesell-Verlag, Düsseldorf)
Burbidge, G. R., Crowne, A. H., Smith, H. E. (1977): Astrophys. J., Suppl. Ser. **33**, 113

Cahn, J. H., Wyatt, S. P. (1978): Astrophys. J. **221**, 163
Cannizzo, J. K., Ghosh, P., Wheeler, J. C. (1982): Astrophys. J. **260**, L83
Cannon, A. J. (1912): Popular Astronomy **20**, Nos. 2, 3 and 4
Cannon, A. J. (1920): Ann. Harvard Obs. **81**, 179
Carson, R., Stothers, R. (1982): Astrophys. J. **259**, 740
Catalano, S., Rodonò, M. (1967): Mem. Soc. Astron. Ital. **38**, 395
Catchpole, R. M. et al. (1979): South African Astron. Obs. Circ. **1**, 61
Celis, S. L. (1981): Astron. Astrophys. **99**, 58
Chanmugam, G., Dulk, G. A. (1981): Astrophys. J. **244**, 569
Chapman, R. D. (1981): Astrophys. J. **248**, 1043
Charles, P. (1980): Sky Telesc. **59**, 188
Charles, P. A. (1982): Observatory **102**, 168
Chentsov, E. L. (1980): Pis'ma v Astron. Zhurn. **6**, 360
Chentsov, E. L. (1981): lecture given at Sonneberg
Chernykh, N. S. (1981): private communication
Chester, T. J. (1979): Astrophys. J. **230**, 167
Chevalier, C. et al. (1980): Astron. Astrophys. **81**, 368
Chevalier, R. A. (1977): Astrophys. Space Sci. **66**, 53
Chevalier, R. A. (1981): Astrophys. J. **246**, 267
Chugainov, P. F. (1966): Inf. Bull. Variable Stars 122
Chugainov, P. F. (1973): Izv. Krymskoj Astrofiz. Obs. **48**, 3
Chugainov, P. F. (1976): Izv. Krymskoj Astrofiz. Obs. **54**, 85
Ciatti, F., Mammano, A., Vittone, A. (1978): Astron. Astrophys. **68**, 251
Clark, D. H., Stephenson, F. R. (1977): *The Historical Supernovae* (Pergamon Press, Oxford)
Clark, F. O. et al. (1981): Astrophys. J. **244**, L99

Clayton, M. L., Feast, M. W. (1969): Mon. Not. R. Astron. Soc. **146**, 411
Cline, T. L. et al. (1982): Astrophys. J. **255**, L 45
Cocke, W. J., Disney, M. J., Taylor, D. J. (1969): IAU Circ. 2128
Cohen, M. (1981): Sky Telesc. **62**, 300
Cohen, M. (1982): Publ. Astron. Soc. Pac. **94**, 266
Cordova, F. A., Jensen, K. A., Nugent, J. J. (1981a): Mon. Not. R. Astron. Soc. **196**, 1
Cordova, F. A., Mason, K. O., Nelson, J. E. (1981b): Astrophys. J. **245**, 609
Cosmovici, C. B. (1974): Astrophys. Space Sci. Libr. **45**
Cousens, A. (1983): Mon. Not. R. Astron. Soc. **203**, 1171
Coutts, C. M., Sawyer-Hogg, H. B. (1969): Publ. David Dunlap Obs. **3**, No. 1
Cowley, A. P., Crampton, D., Hesser, J. E. (1977): Astrophys. Space Sci. Libr. **65**, 54
Cowley, A., Stencel, R. (1973): Astrophys. J. **184**, 687
Cox, A. N., King, D. S., Hodson, W. (1979): Astrophys. J. **228**, 870
Cox, A. N., King, D. S., Tabor, J. E. (1973): Astrophys. J. **184**, 201
Craine, E. R. (1977): *A Handbook of Quasistellar and BL Lacertae Objects* (Parchert Publishing House, Tucson)
Culhane, J. L. (1977): Vistas Astron. **19**, 1

Dautcourt, G. (1976): *Was sind Pulsare?* (What are pulsars?) 2nd ed. (Teubner Verlagsgesellschaft, Leipzig)
Dawson, D. W. (1979): Astrophys. J., Suppl. Ser. **41**, 97
De Groot, M. (1978): Irish Astron. J. **13**, 267
De Loore, C. (1980): Proc. fifth Europ. Reg. Meet., Liège, p. D. 1, 1
Delpino, F. (1981): Coelum **49**, 65
Detre, L. (1969): Mitt. Sternw. Ungar. Akad. Wiss. **6**, 3
Deupree, R. G., Hodson, S. W. (1976): Astrophys. J. **208**, 426
De Vaucouleurs, G. (1978): Astrophys. J. **223**, 730
Dickens, R. J., Carey, J. V. (1967): R. Obs. Bull. Greenwich No. 129, E 340
Diethelm, R. (1981): ESO Messenger **25**, 29
Diethelm, R. (1983): Astron. Astrophys. **124**, 108
Dickinson, D. F. et al. (1978): Astrophys. J. **220**, L 113
Dixon, R. S. (1970): Astrophys. J., Suppl. Ser. **20**, 1
Dokuchaeva, O. D. (1976): Inf. Bull. Variable Stars 1189
Dorschner, J., Gürtler, J. (1980): Sterne **56**, 117
Dorschner, J., Gürtler, J., Fröhlich, H.-E. (1980) Sterne **56**, 245
Dower, R. G., Bradt, H. V., Morgan, E. H. (1982): Astrophys. J. **261**, 228
Duerbeck, H. W. (1977): Astrophys. Space Sci. Libr. **65**, 150
Duerbeck, H. W. (1981): Publ. Astron. Soc. Pac. **93**, 165
Duerbeck, H. W. (1983): IAU Colloq. **80**, Preprint
Dufay, J., Bloch, M., Chalonge, D. (1965): *Novae, Novoides et Supernovae* (Centre National de la Recherche Scientifique, Paris) p. 63
Duldig, M. L., Thomas, R. M., Haynes, R. F. (1980) Proc. Astron. Soc. Aust. **4**, 108
Dupree, A. K. (1981): Astrophys. Space Sci. Libr. **89**, 87
Durison, R. H., Burns, J. O. (1981): Mon. Not. R. Astron. Soc. **195**, 535
Dziembowski, W. (1974): Commun. 20. Colloq. Int. Astrophys. Liège; Mém. Soc. R. Sci. Liège, Sér. 6, **8**, p. 287

Eachus, L. J., Liller, W. (1975): Astrophys. J. **200**, L 61
Eaton, J. A., Hall, D. S. (1979): Astrophys. J. **227**, 907
Eddington, A. S. (1918): Mon. Not. R. Astron. Soc. **79**, 177
Efremov, Yu. M., Kopylov, I. M. (1967): Izv. Krymskoj Astrofiz. Obs. **36**, 240
Eggleton, P. P. (1976): IAU Symp. **73**, 209
Eggleton, P. P. (1983): Astrophys. Space Sci. Libr. **101**, 239
Evans, D. A. (1975): IAU Symp. **67**, 93
Evans, T. L. (1976): Mon. Not. R. Astron. Soc. **174**, 169

Faulkner, D. J. (1977): Proc. Astron. Soc. Aust. **3**, 124
Faulkner, J. (1971): Astrophys. J. **170**, L 99

Feast, M. W. (1974): Proc. ESO Conference on New Large Telescopes Geneva, p. 169
Feast, M. W. (1975): IAU Symp. **67**, 129
Feast, M. W. (1980): Proc. fifth Europ. Reg. Meet., Liège, p. B. 1, 1
Feast, M. W. (1983): Mon. Notes Astron. Soc. S. Afr. **41**, 72
Feast, M. W. et al. (1983): Mon. Not. R. Astron. Soc. **202**, 951
Fernie, J. D., Demers, S. (1966): Astrophys. J. **144**, 440
Ferrari, K. (1950): Mitt. Sternw. Wien **4**, 207
Fetlaar, J. (1923): Rech. astr. Obs. Utrecht **9**, 1
Feuchter, C. A. (1967): Astron. J. **72**, 702
Finlay, E. A., Jones, B. B. (1977): Aust. J. Phys. **26**, 389
Fitch, W. S. (1970): Astrophys. J. **161**, 669
Fitch, W. S. (1976): IAU Colloq. **29**, 185
Fitch, W. S., Szeidl, B. (1976): Astrophys. J. **203**, 616
Flannery, B. P., Ulrich, R. K. (1977): Astrophys. J. **212**, 533
Foy, R., Heck, A., Menessier, M.-O. (1975): Astron. Astrophys. **43**, 175
Frank, J, King, A. R. (1981): Mon. Not. R. Astron. Soc. **195**, 227
Frenk, C. S., White, S. D. M. (1982): Mon. Not. R. Astron. Soc. **198**, 173
Fried, J. W. (1980): Astron. Astrophys. **81**, 182
Friedemann, C., Gürtler, J. (1975): Astron. Nachr. **296**, 125
Friedemann, C., Schmidt, K.-H. (1967): Astron. Nachr. **289**, 223
Fuhrmann, B. (1982): Mitt. Veränderl. Sterne **9**, Heft 4
Fürtig, W. (1975): Inf. Bull. Variable Stars 1071

Gahm, G. F. (1979): Trans. IAU **27 A**, part 2, 121
Gahm, G. F. (1980a): *The Universe in UV Wavelengths: The First Two Years of IUE* (NASA
    Publication)
Gahm, G. F. (1980b): Astrophys. J. **242**, L 163
Gahm, G. F. et al. (1974): Astron. Astrophys. **33**, 399
Gaposchkin, S. (1946): Bull. Harvard Obs. No. 918
Garcia, M.et al. (1980): Astrophys. J. **240**, L 107
Gershberg, R. E. (1970): *Flares of Red Dwarf Stars* (Armagh Obs.) p. 111
Gershberg, R. E., Shakhovskaya, N. I. (1974): Izv. Krymskoj Astrofiz. Obs. **49**, 73
Geßner, H. (1981): Inf. Bull. Variable Stars 1789
Geßner, H. (1981a): Mitt. Veränderliche Sterne **9**, 55
Geßner, H. (1981b): Veröff. Sternw. Sonneberg **9**, Heft 5
Geßner, H. (1982a): Mitt. Veränderliche Sterne **9**, Heft 4
Ghigo, F. D., Cohen, M. L. (1981): Astrophys. J. **245**, 988
Gieseking, F. (1973): Veröff. Astron. Inst. Bonn Nr. 87
Gilmozzi, R., Messi, R., Natali, G. (1981): Astrophys. J. **245**, L 119
Gingold, R. A., Monaghan, J. J. (1979): Proc. Astron. Soc. Aust. **3**, 364
Ginzburg, V. L., Zheleznyakov, V. V. (1975): Annu. Rev. Astron. Astrophys. **13**, 511
Glass, I. S., Feast, M. W. (1982): Mon. Not. R. Astron. Soc. **198**, 199
Gorbatskij, V. G. (1949): Astron. Zh. **26**, 307
Gorbatskij, V. G. (1975): IAU Symp. **67**, 357
Götz, W. (1961): Veröff. Sternw. Sonneberg **5**, Heft 2
Götz, W. (1965): Sterne **41**, 150
Götz, W. (1968): Mitt. Veränderliche Sterne **5**, 1
Götz, W. (1973): Veröff. Sternw. Sonneberg **8**, Heft 3
Götz, W. (1980a): Veröff. Sternw. Sonneberg **9**, Heft 3
Götz, W. (1980b): Veröff. Sternw. Sonneberg **9**, Heft 4
Götz, W. (1981): Veröff. Sternw. Sonneberg **9**, Heft 5
Götz, W., Wenzel, W. (1967): Mitt. Veränderliche Sterne **4**, 71
Graham, J. A. (1972): IAU Colloq. **21**, 120
Graham, J. A. (1974): IAU Symp. **59**, 107
Graham, J. A. (1975): Publ. Astron. Soc. Pac. **87**, 641
Graham, J. A. (1979): IAU Symp. **84**, 195
Graham, J. A., Araya, G. (1971): Astron. J. **76**, 768

Gunther, J., Schweitzer, E. (1982): Bull. AFOEV No. 19, 8
Gurzadyan, G. A. (1980): *Flare Stars* (Pergamon Press, Oxford)
Guthnick, P. (1902): Nova Acta Leopoldina 79; Astron. Nachr. **157**, 1
Guthnick, P., Prager, R. (1915): Astron. Nachr. **201**, 443
Guthnick, P., Prager, R. (1917): Sitzungsber. Preuß. Akad. Wiss., math.-naturwiss. Klasse, 1917, p. 222
Gyldenkerne, K. (1970): Vistas Astron. **12**, 199
Gyldenkerne, K., West, R. M. (1970): IAU Colloq. **6**

Hall, D. S. (1972): Publ. Astron. Soc. Pac. **84**, 323
Hall, D. S. (1976): IAU Colloq. **29**, 287
Hall, D. S. et al. (1979): Sky Telesc. **57**, 132
Hamilton, D., Keel, W., Nixon, J. F. (1978): Sky Telesc. **55**, 372
Hamzaoğlu, E. (1981): Astron. Astrophys. **104**, 65
Hamzaoğlu, E., Keskin, V., Eker, T. (1982): Inf. Bull. Variable Stars 2102
Handbury, M. J., Williams, I. P. (1976): Astrophys. Space Sci. **45**, 439
Hanley, C. M., Shapley, H. (1940): Harvard Obs. Bull. No. 913
Hansen, C. J. (1980): Proc. Workshop on Nonradial and Nonlinear Stellar Pulsations (Springer, Berlin, Heidelberg, New York) p. 445
Harmanec, P. (1982): IAU Symp. **98**, 279
Harmanec, P., Křiž, S. (1975): IAU Symp. **70**, 386
Haro, G. (1968): *Nebulae and Interstellar Matter* (University of Chicago Press, Chicago) p. 157
Haro, G., Morgan, W. W. (1953): Astrophys. J. **118**, 16
Harris, W. E. (1976): Astron. J. **81**, 1095
Hartmann, L., Rosner, R. (1979): Astrophys. J. **230**, 802
Haynes, R. F., Lerche, I., Murdin, P. (1980): Astron. Astrophys. **87**, 299
Hazard, C., Mitton, S. (1979): *Active Galactic Nuclei* (Cambridge University Press, Cambridge and New York)
Hazlehurst, J. (1976): IAU Symp. **73**, 323
Henize, K. G. (1961): Publ. Astron. Soc. Pac. **73**, 159
Herbig, G. H. (1958a): Astrophys. J. **127**, 312
Herbig, G. H. (1958b): Astrophys. J. **128**, 259
Herbig, G. H. (1960): Astrophys. J., Suppl. Ser. **4**, 337
Herbig, G. H. (1962): Adv. Astron. Astrophys. **1**, 47
Herbig, G. H. (1969): Contrib. Lick Obs. No. 282
Herbig, G. H. (1973): Astrophys. J. **182**, 129
Herbig, G. H. (1977): Astrophys. J. **217**, 693
Herbig, G. H. (1981): *Recent Advances in Observational Astronomy*, eds. H. L. Johnson, C. Allen (Univerisdad Nac. Autón. México, México) p. 19
Herbig. G. H., Rao, N. K. (1972): Astrophys. J. **174**, 401
Hertzsprung, E. (1926): Bull. Astron. Inst. Netherlands **3**, 115
Hewitt, A., Burbidge, G. (1980): Astrophys. J., Suppl. Ser **43**, 57
Hill, P. W. et al. (1981): Mon. Not. R. Astron. Soc. **197**, 81
Hillebrand, W. (1982): Sterne Weltraum **21**, 406
Hoffmeister, C. (1929): Astron. Nachr. **236**, 233
Hoffmeister, C. (1933): Astron. Nachr. **250**, 397
Hoffmeister, C. (1934): Astron. Nachr. **253**, 91
Hoffmeister, C. (1944): Astron. Nachr. **274**, 232
Hoffmeister, C. (1949): Astron. Nachr. **278**, 24
Hoffmeister, C. (1955): Astron. Nachr. **282**, 257
Hoffmeister, C. (1962a): Kleine Veröff. Remeis-Sternw. Bamberg **3**, 105
Hoffmeister, C. (1962b): Z. Astrophys. **55**, 46; Astron. Nachr. **287**, 55
Hoffmeister, C. (1963): Astron. Nachr. **287**, 169
Hoffmeister, C. (1964): Astron. Nachr. **289**, 49 and Inf. Bull. Variable Stars 67
Hoffmeister, C. (1965): Veröff. Sternw. Sonneberg **6**, Heft 3
Hoffmeister, C. (1970): Veränderliche Sterne, 1st ed. (Verlag J. A. Barth, Leipzig)
Hofmeister, E. (1965): *Delta-Cephei-Sterne vom Standpunkt der Sternentwicklung*, München

Hofmeister, E., Kippenhahn, R., Weigert, A. (1964) Z. Astrophys. **59**, 215 and 242
Hofmeister, E., Kippenhahn, R., Weigert, A. (1965): Z. Astrophys. **60**, 57
Hopp, U., Witzigmann, S., Geyer, E. H. (1982): Inf. Bull. Variable Stars 2148
Howarth, I. D., Wilson, R. (1981): Astrophys. Space Sci. Libr. **89**, 481
Howarth, I. D., Wilson, R. (1983): Mon. Not. R. Astron. Soc. **202**, 347
Hoyle, F., Wickramasinghe, N. C. (1962): Mon. Not. R. Astron. Soc. **124**, 117
Hudec, R. (1981a): Bull. Astron. Inst. Czech. **32**, 93
Hudec, R. (1981b): Bull. Astron. Inst. Czech. **32**, 108
Hudec, R., Meinunger, L. (1977): Mitt. Veränderliche Sterne **7**, 194
Hudec, R., Wenzel, W. (1976): Bull. Astron. Inst. Czech. **27**, 325
Humphreys, R. M. (1978): Astrophys. J. **219**, 445 and IAU Symp. **84**, 93
Hutchings, J. B. (1977): Highlights of Astron. **4**, I, 129
Huth, H. (1966): Sterne **42**, 129
Huth, H., Wenzel, W. (1981): *Bibliographic Catalogue of Variable Stars* (Centre de Données
    Stellaires, Strasbourg)

Iben, I. (1974) IAU Symp. **59**, 3
Ikaunieks, J. (1963): Trans. Astrophys. Lab. Riga **9**, 33
Ikaunieks, J. (1971): Trans. Riga Obs. **12**
Ilovaisky, S. A., Chevalier, C. (1977): Astrophys. Space Sci. Libr. **65**, 149

Jakate, S. M. (1979): Astron. J. **84**, 1042
Jarrett, A. H., Gibson, J. B. (1975): Inf. Bull. Variable Stars 979
Jensch, A. (1934): Astron. Nachr. **253**, 91
Jensch, A. (1936): Unterrichtsblätter für Mathematik und Naturwissenschaften, p. 253
Jerzykiewicz, M. (1978): Acta Astron. **28**, 465
Jerzykiewicz, M., Sterken, C. (1979): IAU Colloq. **46**, 474
Jerzykiewicz, M., Wenzel, W. (1977): Acta Astron. **27**, 35
Joy, A. H. (1942): Astrophys. J. **96**, 344
Joy, A. H. (1945): Astrophys. J. **102**, 168
Joy, A. H. (1952): Astrophys. J. **115**, 24
Jurcsik, J., Szabados, L. (1979): Inf. Bull. Variable Stars 1722

Kafatos, M., Michalitsanos, A. G., Vardya, M. S. (1977): Astrophys. J. **216**, 526
Kahn, S. M. et al. (1981): Astrophys. J. **250**, 733
Kaler, J. B. (1981): Astrophys. J. **245**, 568
Kamper, K., van den Bergh, S. (1976): Sky Telesc. **51**, 236
Katz, J. I. (1975): Astrophys. J. **200**, 298
Keenan, P. C. (1966): Astrophys. J., Suppl. Ser. **13**, 333
Kellermann, K. I. (1980): Ann. New York Akad. Sci. **336**, 1; Green Bank Repr. B No. 509; see
    also Green Bank Repr. B No. 478, and *Galactic and Extra-Galactic Radio Astronomy* (Sprin-
    ger, Berlin, Heidelberg, New York 1974) Chap. 12
Kellermann, K. I., Pauliny-Toth, I. I. K. (1981): Annu. Rev. Astron. Astrophys. **19**, 373; Green
    Bank Repr. B No. 524
Kholopov, P. N. (1951): Perem. Zvezdy, Byull. **8**, 83
Kholopov, P. N. (1956): Perem. Zvezdy, Byull. **11**, 325
Kholopov, P. N. (1981): Perem. Zvezdy, Byull. **21**, 465
Kholopov, P. N. (1982): *New Catalogue of Suspected Variable Stars* (Nauka, Moskva 1982)
    (NSV)
Kholopov, P. N. et al. (1981): 66th name list of variable stars, Inf. Bull. Variable Stars 2042
Kiang, T. (1962): Observatory **82**, 57
Kilkenny, D. (1982): Mon. Not. R. Astron. Soc. **200**, 1019
Kilkenny, D., Flanagan, C. (1983): Mon. Not. R. Astron. Soc. **203**, 19
King, I. R. (1977): Highlights of Astron. **4**, II, 41
Kinman, T. D. (1959): Mon. Not. R. Astron. Soc. **119**, 559
Kinman, T. D. (1964): Astrophys. J., Suppl. Ser. **11**, 999
Kinman, T. D. (1975): IAU Symp. **67**, 573

Kinman, T. D., Wirtanen, C. A., Janes, K. A. (1964): Astrophys. J., Suppl. Ser. **11**, 223
Kinman, T. D., Wirtanen, C. A., Janes, K. A. (1966): Astrophys. J., Suppl. Ser. **13**, 379
Kippenhahn, R. (1973): Sterne Weltraum **12**, 133
Kippenhahn, R., Kohl, K., Weigert, A. (1967): Z. Astrophys. **66**, 58
Kippenhahn, R., Thomas, H.-C. (1978): Astron. Astrophys. **63**, 265
Kippenhahn, R., Weigert, A. (1964): Sterne Weltraum **3**, 173
Kippenhahn, R., Weigert, A. (1965): Sterne Weltraum **4**, 148
Kippenhahn, R., Weigert, A. (1967): Z. Astrophys. **65**, 251; see also Sterne Weltraum **6**, 176
Kirshner, R. P. (1974): Highlights of Astron. **3**, 533
Klebesadel, R. et al. (1982): Astrophys. J. **259**, L 51
Kohoutek, L. (1982): Inf. Bull. Variable Stars 2113
Kopal, Z. (1965): Adv. Astron. Astrophys. **3**, 89
Kopal, Z. (1978): Astrophys. Space Sci. Libr. **68**
Kopal, Z. (1979): Astrophys. Space Sci. Libr. **77**
Kopylov, I. M. (1957): IAU Symp. **3**, 71
Kraft, R. P. (1958): Astrophys. J. **127**, 625
Kraft, R. P. (1959): Astrophys. J. **130**, 110
Kraft, R. P. (1974): Sky Telesc. **48**, 18
Kraft, R. P., Luyten, W. J. (1965): Astrophys. J. **142**, 1041
Kraft, R. P., Schmidt, M. (1963): Astrophys. J. **137**, 249
Krautter, J. (1978): Sterne Weltraum **17**, 56
Kreiner, J. M., Tremko, J. (1978): Inf. Bull. Variable Stars 1446
Krelowski, J. (1975): IAU Symp. **67**, 149
Kron, G. E. (1952): Astrophys. J. **115**, 301
Krzemiński, W., Serkowski, K. (1977): Astrophys. J. **216**, L 45
Kuhi, L. (1964): Astrophys. J. **140**, 1409
Kukarkin, B. V. (1949): *Issledovanie stroeniya i rasvitiya zvezdnykh sistem na osnove izlucheniya peremennykh zvezd* (Gos. Izd. Tekhn.-Teor. Lit., Moskva-Leningrad); German edition: *Erforschung der Struktur und Entwicklung der Sternsysteme auf der Grundlage des Studiums der Veränderlichen Sterne* (Akademie-Verlag, Berlin 1954)
Kukarkin, B. V. (1972): IAU Colloq. **21**, 9
Kukarkin, B. V. (1975): IAU Symp. **67**, 511
Kukarkin, B. V., Parenago, P. P. (1934): Perem. Zvezdy, Byull. **4**, 251
Kukarkin, B. V. et al. (1951): *Katalog zvezd, zapodozrennykh v peremennosti* (Izd. Akad. Nauk SSSR, Moskva) (CSVS 1951)
Kukarkin, B. V. et al. (1965): *The Second Catalogue of Suspected Variable Stars* (Astronomical Council and Sternberg Astron. Institute, Moskva (CSVS 1965)
Kukarkin, B. V. et al. (1969): *General Catalogue of Variable Stars*, 3ed. (Astronomical Council and Sternberg Astron. Institute, Moskva) (GCVS)
Kukarkin, B. V. et al. (1971): GCVS, Suppl. 1
Kukarkin, B. V. et al. (1974): GCVS, Suppl. 2
Kukarkin, B. V. et al. (1976): GCVS, Suppl. 3
Kundt, W. (1981): Naturwissenschaften **68**, 63
Kundt, W. (1982): Sterne Weltraum **21**, 66
Kunkel, W. E. (1975): IAU Symp. **67**, 42
Kurochkin, N. E. (1960): Astron. Tsirk. Nos. 210 and 212
Kurtz, D. W. (1979): Mon. Notes Astron. Soc. South. Afr. **38**, 36
Kvíz, Z. (1956a): Bull. Astron. Inst. Czech. **9**, 70
Kvíz, Z. (1956b): Contr. Astron. Inst. Brno **1**, No. 14
Kvíz, Z. (1959): Bull. Astron. Inst. Czech. **11**, 71
Kwee, K. K. (1968) Bull. Astron. Inst. Netherlands **19**, 260
Kwok, S., Purton, C. R. (1979): Astrophys. J. **229**, 187

Lafler, J., Kinman, T. D. (1964): Astrophys. J., Suppl. Ser. **11**, 216
Lamb, D. Q., Van Horn, H. M. (1975): Astrophys. J. **200**, 306
Lamla, E. (1965): Landolt-Börnstein, Zahlenwerte und Funktionen NS, Gr. VI, Bd. **4**, 322 (Springer, Berlin, Heidelberg, New York)

Landolt, A. U. (1968): Astrophys. J. **153**, 151
Laustsen, S. (1980): Rep. first ESO/ESA Workshop, Geneva, p. 39
Lawrence, A. et al. (1983): Astrophys. J. **271**, 793
Leavitt, H. A. (1908): Ann. Harvard Obs. **60**, 87
Leavitt, H. A. (1912): Circ. Harvard Obs. No. 173
Le Contel, J.-M. (1981): Proc. Workshop Puls. B Stars, Nice Obs., p. 45
Ledoux, P. (1951): Astrophys. J. **114**, 373
Ledoux, P., Walraven, Th. (1958): Handbuch der Physik **51**, 384 (Springer, Berlin, Göttingen, Heidelberg)
Lewin, W. H. G., van Paradijs, J. (1979): Sky Telesc. **57**, 446
Liebert, J., Stockman, H. S. (1983): Prepr. Steward Obs. No. 441
Lightman, A. P. (1976): Sky Telesc. **52**, 243
Liller, W. (1977): Sky Telesc. **53**, 351
Livio, M., Bath, G. T. (1982): Astron. Astrophys. **116**, 286
Livio, M., Shaviv, G. (1983): IAU Colloq. **72**
Loreta, E. (1934): Astron. Nachr. **254**, 151
Lü, P. K. (1977): Astron. J. **82**, 773
Lub, J. (1977): Astron. Astrophys., Suppl. Ser. **29**, 345
Ludendorff, H. (1928): Handbuch der Astrophysik **6**, 99 (Springer, Berlin)
Luthardt, R. (1983): Inf. Bull. Variable Stars 2360
Lynas-Gray, A. E. (1981): Irish Astron. J. **15**, 42
Lynds, R. et al. (1969): IAU Circ. 2129
Lyuty, V. M., Pronik, V. I. (1975): IAU Symp. **67**, 591

Maeder, A. (1980): Astron. Astrophys. **90**, 311
Maeder, A. (1981): Astron. Astrophys. **99**, 97
Maffei, P. (1967): Astrophys. J. **147**, 802
Mallama, A. D., Trimble, V. L. (1978): Q. J. R. Astron. Soc. **19**, 430
Mammano, A., Ciatti, F. (1975): Astron. Astrophys. **39**, 405
Manchester, R. N., Taylor, J. H. (1977): *Pulsars* (Freeman, San Francisco)
Manchester, R. N., Taylor, J. H. (1981): Astron. J. **86**, 1953
Margon, B., Grandi, S. A., Downes, R. A. (1980): Astrophys. J. **241**, 306
Marino, B. F. (1980): J. R. Astron. Soc. New Zealand **28**, 158
Marino, B. F., Williams, H. O. (1983): Inf. Bull. Variable Stars 2266
Markova, L. T., Fomin, S. K. (1975): Astron. Tsirk. No. 856
Martynov, D. Ya. (1971): in Tsesevich, *Zatmennye Peremennye Zvezdy* (Izdatel'stvo Nauka, Moskva)
Matese, J. J., Whitmire, D. P. (1983): Astron. Astrophys. **117**, L 7
Mauder, H. (1981): ESO Messenger 24, 13
Mayall, M. W. (1949): Astron. J. **54**, 191
Mayall, M. W. (1960): J. R. Astron. Soc. Can. **54**, 194
Mayor, M., Acker, A. (1980): Astron. Astrophys. **92**, 1
McGraw, J. T. (1978): cited without reference in Pettersen, B. R. (1980)
McGraw, J. T., Starrfield, S. G., Angel, J. R. P. (1979): Smithson. Astrophys. Obs. Spec. Rep. 385, 125
McLaughlin, D. B. (1945): Publ. Astron. Soc. Paci. **57**, 69
McLaughlin, D. B. (1960): Stars and Stellar Syst. **6**, 585
McLaughlin, D. B. (1965): *Novae, Novoides et Supernovae* (Centre National de la Recherche Scientifique, Paris) p. 1
Meinunger, I. (1976): Astron. Nachr. **297**, 23
Meinunger, I. (1977): Astron. Nachr. **298**, 171
Meinunger, L. (1971): Mitt. Veränderliche Sterne **5**, 177
Meinunger, L. (1979): Mitt. Veränderliche Sterne **8**, 105
Meinunger, L. (1981): Mitt. Veränderliche Sterne **9**, 67
Meinunger, L. (1982): personal communication
Meinunger, L., Wenzel, W. (1971): Mitt. Veränderliche Sterne **5**, 170
Mendez, R., Gathier, R., Niemala, V. (1982): Astron. Astrophys. **116**, L 1

Mennesier, M. O. (1981): Astron. Astrophys. **93**, 325
Merrill, P. W. (1952): Astrophys. J. **115**, 145
Merrill, P. W. (1959): Sky Telesc. **18**, 490
Meyer, F., Meyer-Hofmeister, E. (1979): Astron. Astrophys. **78**, 167
Meyer, F., Meyer-Hofmeister, E. (1982a): Astron. Astrophys. **104**, L 10
Meyer, F., Meyer-Hofmeister, E. (1982b): Astron. Astrophys. **106**, 34
Meyer, F., Meyer-Hofmeister, E. (1983): Astron. Astrophys. **121**, 29
Mikolajewska, J., Mikolajewski, M. (1980): Inf. Bull. Variable Stars 1846
Millis, R. L. (1973): Publ. Astron. Soc. Pac. **85**, 410
Mitchell, S. A. (1935): Publ. Leander McCormick Obs. **6**, 201
Mitrofanov, I. G. (1978): Pis'ma Astron. Zh. **4**, 219
Miyaji, S. (1983): Astrophys. Space Sci. Libr. **101**, 263
Moffett, Th. J. (1974): Astrophys. J., Suppl. Ser. **29**, 1
Möllenhoff, C., Schaifers, K. (1978): Astron. Astrophys. **64**, 253; see also Sterne Weltraum **17**, 336
Molteni, D. et al. (1980): Astron. Astrophys. **87**, 88
Mumford, G. S. (1962): Sky Telesc. **23**, 135
Mumford, G. S. (1963): Sky Telesc. **26**, 190
Mustel, E. R. (1974): Highlights of Astron. **3**, 545

Nather, R. E. (1973): Vistas Astron. **15**, 91
Nather, R. E., Robinson, E. L. (1974): Astrophys. J. **190**, 637
Nather, R. E. et al. (1977): Astrophys. J. **211**, L 125
Nather, R. E, Robinson, E. L., Stother, R. J. (1981): Astrophys. J. **244**, 269
Nather, Warner, MacFarlane (1969): IAU Circ. 2129
Nelson, B., Young, A. (1970): Publ. Astron. Soc. Pac. **82**, 699
Nelson, B., Young, A. (1976): IAU Symp. **73**, 141
North, P., Rufener, F. (1981): Inf. Bull. Variable Stars 2036
Nugis, T., Kolka, I., Luud, L. (1978): IAU Symp. **83**, 39

O'Keefe, J. A. (1939): Astrophys. J. **90**, 294
Oort, J. H. (1965): Sterne **41**, 178
Oort, J. H., Plaut, L. (1975): Astron. Astrophys. **41**, 71
Oosterhoff, P. Th. (1941): Ann. Sternw. Leiden **17**, 4
Oosterhoff, P. Th. (1957): Bull. Astron. Inst. Netherlands **13**, 317
Oskanyan, V. (1964): *The UV Ceti Variable Stars* (Publ. Obs. Astron. Beograd)
Oskanyan, V. S. et al. (1977): Astrophys. J. **214**, 430
Osvalds, V., Risley, A. M. (1961): Publ. Leander McCormick Obs. **11**, part XXI
Osváth, I. (1957): Mitt. Sternw. Ungar. Akad. Wiss. Budapest Nr. 42
Ott, H.-A. (1979): Sterne Weltraum **18**, 206
Ourassine, L. A., Ourassina, I. A. (1975): Inf. Bull. Variable Stars 973
Overbye, D. (1979): Sky Telesc. **58**, 510
Ozernoy, L. M., Usov, V. V. (1977): Astron. Astrophys. **56**, 163

Pacini, F., Salvati, M. (1981): Astrophys. J. **245**, L 107
Paczynski, B. (1976): IAU Symp. **73**, 75
Paczynski, B. (1980): Acta Astron. **30**, 113
Paczynski, B. (1981): Acta Astron. **31**, 1
Paczynski, B., Rudak, B. (1980): Astron. Astrophys. **82**, 349
Parenago, P. P. (1953): Tr. vtorogo soveshchaniya po voprosam kosmogonii (Akademiya Nauk SSSR, Moskva) p. 334
Parenago, P. P. (1954) Tr. Gos. Astron. Inst. Shternberga **25**, 225
Parenago, P. P. (1957): Mitt. Sternw. Ungar. Akad. Wiss. Budapest Nr. 42, 53
Patkós, L. (1981): Astrophys. Lett. **22**, 1
Patterson, J. (1979): Astrophys. J. **233**, L 13
Patterson, J. (1981): Astrophys. J., Suppl. Ser. **45**, 517
Pavel, F. (1949): Astron. Nachr. **278**, 57

Pavlovskaya, E. D. (1957): Astron. Zh. **34**, 956
Payne-Gaposchkin, C. (1954): *Variable Stars and Galactic Structure* (The Athlone Press, London)
Payne-Gaposchkin, C. (1957): *The Galactic Novae* (North-Holland Publishing Company, Amsterdam) p. 98
Payne-Gaposchkin, C. (1958): Handbuch der Physik **51**, 753 (Springer, Berlin, Göttingen, Heidelberg)
Payne-Gaposchkin, C. (1963): Astrophys. J. **138**, 320
Payne-Gaposchkin, C. (1971): Smithson. Contr. Astrophys. No. 13
Payne-Gaposchkin, C. (1977a): Astron. J. **82**, 665
Payne-Gaposchkin, C. (1977b): Astrophys. Space Sci. Libr. **65**, 3
Pedersen, H. (1979): ESO Messenger No. 18, 34
Pedersen, H. et al. (1983): ESO Sci. Prepr. No. 247
Pedersen, H. et al. (1983a): Astrophys. J. **263**, 325
Pedersen, H. et al. (1983b): Astrophys. J. **263**, 340
Pel, J. W. (1976): Astron. Astrophys., Suppl. Ser. **24**, 413
Pel, J. W., Lub, J. (1978): IAU Symp. **80**, 229
Percy, J. R. (1981): Proc. Workshop Puls. B Stars, Nice Obs., p. 277
Perek, L. (1951): Contrib. Astron. Inst. Brno **1**, No. 8
Perek, L., Kohoutek, L. (1967): *Catalogue of Galactic Planetary Nebulae* (Academic Publishing House, Prague)
Persi, P., Ferrari Toniolo, M. (1980): Mem. Soc. Astron. Ital. **51**, 695
Petersen, J. O. (1973): Astron. Astrophys. **27**, 89
Petersen, J. O. (1976): IAU Colloq. **29**, 195
Petit, M. (1960): Ann. Astrophys. **23**, 713
Pettersen, B. R. (1980): Astron. Tidsskr. **13**, 173
Plaut, L. (1963): Bull. Astron. Inst. Netherlands **17**, 81
Plaut, L. (1965a): Stars and Stellar Syst. **5**, Chap. 13
Plaut, L. (1965b): Stars and Stellar Syst. **5**, Chap. 14
Plaut, L. (1966): Bull. Astron. Inst. Netherlands, Suppl. Ser. **1**, 105
Plaut, L. (1968a): Bull. Astron. Inst. Netherlands, Suppl. Ser. **2**, 293
Plaut, L. (1968b): Bull. Astron. Inst. Netherlands, Suppl. Ser. **3**, 1
Plaut, L. (1970): Astron. Astrophys. **8**, 341
Plaut, L., Borgman, J. (1954): Observatory **74**, 181
Pollock, J. T., Hall, D. L. (1974): Astron. Astrophys. **30**, 41
Popova, M. (1975): IAU Symp. **67**, 223
Poveda, A. (1964): Nature **202**, 1319
Prager, R. (1932): Kleinere Veröff. Sternw. Berlin-Babelsberg 12
Prager, R. (1940): Bull. Harvard Obs. No. 912
Preston, G. W. (1959): Astrophys. J. **130**, 507
Prialnik, D., Shara, M. M., Shaviv, G. (1978): Astron. Astrophys. **62**, 339
Proust, D., Ochsenbein, F., Pettersen, B. R. (1981): Astron. Astrophys., Suppl. Ser. **44**, 179
Pskowski, Ju. P. (1978): *Novae und Supernovae* (Teubner-Verlagsgesellschaft, Leipzig)
Pugach, A. F. (1977): Inf. Bull. Variable Stars 1277
Pustil'nik, S. A. (1976): Commun. Spec. Astrophys. Obs. USSR AS No. 18, 5

Racine, R. (1968): Astron. J. **73**, 588
Rappaport, S., Joss, P. C., Webbink, R. F. (1982): Astrophys. J. **254**, 616
Rappaport, S., Van den Heuvel, E. P. J. (1982): IAU Symp. **98**, 327
Rees, M. J., Stoneham, R. J. (ed.) (1982): *Supernovae* (Reidel Publ. Company, Dordrecht)
Reimers, D. (1975): Commun. 19. Colloq. Int. Astrophys. Liège; Mém. Soc. R. Sci. Liège, Sér. 6, **8**, p. 369
Reimers, D. (1977) Astron. Astrophys. **61**, 217 and Mitt. Astron. Ges. Nr. **43**, 70
Richter, G. (1960): Astron. Nachr. **285**, 274
Richter, G. (1967a): Veröff. Sternw. Sonneberg **7**, Nr. 3
Richter, G. A. (1967b): Sterne **43**, 38
Richter, G. A., Börngen, F. (1981): Astrophys. Lett. **21**, 101

Richter, G., Schaifers, K., Wenzel, W. (1961): Mitt. Veränderliche Sterne **1**, 526
Richter, G. A., Meinunger, I. (1962): Astron. Nachr. **294**, 39
Richter, G. A. et al. (1981): Astron. Nachr. **302**, 211
Ritter, H. (1976): IAU Symp. **73**, 205
Ritter, H. (1980): ESO Messenger No. 21, 16
Ritter, H. (1982): Publ. Max-Planck-Inst. Astrophys. Garching/München No. 22
Ritter, H. (1983): Publ. Max-Planck-Inst. Astrophys. Garching/München No. 51
Robertson, B. S. C., Warren, P. R., Bywater, R. A. (1976): Inf. Bull. Variable Stars 1173
Robertson, B. S. C., Feast, M. W. (1981): Mon. Not. R. Astron. Soc. **196**, 111
Robinson, E. L. (1975): Astron. J. **80**, 515
Robinson, E. L. (1976 a): Astrophys. J. **203**, 485
Robinson, E. L. (1976 b): Annu. Rev. Astron. Astrophys. **14**, 119
Robinson, E. L. (1983): Astrophys. Space Sci. Libr. **101**, 1
Robinson, E. L., McGraw, J. T. (1976): cited without reference in Hansen, C. J. (1980)
Rodonò, M. (1980): Mem. Soc. Astron. Ital. **51**, 623
Rodonò, M. (1981): *Photometric and Spectroscopic Binary Systems* eds. Carling, E. B., Kopal, Z. (Reidel Publ. Comp., Dordrecht) p. 285
Rosenberg, H. (1906): Nova Acta Leopoldina **85**, Nr. 2; compare Geschichte und Literatur Veränderl. Sterne I, **2**, 224
Rosino, L. (1972): IAU Colloq. **21**, 51
Rosino, L. (1978): Vistas Astron. **22**, 39
Rosner, R., Tucker, W. H., Vaiana, G. S. (1978): Astrophys. J. **220**, 643
Rößiger, S. (1979): Sterne **55**, 76
Rößiger, S. (1982): Sterne **58**, 147
Rößiger, S., Wenzel, W. (1972): Astron. Nachr. **294**, 29
Rößiger, S., Wenzel, W. (1973): Astron. Nachr. **295**, 47
Rucinski, S. M. (1981): Acta Astron. **31**, 37
Rucinski, S. M. (1982): Publ. Max-Planck-Inst. Astrophys. Garching/München No. 38
Russell, H. N. (1912): Astrophys. J. **35**, 315 and **36**, 54
Rydgren, A. E., Vrba, J. (1983): Astron. J. **88**, 1027

Saha, A. (1982): Bull. Am. Astron. Soc. **14**, 886
Sahade, J. (1976): Commun. 20. Colloq. Int. Anstrophys. Liège; Mém. Soc. R. Sci. Liège, Sér. 6, **8**, p. 303
Sahade, J. (1980): *The Wolf-Rayet Stars* (Collège de France, Paris)
Sahade, J., Wood, F. B. (1978): *Interacting Binary Stars* (Pergamon Press, Oxford)
Sandage, A., Tammann, G. A. (1969): Astrophys. J. **157**, 683
Sanders, W. T., Cassinelli, J. P., van der Hucht, K. A. (1981): Wisconsin Astrophys. No. 143
Sanford, R. (1949): Astrophys. J. **109**, 208
Sawyer-Hogg, H. (1973): Publ. David Dunlap Obs. **3**, No. 6
Scalo, J. M. (1980): Astrophys. Space Sci. Libr. **88**, 78
Schaefer, B. E. (1981): Nature **294**, 722
Schaefer, B. E. (1983): Inf. Bull. Variable Stars 2281
Schaefer, B., Bradt, H. (1982): IAU Circ. 3752
Schatzman, E. (1950): Ann. Astrophys. **13**, 384
Schatzman, E. (1951): Ann. Astrophys. **14**, 294
Schiller, K. (1923): *Einführung in das Studium der veränderlichen Sterne* (Verlag J. A. Barth, Leipzig)
Schlickeiser, R. (1981): Astron. Astrophys. **94**, 229
Schmidt, E. G. (1972): Astrophys. J. **176**, 165
Schmidt, M. (1956): Bull. Astron. Inst. Netherlands **13**, 15
Schneller, H. (1949): Veröff. Sternw. Sonneberg **1**, 355
Schneller, H. (1952): Geschichte u. Literatur Veränderl. Sterne II, **3**
Schneller, H. (1960): Kleine Veröff. Remeis-Sternw. Bamberg Nr. 27, 32
Schneller, H. (1965): Mitt. Veränderliche Sterne **2**, 86
Schönberner, D. (1977): Astron. Astrophys. **57**, 437
Schorn, R. A. (1981): Sky Telesc. **62**, 100

Schramm, D. N. (1977): Astrophys. Space Sci. Libr. **66**
Schwarzschild, M., Härm, R. (1959): Astrophys. J. **129**, 637
Schwartz, D. A et al. (1981): Mon. Not. R. Astron. Soc. **196**, 95
Scovil, C. E. (1980): *The AAVSO Variable Star Atlas* (Sky Publishing Co., Cambride, Mass.)
Seares, F. H. (1925): Contr. Mt. Wilson Obs. **13**, 145
Shapley, H. (1914): Astrophys. J. **40**, 448
Shapley, H. (1915): Princeton Contr. No. 3
Shapley, H. (1916): Astron. J. **43**, 217
Shara, M. M. (1982): Astrophys. J. **261**, 649
Sharov, A. S. (1975): IAU Symp. **67**, 275
Sharov, A. S., Alksnis, A. K. (1975): Astron. Tsirk. No. 869
Sharov, A. S., Lyuty, V. M. (1976): IAU Symp. **70**, 107
Sherwood, V. E., Plaut, L. (1975): IAU Symp. **67**
Shklovskij, I. S. (1976): *Sverkhnovye zvezdy* (Izd. Nauka, Moskva)
Shklovskij, I. S. (1978): Astron Zh. **55**, 726
Shobbrook, R. R., Stobie, R. S. (1976): Mon. Not. R. Astron. Soc. **174**, 401
Slowak, M. H. (1980): Publ. Astron. Soc. Pac. **92**, 550
Smak, J. (1971): IAU Colloq. **15**, 248
Smak, J. (1982a): Commun. Konkoly Obs. Budapest No. 83
Smak, J. (1982b): Acta Astron. **32**, 199
Smak, J. (1982c): Acta Astron. **32**, 213
Smith, F. G. (1977): *Pulsars* (University Press, Cambridge, London, New York, Melbourne)
Smith, H. A. (1981): Astron. J. **86**, 998
Smith, H. J. (1955): Astron. J. **60**, 179
Smith, H. J. (1965): *Quasi-Stellar Sources and Gravitational Collapse* ed. I. Robinson et al. (University of Chicago Press, Chicago 1964) p. 221
Starrfield, S., Sparks, W. M., Truran, J. W. (1974): Astrophys. J., Suppl. Ser. **28**, 247 and Astrophys. J. **192**, 647
Starrfield, S., Sparks, W. M., Truran, J. W. (1976): IAU Symp. **73**, 155
Starrfield, S. et al. (1981): Astrophys. J. **243**, L 27
Starrfield, S. et al. (1983a): Astrophys. J. **268**, L 27
Starrfield, S. et al. (1983b): Sky Telesc. **65**, 506
Stebbins, J., Huffer, C. M. (1930): Washburn Obs. Publ. **25**, part 3, 143
Stein, W. A., O'Dell, S. L., Strittmatter, P. A. (1976): Annu. Rev. Astron. Astrophys. **15**, 173
Stellingwerf, R. F. (1975): Astrophys. J. **199**, 705
Stephenson, C. B. (1967): Publ. Astron. Soc. Pac. **79**, 584
Stephenson, C. B., Herr, R. B. (1963): Publ. Astron. Soc. Pac. **75**, 444
Stepinski, T. (1980): Acta Astron. **30**, 414
Sterken, C., Jerzykiewicz, M. (1980): Proc. Workshop on Nonradial and Nonlinear Stellar Pulsation (Springer, Berlin, Heidelberg, New York) p. 114
Sterne, T. E. (1934): Harvard Obs. Circ. Nos. 386 and 387; Popular Astron. **42**, No. 10; Harvard Repr. No. 107
Stobie, R. S. (1980): Proc. fifth Europ. Reg. Meet., Liège, p. B. 2, 1
Stothers, R. (1977): Astrophys. J. **213**, 791
Stothers, R., Chin, Ch.-W. (1983): Astrophys. J. **264**, 583
Strittmatter, P. A. (1976): *The Physics of Non-Thermal Radio Souces* ed. Setti, G. (Reidel Publ. Company, Dordrecht)
Strom, S. E. (1977): IAU Symp. **75**, 190
Struve, O. (1947): Publ. Astron. Soc. Pac. **59**, 192
Struve, O. (1953): Sky Telesc. **12**, 99
Struve, O. (1954): Sky Telesc. **13**, 368
Struve, O. (1957): Sky Telesc. **16**, 418
Struve, O. (1962): *Astronomie – Einführung in ihre Grundlagen* (Verlag W. de Gruyter & Co., Berlin)
Struve, O., Herbig, G., Horak, H. (1950): Astrophys. J. **112**, 216
Szeidl, B. (1965): Mitt. Sternw. Ungar. Akad. Wiss. Budapest **5**, 265

Szeidl, B. (1975): IAU Symp. **67**, 545
Szeidl, B. (1976): IAU Colloq. **29**, 134

Taam, R. E. (1980): Astrophys. J. **242**, 749
Tapia, S. (1977): Astrophys. J. **212**, L 125
Thackeray, A. D. (1953): Mon. Not. R. Astron. Soc. **113**, 237
Thackeray, A. D. (1958): Mon. Not. R. Astron. Soc. **118**, 117
Thackeray, A. D. (1967): Mon. Not. R. Astron. Soc. **135**, 51
Thiessen, G. (1956): Zeitschr. Astrophys. **39**, 36.
Thomas, H. (1932): Veröff. Sternw. Babelsberg **9**, Nr. 4
Tomkin, J., Lambert, D. L. (1978): Astrophys. J. **222**, L 119
Torres, C. A. O., Ferraz-Mello, S. (1973): Astron. Astrophys. **27**, 231
Trimble, V. (1968): Astron. J. **73**, 535 and 657
Trimble, V. (1972): Mon. Not. R. Astron. Soc. **156**, 411
Truran, J. W. (1980): Illinois Astron. Prepr. 80–12 and 80–41
Tsesevich, V. P. (1971): *Zatmennye peremennye zvezdy* (Izdatel'stvo Nauka, Moskva); English
    translation (1973): *Eclipsing Variable Stars* (Halsted Press, New York, Toronto)
Tsesevich, V. P., Kazanasmas, M. S. (1963): *Atlas poiskovykh kart* (Odesskaya Astron. Observa-
    toriya)
Tsesevich, V. P., Kazanasmas, M. S. (1971): *Atlas poiskovykh kart peremennykh zvezd* (Odessa)
    (Izdatel'stvo Nauka, Moskva)
Turner, H. H. (1920): Mon. Not. R. Astron. Soc. **80**, 279

Underhill, A. B. (1966): Astrophys. Space Sci. Libr. **6**, 237
Usher, P. D. (1972): Astrophys. J. **172**, L 25

Van Agt, S. (1973): IAU Colloq. **21**, 35
Van Agt, S. (1978): Publ. David Dunlap Obs. **3**, No. 7
Van Agt, S. (1980): Rep. first ESO/ESA Workshop, Geneva, p. 33
Van de Kamp, P. (1978): Sky Telesc. **56**, 397
Van den Bergh, S. (1974): Highlights of Astron. **3**, 559
Van den Bergh, S. (1978): Vistas Astron. **22**, 307; Contr. Dominion Astrophys. Obs. No. 360
Van den Bergh, S., Herbst, E., Pritchet, C. (1973): Astron. J. **78**, 375
Van Gent, H. (1933): Bull. Astron. Inst. Netherlands **7**, 21
Van Herk, G. (1965): Bull. Astron. Inst. Netherlands **18**, 71
Van Paradijs, J. (1981): ESO Messenger No. 23; see also Astron. Astrophys. **103**, 140 and Proc.
    fifth Europ. Reg. Meet. Liège, p. GL. 3, 1
Van Schewick, H. (1937): Astron. Nachr. **262**, 97
Vogt, N. (1980): Astron. Astrophys. **88**, 66
Vogt, N. (1981): Astrophys. J. **252**, 653
Vogt, N. (1983): Sterne Weltraum **22**, 123; 278 and 404
Vogt, N., Bateson, F. M. (1981): ESO Prepr. No. 161
Vogt, N. et al. (1981): Astron. Astrophys. **94**, L 29

Wachmann, A. A. (1961): Astron. Abh. Sternw. Hamburg **6**, 4
Walker, A. R. (1976): Mon. Not. R. Astron. Soc. **179**, 587
Walker, H. J., Kilkenny, D. (1980): Mon. Not. R. Astron. Soc. **190**, 299
Walker, H. J., Schönberner, D. (1981): Astron. Astrophys. **97**, 291
Walker, M. F. (1954): Publ. Astron. Soc. Pac. **66**, 230
Walker, M. F. (1963a): Mitt. Veränderliche Sterne **2**, 17
Walker, M. F. (1963b): Astrophys. J. **138**, 313
Walker, M. F. (1978): Astrophys. J. **224**, 546
Walker, M. F. (1980): Publ. Astron. Soc. Pac. **92**, 66
Walker, M. F., Bell, M. (1980): Astrophys. J. **237**, 89
Wallace, P. T. et al. (1977): Nature **266**, 692
Wallerstein, G. (1958): Astrophys. J. **127**, 588
Wallerstein, G. (1959): Astrophys. J. **130**, 564

Wallerstein, G., Crampton, D. (1967): Astrophys. J. **149**, 225
Wallerstein, G., Greenstein, J. L. (1980): Publ. Astron. Soc. Pac. **92**, 275
Walter, F. M., Kuhi, L. V. (1981): Astrophys. J. **250**, 254
Wamsteker, W. (1979): ESO Messenger No. 18, 31
Warner, B. (1973): Mon. Not. R. Astron. Soc. **162**, 189
Warner, B. (1974): Mon. Not. R. Astron. Soc. **167**, 61 P
Warner, B. (1974a): Mon. Notes Astron. Soc. S. Afr. **33**, 21
Warner, B. (1976): IAU Symp. **73**, 85
Warner, B., Cropper, M. (1983): Mon. Not. R. Astron. Soc. **203**, 909
Warner, B., Nather, E. N. (1972): Sky Telesc. **43**, 82
Warner, B., Robinson, E. L. (1972): Mon. Not. R. Astron. Soc. **159**, 101
Watson, M. G., Warwick, R. S., Corbet, R. H. D. (1982): Mon. Not. R. Astron. Soc. **199**, 915
Weaver, H. (1974): Highlights of Astron. **3**, 509
Webbink, R. F. (1978): Publ. Astron. Soc. Pac. **90**, 57
Webster, B. L., Allen, D. A. (1975) Mon. Not. R. Astron. Soc. **171**, 171
Weedman, D. W. (1977): Vistas Astron. **21**, 55 and Annu. Rev. Astron. Astrophys. **15**, 69
Wehlau, A. (1964): Sky Telesc. **27**, 147
Weiler, K. W., Johnston, K. J. (1980): Mon. Not. R. Astron. Soc. **190**, 269
Weiss, W. W., Jenkner, H., Wood, H. J. (1976): IAU Colloq. **32**
Welter, G. L., Worden, S. P. (1980): Astrophys. J. **242**, 673
Wenzel, W. (1961): Veröff. Sternw. Sonneberg **5**, Nr. 1
Wenzel, W. (1962): Mitt. Veränderliche Sterne, Suppl. 2
Wenzel, W. (1963): Mitt. Veränderliche Sterne, **1**, 730
Wenzel, W. (1969): Mitt. Veränderliche Sterne, **5**, 75
Wenzel, W. (1975): Astron. Nachr. **296**, 183
Wenzel, W. (1976): Inf. Bull. Variable Stars 1222
Wenzel, W. (1980): Mitt. Veränderliche Sterne **8**, 141
Wenzel, W. (1980a): Mitt. Veränderliche Sterne **8**, 182
Wenzel, W. (1981): Bull. Inf. Cent. Données Stellaires No. 20, 105
Wenzel, W., Dorschner, J., Friedemann, Ch. (1971): Astron. Nachr. **292**, 221
Wenzel, W., Fürtig, W. (1967): Sterne **43**, 19
Wenzel, W, Geßner, H. (1975): Mitt. Veränderliche Sterne **7**, 23
Wenzel, W., Meinunger, I. (1978): Astron. Nachr. **299**, 239
Wheeler, J. C. (1980): Proc. Texas Workshop on Type I Supernovae, McDonald Observatory,
    Austin
Wheeler, J. C. (1981): Rep. Prog. Phys. **44**, 85; Austin Repr. No. 944
White, N. E., Swank, J. H. (1982): Astrophys. J. **253**, L 61
Whitney, C. A. (1978): Astrophys. J. **220**, 245
Wickramasinghe, D. T. (1982): Proc. Astron. Soc. Aust. **4**, 328
Wilkens, H. (1964): Mitt. Veränderliche Sterne **2**, 101
Willmore, A. P. (1977): Highlights of Astron. **4**, I, 87
Willson, L. A. (1980): Bull. Am. Astron. Soc. **12**, 805
Willson, L. A. (1981): Astrophys. Space Sci. Libr. **89**, 353
Willson, L. A., Garnavich, P., Mattei, J. A. (1981): Inf. Bull. Variable Stars 1961
Wilson, R. E. (1942): Astrophys. J. **96**, 371
Wilson, R. E., Fox, R. K. (1981): Astron. J. **86**, 1259
Wilson, jr., R. H. (1937): Publ. Astron. Soc. Pac. **49**, 202
Winget, D. E., Van Horn, H. M. (1982): Sky Telesc. **64**, 216
Wittmann, A. (1974): Sterne Weltraum **13**, 269
Wolf, M. (1914): Astron. Nachr. **198**, 371
Wolf, M. (1916): Astron. Nachr. **202**, 415
Wood, F. B. (1950): Astrophys. J. **112**, 196
Wood, P. R. (1974): IAU Symp. **59**, 101
Wood, P. R. (1975): IAU Colloq. **29**, 69
Wood, P. R. (1979): Astrophys. J. **227**, 220
Wood, P. R., Cahn, J. H. (1977): Astrophys. J. **211**, 499
Wood, P. R., Zarro, D. M. (1981): Astrophys. J. **247**, 247

Woolley, R. (1966): Observatory **86**, 76
Woolley, R., Savage, A. (1971): R. Obs. Bull. Greenwich No. 170

Yahel, R. Z. (1980): Astron. Astrophys. **90**, 26
Yamashita, Y., Maehara, H., Norimoto, Y. (1978): Publ. Astron. Soc. Japan **30**, 219; Tokyo Astron. Obs. Rep. No. 532
Young, P. J. et al. (1976): Astrophys. J. **209**, 882

Zhilyaev, B. E. et al. (1978): *Zvezdy tipa R Severnoj Korony* (Naukova Dumka, Kiev)
Zombeck, M. V. (1980): Smithson. Astrophys. Obs. Spec. Rep. No. 386

Star Index

To avoid confusion with star designations, the page numbers are given in italics.

Andromeda
Z   *81, 123ff., 129f.*
RX   *111*
SW   *44*
AC   *37f., 44*
AE   *176*
AF   *176*
CC   *54*
ζ *202*
λ *286*
o *178*

Apus
S   *186f.*
MY   *77*

Aquarius
R   *2, 126f., 129, 130*
Z   *55, 68ff.*
VY   *87*
AE   *109, 111, 116*
BS   *52*
BV   *43*
CY   *50ff., 54*

Aquila
R   *60f., 62*
U   *186*
UU   *111*
CM   *126*
V 603 (Nova 1918)   *82, 89, 93f., 111*
V 605   *186*
V 672   *42*
V 976   *181*
V 1208   *53*
V 1229 (Nova 1970)   *116*
V 1333   *135, 143ff.*
V 1343 (SS 433)   *141ff.*
η *2, 26f., 31, 33f., 286*
28 (V 1208)   *53*

Ara
V   *71*
AE   *181*
V 801   *138f.*

Aries
V   *71*
W   *96*
RV   *52, 54*
SX   *195*
TT   *101, 111*
UX   *192*

Auriga
R   *59*
T   *82, 111*
RT   *286*
RW   *156f., 160f., 163f., 169, 265*
SS   *111*
UV   *129, 131*
CO   *37*
KN   *271*
KR   *111*
V 363 (Lanning 10)   *111*
β *204*
ε *2, 210, 217f., 286*
ζ *202, 218f., 286*

Bootes
R   *59*
S   *59*
V   *60*
RS   *43*
SS   *192*
TV   *44*

Camelopardalis
X   *59*
Z   *99ff., 111, 113, 116*
RU   *32f.*
TW   *69*
XX   *186*

Cancer
R   *2*
RW   *44*
SY   *111, 113*
TT   *44*
VZ   *52, 54*
YZ   *99*

AC   *111*
BN   *50*

Canes Venatici
Z   *44*
RS   *191f.*
TT   *71*
TX   *126*
AM   *81, 89, 103f., 108f., 111, 114, 122, 210*
BG   *77*
α_2 *195*

Canis Major
Z   *176, 180*
UW   *286*
EW   *286*
FW   *286*
α (Sirius)   *220*
β *73*

Canis Minor
AD   *50*
BG (3 A 0729 + 103)   *111*

Capricornus
TW   *49*

Carina
R   *286*
S   *286*
Y   *38, 39*
RT   *126*
AG   *175f.*
GL   *207*
GZ   *38*
HH   *207*
HR   *175f.*
OY   *111, 114f.*
QU   *111*
V 351 (Nova 1970)   *85, 96f.*
V 395 (2 S 0921 − 630)   *135*
η *176ff., 286*
l   *286*

MIX
Papier aus verantwortungsvollen Quellen
Paper from responsible sources
FSC® C105338

If you have any concerns about our products,
you can contact us on
ProductSafety@springernature.com

In case Publisher is established outside the EU,
the EU authorized representative is:
**Springer Nature Customer Service Center GmbH
Europaplatz 3, 69115 Heidelberg, Germany**

Printed by Libri Plureos GmbH
in Hamburg, Germany